Springer Series in Computational Neuroscience

Volume 2

For other titles published in this series, go to
www.springer.com/series/8164

Jose Luis Perez Velazquez · Richard Wennberg
Editors

Coordinated Activity in the Brain

Measurements and Relevance to Brain Function and Behavior

Springer

Editors
Jose Luis Perez Velazquez
Department of Paediatrics
Institute of Medical Science
University of Toronto
Toronto ON M5G 1X8, Canada
jose-luis.perez-velazquez@sickkids.ca

Richard Wennberg
University Health Network
Toronto Western Hospital
University of Toronto, ON M5T 2S8
Canada
r.wennberg@utoronto.ca

ISBN 978-0-387-93796-0 e-ISBN 978-0-387-93797-7
DOI 10.1007/978-0-387-93797-7
Springer Dordrecht Heidelberg London New York

Library of Congress Control Number: 2009921198

Printed on acid-free paper

Springer is part of Springer Science+Business Media (www.springer.com)

Preface

The Numerous Aspects of Coordinated Activity in the Nervous System

It is reasonable to say that the ultimate goal of neuroscience is to understand how nervous systems, and brains in particular, process information. Whether it be a nematode or a primate nervous system, the foundation of information processing lies in the interactions between units, which in the case of nervous systems means neurons and glial cells. Each cell's activity is meaningful only with respect to what other cells are doing. Hence, these multiple cellular interactions imply that some sort of coordinated activity must exist at some level. Synchronization of the cellular and network activities is one such aspect, and because of this, measures of synchronized activity are fundamental to the understanding of neuronal information processing. The search for methods and conceptual frameworks to capture coordinated activities in nervous systems has had a long history, as expressed by Varela and colleagues: "...since the time of Sherrington and Pavlov the understanding of global distributed properties has been an El Dorado of neuroscience, one that is difficult to reach. The reasons for these difficulties have been both technical and conceptual" (F.J. Varela, E. Thompson, E. Rosch, *The Embodied Mind*, MIT Press, 1991). This book contains contributions that address some of those technical and conceptual matters that continue to hinder, to various extents, the comprehension of coordinated activity in the brain.

It is generally accepted that synchronization of neuronal activity supports the temporal correlations in cell action potential firing that seem to be necessary for neuronal networks to activate one another (information processing relies on the stable propagation of action potentials from network to network). However, the functional role of the synchronization of cellular activity upon cognition and brain pathologies is still debated. From the "binding problem" to the extent of pathological synchrony in epilepsy, different perspectives have been advanced regarding the possible roles of the synchronization that may be seen and measured using a variety of methods. It is important to note that some of these methods will also record activity from glial cells, and that therefore the measured synchrony will reflect neuronal and glial contributions, despite the fact that most studies of nervous system synchrony focus exclusively on neurons. There may be a coming danger that, with the advent of more powerful recording methods and sophisticated analytical techniques, fast data processing may lead to misinterpretations in the physiological account of synchrony measures. Since the analysis of brain coordinated activity will be of fundamental importance in the years to come, careful consideration of the limitations in the recording and analytical methodologies is fundamental for enthusiasts and apprentices in the field. This book is designed to serve that purpose, to illustrate novel developments and basic methodologies, along with some of their constraints, in the assessment of neuronal correlated activity.

A general introduction to the field along with some historical facts is presented in chapters "Correlations of Cellular Activities in the Nervous System: Physiological and Methodological Considerations" and "Synchronization Between Sources: Emerging Methods for Understanding Large-Scale Functional Networks in the Human Brain". Classical analysis methods such as phase synchronization are described in chapters "Correlations of Cellular Activities in the Nervous System: Physiological and Methodological Considerations", "Synchronization Between Sources: Emerging Methods for Understanding Large-Scale Functional Networks in the Human Brain", and "From Synchronization to Networks: Assessment of Functional Connectivity in the Brain", and other methods based on spectral analysis are detailed in various chapters. Chapter "The Size of Neuronal Assemblies, Their Frequency of Synchronization and Their Cognitive Function" in particular focuses on coherence and power spectral analyses. Chapter "Synchronization Between Sources: Emerging Methods for Understanding Large-Scale Functional Networks in the Human Brain" devotes sections to the need for source localization when analysing synchrony, using methods such as independent component and beamformer analysis. This is of importance because the data obtained are coming from sensors (usually electroencephalographic or magnetoencephalographic), whereas the synchrony that is actually sought is that between brain sources. Along these lines, chapter "Correlations of Cellular Activities in the Nervous System: Physiological and Methodological Considerations" demonstrates that it is possible to measure (spurious) synchrony between sensor data signals in the setting of completely desynchronized neural sources. Wavelet analysis, another popular analytical method used to assess synchronization, is described in chapters "Synchronization Between Sources: Emerging Methods for Understanding Large-Scale Functional Networks in the Human Brain", "Denoising and Averaging Techniques for Electrophysiological Data", and "Time–Frequency Methods and Brain Rhythm Signal Processing" along with more details on Fourier analysis. A clear and intuitive comparison between wavelet and Fourier decomposition is presented in chapter "Denoising and Averaging Techniques for Electrophysiological Data". Cluster analysis of phase differences is detailed in chapters "Detection of Phase Synchronization in Multivariate Single Brain Signals by a Clustering Approach" and "Denoising and Averaging Techniques for Electrophysiological Data". Methods to determine not only the synchronization between brain areas but also the directionality of coupling are detailed in chapters "Approaches to the Detection of Direct Directed Interactions in Neuronal Networks" and "Synchrony in Neural Networks Underlying Seizure Generation in Human Partial Epilepsies", particularly as applied to studies of epilepsy. The important new developments in graph theory as applied to the study of coordinated brain activity are described in chapters "From Synchronization to Networks: Assessment of Functional Connectivity in the Brain" and "Complex Network Modeling: A New Approach to Neurosciences". Chapter "Complex Network Modeling: A New Approach to Neurosciences" presents a more technical description of graph theory, with many mathematical details. Graph theory applied to the study of brain functional connectivity is an emerging area of research that combines synchronization analysis with complex systems theory, and can serve to identify universal principles and statistical properties common to disparate systems. An historical introduction to the field is presented in a section of chapter "From Synchronization to Networks: Assessment of Functional Connectivity in the Brain", and examples of graph theory applied to the study of neural functional connectivity are discussed in chapters "From Synchronization to Networks: Assessment of Functional Connectivity in the Brain" and "Complex Network Modeling: A New Approach to Neurosciences".

Variability in electrophysiological recordings is a common source of distress for the experimentalist, and the authors of chapter "Denoising and Averaging Techniques for Electrophysiological Data" discuss that what is normally considered a sensory-evoked or event-related potential may be a modulation of the ongoing, spontaneous brain activity, and techniques are described to improve noisy data, specifically addressing the important question of the variability present in single trials. These queries about variability in the signals are closely related to the debate on whether a resting baseline in brain activity really exists. Chapters "Detection of Phase Synchronization in Multivariate Single Brain Signals by a Clustering Approach" and "Denoising and Averaging Techniques for Electrophysiological Data" address these questions of what can be considered baseline for event-related potentials and single trials. Fluctuations in whole-head recordings are related to fluctuations in the activity of individual cells, a topic that is addressed in chapter "Dissection of Synchronous Population Discharges in Vitro", with elegant descriptions of in vitro experiments combining local field potentials, multiunit cellular activity and intracellular recordings, designed to address the cellular and network mechanisms underlying the generation of synchronous population activity. A theoretical framework for understanding the emergence of coherent activity in nervous systems is presented in chapter "The Phase Oscillator Approximation in Neuroscience: an Analytical Framework to Study Coherent Activity in Neural Networks", focusing on coupled oscillator formalisms and the reduction to a phase model. In the end, the correlated activity estimated from electrophysiological recordings represents a summation of synaptic potentials and ion movements, and therefore a close consideration of the nature of these events, in terms of the number of active synapses/cells and cell assemblies resulting from the functional contacts, is fundamental. In this regard, chapter "The Size of Neuronal Assemblies, Their Frequency of Synchronization and Their Cognitive Function" presents some estimations on the number of active cells needed to produce field potentials or electroencephalographic signals, as well as a discussion of the relation between the frequencies at which synchronization takes place and the size of the cell assemblies.

Several chapters make reference to particular software and code sources used to perform specific analyses, including the following: BESA ("Synchronization Between Sources: Emerging Methods for Understanding Large-Scale Functional

Networks in the Human Brain"), EEGLAB ("Synchronization Between Sources: Emerging Methods for Understanding Large-Scale Functional Networks in the Human Brain"), MATLAB ("Correlations of Cellular Activities in the Nervous System: Physiological and Methodological Considerations", "Synchronization Between Sources: Emerging Methods for Understanding Large-Scale Functional Networks in the Human Brain", and "The Size of Neuronal Assemblies, Their Frequency of Synchronization and Their Cognitive Function"), BioSig ("Synchronization Between Sources: Emerging Methods for Understanding Large-Scale Functional Networks in the Human Brain"), Spikoscope and Labview ("Dissection of Synchronous Population Discharges in Vitro"). Available as a free download on the book's website at http://www.springer.com/978-0-387-93796-0 are specific codes for the following purposes: MATLAB code to estimate power and coherence spectra ("The Size of Neuronal Assemblies, Their Frequency of Synchronization and Their Cognitive Function") as well as MATLAB code (software package FDMa) containing scripts and data files to compute the simulations and application to tremor signals as described in chapter "Approaches to the Detection of Direct Directed Interactions in Neuronal Networks".

Finally, we would like to express our profound gratitude to all of the contributing authors who heeded our requests to provide chapters for this book. Their shared enthusiasm and dedication to progress in this multidisciplinary area, combined with the inspiring breadth and depth of their expertise, is something we hope will serve to stimulate much further thought and experimentation into the coordinated activity of the brain.

Toronto, Ontario
September 2008

Jose Luis Perez Velazquez
Richard Wennberg

Contents

Supplemental electronic material can be found at
http://www.springer.com/978-0-387-93796-0

Contributors

Fabrice Bartolomei INSERM, U751; Faculté de Médecine, Aix Marseille Université; Assistance Publique – Hôpitaux de Marseille, Hôpital de la Timone, Service de Neurophysiologie Clinique, Marseille F-13000, France, fbartolo@medecine.univ-mrs.fr, Fabrice.Bartolomei@medecine.univ-mrs.fr

Ivan Cohen INSERM UMRS975/CNRS UMR7225, CHU Pitié-Salpêtrière, 105 boulevard de l'hôpital, 7501 Paris, France, coheni@upmc.fr

Luis Garcia Domínguez Hospital for Sick Children, University of Toronto, Toronto, ON, Canada, l.garcia.d@gmail.com

Sam M. Doesburg Department of Psychology, University of British Columbia, Vancouver, BC, Canada, sam@psych.ubc.ca

Ramon Guevara Erra Hospital for Sick Children, University of Toronto, Toronto, ON, Canada, guevara.erra@gmail.com

Roberto F. Galán Department of Neurosciences, School of Medicine, Case Western Reserve University, 10900 Euclid Avenue, Cleveland, OH 44106, USA, rfgalan@case.edu

Jesse Gillis The Centre for High-Throughput Biology, 177, Michael Smith Laboratories, 2185 East Mall, University of British Columbia, Vancouver, BC V6T1Z4, Canada, jesse.gillis@gmail.com

Kathrin Henschel FDM, Freiburg Center for Data Analysis and Modeling, University of Freiburg, Eckerstr. 1, 79104 Freiburg, Germany, kathrin.henschel@fdm.uni-freiburg.de

J. Michael Herrmann Bernstein Center for Computational Neuroscience Göttingen, Bunsenstrasse 10, 37073 Göttingen, Germany; School of Informatics, Institute of Perception, Action and Behaviour, University of Edinburgh, 11 Crichton Street Edinburgh, EH8 9AB, UK, michael@chaos.gwdg.de, michael@nld.ds.mpg.de

Axel Hutt INRIA CR Nancy – Grand Est, CS20101, Villers-ls-Nancy Cedex 54603, France, axel.hutt@loria.fr

Matthias Ihrke Bernstein Center for Computational Neuroscience Göttingen; Max-Planck Institut for Dynamics and Self-Organization, Bunsenstrasse 10, 37073 Göttingen, Germany, mihrke@gwdg.de

Michael Jachan FDM, Freiburg Center for Data Analysis and Modeling, University of Freiburg, Eckerstr. 1, 79104 Freiburg, Germany, michael.jachan@fdm.uni-freiburg.de

Liset Menendez de la Prida Instituto Cajal – CSIC, Ave. Doctor Arce, 37, Madrid 28002, Spain, lmprida@cajal.csic.es

Yamir Moreno Department of Theoretical Physics, Institute for Biocomputation and Physics of Complex Systems (BIFI), University of Zaragoza, Zaragoza 50009, Spain, yamir.moreno@gmail.com

Matthias H.J. Munk Max-Planck-Institute for Biological Cybernetics, Spemannstr. 38, 72076 Tuebingen, Germany, matthias.munk@tuebingen.mpg.de

Jakob Nawrath FDM, Freiburg Center for Data Analysis and Modeling, University of Freiburg, Eckerstr. 1, 79104 Freiburg, Germany, nawrath@fdm.uni-freiburg.de

Johannes Sarnthein Neurochirurgie, UniversitätsSpital Zürich, Zürich 8091, Switzerland, johannes.sarnthein@usz.ch

Ariane Schad FDM, Freiburg Center for Data Analysis and Modeling, University of Freiburg, Eckerstr. 1, 79104 Freiburg, Germany, schad@fdm.uni-freiburg.de

Björn Schelter FDM, Freiburg Center for Data Analysis and Modeling, University of Freiburg, Eckerstr. 1, 79104 Freiburg, Germany, schelter@fdm.uni-freiburg.de

Hecke Schrobsdorff Bernstein Center for Computational Neuroscience Göttingen; Max-Planck Institut for Dynamics and Self-Organization, Bunsenstrasse 10, 37073 Göttingen, Germany, hecke@nld.ds.mpg.de

Linda Spindeler FDM, Freiburg Center for Data Analysis and Modeling, University of Freiburg, Eckerstr. 1, 79104 Freiburg, Germany, spindeler@fdm.uni-freiburg.de

Cornelis J. Stam Department of Clinical Neurophysiology, VU University Medical Center, Amsterdam, The Netherlands, CJ.Stam@VUmc.nl

Jens Timmer FDM, Freiburg Center for Data Analysis and Modeling, University of Freiburg, Eckerstr. 1, 79104 Freiburg, Germany, jeti@fdm.uni-freiburg.de

Jose Luis Perez Velazquez Neurosciences and Mental Health Programme, Department of Pediatrics and Institute of Medical Science, Brain & Behaviour Centre, Hospital for Sick Children, University of Toronto, Toronto, ON, Canada, jose-luis.perez-velazquez@sickkids.ca

Astrid von Stein Neurochirurgie, UniversitätsSpital Zürich, Zürich, Switzerland, astrid.vonstein@ini.phys.ethz.ch

Lawrence M. Ward Department of Psychology; The Brain Research Centre, University of British Columbia, Vancouver, BC, Canada, lward@psych.ubc.ca

Fabrice Wendling INSERM, U642, Rennes; Université de Rennes 1, LTSI, Rennes F-35000, France, fabrice.wendling@univ-rennes1.fr

Richard Wennberg Clinical Neurophysiology Laboratory, Krembil Neuroscience Centre, Toronto Western Hospital, University Health Network, University of Toronto, Toronto, ON, Canada, r.wennberg@utoronto.ca

Correlations of Cellular Activities in the Nervous System: Physiological and Methodological Considerations

Jose Luis Perez Velazquez, Ramon Guevara Erra, Richard Wennberg, and Luis Garcia Dominguez

Abstract Rhythms (cycles, oscillations) and synchronization processes pervade all aspects of the living and nonliving, from circadian rhythms to individual habits and traits (Glass, 2001; Pikovsky et al., 2001). The distributed nature of cognition implies that distinct brain areas must somehow coordinate their activity: cognition, whether in the form of perception or motor actions, results from the integrated spatiotemporal coordinated activity of cell populations in the nervous system, including both glial and neuronal elements. Because the system's collective behavior is difficult, if not impossible, to deduce from its individual components, the determination of brain-coordinated activity at adequate levels of description becomes fundamental to the understanding of nervous system function and its relation to behavior. Some questions arise: what variables are best suited to measure these correlations in activity? What order parameters should be used to characterize coordinated activity among cellular populations and how do these activity patterns relate to behavioral responses? What level of description should be used? This chapter summarizes some physiological and methodological considerations to aid the science practitioner in the study of correlated activity patterns in nervous systems.

1 Introduction

The first description of synchronization is attributed to Christiaan Huygens in 1665, who, while sick at home, spent time scrutinizing the relation between the periodic motion of two pendulum clocks (Huygens, 1673). Since then, synchronization phenomena have been the subject of enormous interest among scientists in a variety of disciplines. The mathematical definition of synchronization, entrainment (or one to

J.L. Perez Velazquez (✉)
Neurosciences and Mental Health Program, Division of Neurology, Department of Paediatrics and Institute of Medical Science, Brain and Behaviour Centre, Hospital for Sick Children, University of Toronto, Toronto, ON, Canada
e-mail: jose-luis.perez-velazquez@sickkids.ca

J.L. Perez Velazquez, R. Wennberg (eds.), *Coordinated Activity in the Brain*, Springer Series in Computational Neuroscience 2, DOI 10.1007/978-0-387-93797-7_1,

one frequency locking) with phase locking (Hoppensteadt and Izhikevich, 1997), and the equivalent, more physical, definition of synchronization as the adjustment of rhythms between weakly coupled oscillating systems (for a comprehensive description of synchronization, the monograph by Pikovsky et al., 2001 is strongly recommended), are somewhat relaxed when used in the context of biology. Since in a synchronized regime the phases of the oscillators remain bounded, synchronization has also been defined in the field of signal processing as the constancy of the instantaneous phase difference between two oscillators. Bear in mind that this definition, however, is only operative and does not entail any particular physical or mathematical model by which this regime is reached. The strict constancy of a phase difference that can be observed in mathematical simulations is rarely found in measurements of biological variables due to noise and nonstationarity, among other factors. Consequently the somewhat precise physicomathematical description of a synchronized state must be substituted by a statistical one in which phase differences, for example, do not need to remain constant but may fluctuate at a significantly low variance. After these considerations, skeptics may reasonably ask whether one can expect to find synchronization in the mostly noisy and stochastic environment of brain activity. A look at any normal electrophysiological brain recording reveals an apparently noisy signal with little trace of robust, long duration rhythms. Nonetheless, the answer to the skeptic's query is yes, we can expect synchronization to occur, as synchronization in stochastic systems does take place and has been studied for a long time (the book by Anishchenko et al., 2002, provides a clear account of the ideas, methods, and mechanisms underlying stochastic synchronization). This may seem paradoxical, nevertheless it is known that two coupled systems governed by chaotic dynamics can synchronize their chaotic trajectories, even though the coupled systems remain chaotic, only their trajectories are locked to each other. Indeed, these chaotic networks can also split into subsets of synchronized units (Kestler et al., 2008), which may tell us something when translated to neuronal activity patterns. Moreover, synchronization also occurs in the presence of noise, although in this setting it is limited to relatively short time intervals as already described in the seminal work of Stratonovich (1963). Chaotic phase synchronization has been studied (Pikovsky et al., 1997) and characterized using diffusion coefficients (Fujisaka et al., 2005). The study of synchronization of chaotic oscillators requires more general definitions of the concept, and thus the notions of generalized, phase, and full synchronization arose (Afraimovich et al., 1986; Pecora et al., 1997; Anishchenko and Vadivasova, 2002). A very general description of synchronization is the adjustment of some sort of relation in the characteristic times (periods for example) or oscillating phases between two interacting systems.

For the purposes of brain research, perhaps an even more general description of coordinated activity is needed. It could be generally described as: any relation between neurophysiological observations—measured by any reliable method that represents cellular activities—in separate brain areas, which may or may not be directly interacting. Thus, the term synchronization can be used to describe statistical constancy of phase differences between two signals and, in a wider sense, to describe all sorts of correspondences among signal parameters. We believe that a defined need for direct interaction can be relaxed for these purposes, as rationalized below. This very general concept of brain-coordinated activity includes the classical description of synchronization in the case of directly interacting areas, as well as instances where one brain region sends synchronous inputs to separate areas, in which case a correlated activity in these receiving areas may be detected, but is not dependent on a direct connectivity among them.

In general, for all these reasons, we recommend the use of the term "coordination", rather than "synchronization", in neuroscience. This coordinated activity of cellular populations would include both glial and neuronal elements, so that glial contributions to the establishment of ionic gradients and thus potential differences are not forgotten, notwithstanding the fact that the consequences of glial activities occur on slower timescales than those of neuronal activities, in general. The approach called "coordination dynamics" in brain research (Kelso, 1995; Kelso and Engstrøm, 2006) may be particularly useful for the comprehension of essential features of brain function and their relation to behavior.

2 Neurophysiological Bases of the Correlation of Activities Among Brain Areas

The current surge of interest in the idea that coordinated activity in brain cellular networks is fundamental for information processing originated with proposals already advanced by the Russian psychologist Luria suggesting that the dynamic interplay between brain areas is the essence of brain function (Luria, 1966), followed later by the ideas of Livanov, who had already articulated many of these issues with his theory of the spatiotemporal organization of brain function (Livanov, 1977), von der Malsburg (1981), and others (Damasio, 1989; Singer and Gray, 1995). These proposals were substantiated by early empirical evidence of correlations in the activity between cortical cells (Li, 1959), as well as by a wide variety of more recent experiments. Synchrony is today invoked to bear on aspects of neurophysiology ranging from perception and cognition (Nunez, 2000; Eckhorn, 2000; Varela et al., 2001; Bressler and Kelso, 2001; Engel et al., 2001) to consciousness (John, 2002; Singer, 2006). Consider, for instance, the somewhat trivial behavior of reaching for and grasping an object. Sensory information about the spatial properties (location, shape) of the object has to be integrated with the spatial properties of the effector systems (arm, hand) using some sort of self-centred coordinate reference frame. Neuroimaging studies indicate that parietal and frontal areas of the brain neocortex compute such sensorimotor transformations and are particularly involved in this grasping task (Binkofski et al., 1999). Without some sort of coordinated cellular activity amongst those brain regions, the accomplishment of this task would not be easy. Hence, it is currently thought that cognitive processes involve not only serial stages of sensory signal processing (Harris, 2005), but also massive parallel information processing circuitries, and that therefore it is the coordinated activity of brain cell ensembles that are the source (and manifestation) of cognitive processes (Bressler and Kelso, 2001; Kelso, 1995; Varela et al., 2001; Friston, 2001; Dehaene et al., 2006; Singer, 2006; Jirsa, 2004). As L. Swanson remarks "the profound question is not 'what is the brain'? but rather 'what are the basic parts of the nervous system and how are they interconnected functionally'?" (Swanson, 2000).

Of note is the extraordinary tendency of brain cell ensembles to synchronize. Even brain slices maintained in vitro are prone to show synchronized rhythms. It does not require much effort—just raise potassium or reduce magnesium in the extracellular medium surrounding hippocampal slices, and then stick a recording electrode in the CA1 area of the hippocampus—synchronized rhythmic activity will be present in the majority of recordings. Most remarkably, even in the absence of synaptic transmission (blocked by reducing extracellular calcium) the circuitries of hippocampal slices synchronize their firing activity (Taylor and Dudek, 1984). Equally unexpected is the observation, in computer simulations, that uncoupled neurons will fire in synchrony even if their input consists of fluctuations, albeit containing some degree of correlation (Galán et al., 2006). The nature of the neuronal input–output relation, studied using a nonlinear transfer function, was elucidated to some extent, by showing the results of distinct input frequencies in producing synchronous output activity (Menendez de la Prida and Sanchez-Andres, 1999). A tendency to phase locking has also been found in a variety of in vivo experiments using EEG or MEG recordings in humans (Friston et al., 1997). The fact that phase locking among brain areas does not seem to be completely in phase, but is more or less distributed, suggests the phenomenon of relative coordination (Kelso, 1995). Thus, neurons may be so prone to synchronize their activity that, as Erb and Aertsen point out "the question might be not so much how the brain functions by virtue of oscillations, as most researchers working on cortical oscillations seem to assume, but rather how it manages to do so in spite of them" (Erb and Aertsen, 1992). Too much synchronization impedes efficient information processing, and hence this is a "problem" for neural circuits and a difficulty in the understanding of how brains integrate information. Without doubt, there must be *functional integration* in the activity of neuronal networks but, simultaneously, *segregation* is needed so that cooperative interactions among discrete brain areas are maintained (Tononi et al., 1998a). Methods and conceptual frameworks to determine the interactions that result in the *functional connectivity* between networks (Friston, 2001; Tononi et al., 1998b) have been developed during past decades, from the classical cross-correlation analysis (Haken, 2002) of brain activity to the more modern mutual or nonlinear interdependence (Schiff et al., 1996; Breakspear and Terry, 2002) and phase synchronization (Mormann et al., 2000; Varela et al., 2001; Pikovsky et al., 2001; Breakspear et al., 2004), described in the next section.

3 On the Measurement of Correlations in Nervous System Activity

Extracting information about synchronization is not a trivial task and has to be evaluated carefully, giving thought to the properties of the systems, the recording methods and the analytical techniques used. As shown in some systems with nonlinear dynamics and complicated synchronization sets, some dynamical limitations to the detection of synchrony are intrinsic to the systems and all efforts to improve them are condemned to failure (So et al., 2002). On a more optimistic note, let us comment on some analytical methods that have been employed for the assessment of coordinated activity in nervous systems. We will describe the essence of some of these, and provide extensive references for the interested readers who wish to pursue details. Methods that are of general application take advantage of the several representations that the dynamics of physical systems have: one of these is as a geometric object in a particular multidimensional state space (an attractor for instance). A geometric representation (an early perspective on geometric representations of neural activity can be found in Schild, 1984) allows for the analysis of the correlated (synchronized) activity between two systems using statistical properties of the possible mapping function relating the dynamics of the systems (Pecora et al., 1995), or analysis using other methods like mutual nonlinear prediction (Schiff et al., 1996), multivariate recurrence plots (Romano et al., 2004), or recurrences of the trajectories in state space (Romano et al., 2005), to mention a few.

A more specific method, particularly for signals with oscillatory activity, is the assessment of phase synchronization, which tends to be a current favorite, as phase relationships in brain signals embody time windows for selective neuronal communication (Fries, 2005) and thus phase could be, in principle, an important order parameter to characterize brain activity (Kelso, 1995). The concept of order parameters is widely used in physics in studies of phase transitions, with order parameters defined as those parameters that are able to capture the macroscopic behavior resulting from microscopic fluctuations. Other studies have used as an order parameter the amplitude of the mean field, to study the onset of synchronization (Popovych et al., 2008). If phase synchronization is studied, the phase in the experimental recordings has to be derived first. However, to achieve this without ambiguity, regular and periodic sustained activity would be optimal, which is a very uncommon situation in neuronal recordings. Thus we encounter a first problem: how to characterize the phase in the noisy and nonperiodic electrophysiological recordings we obtain from the brain? A recommended starting point is to consider what the recordings to be used in a specific synchrony analysis actually represent. The physiological interpretations of specific recordings, either local field potentials, electroencephalography (EEG), or magnetoencephalography (MEG) have been addressed elsewhere (Nunez, 2000; Varela et al., 2001; Perez Velazquez and Wennberg, 2004; Perez Velazquez, 2005). Suffice to say that these recording methods detect macroscopic signals derived mostly from synaptic potentials. From these recordings, time series (normally voltage values) are obtained, and to compute phase synchronization it is necessary to extract the phases. Winfree's classical monograph (1980) devotes several pages to the explanation of the concept of phase in biological observations. Those activities in which the period is well defined, such as alpha (8–13 Hz) rhythms or absence (3 Hz) seizures, seem adequate to extract the phase of the oscillation, however, most brain recordings are nonperiodic and noisy, and therefore methods to extract the instantaneous phase (and amplitude) from these fluctuating time series have been derived. The most commonly used methods in neuroscience are wavelet decompositions and the Hilbert transform (Varela et al., 2001). These techniques are described in Pikovsky et al. (2001), and have been applied extensively to brain synchrony analyses (Lachaux et al., 2000; Quiroga et al., 2002; Breakspear et al., 2004; Chavez et al., 2005; Trujillo et al., 2005; Garcia Dominguez et al., 2005; Perez Velazquez et al., 2007a, b, c). Interestingly, methods to derive phase synchrony without the need to actually measure the oscillation phase have also been derived (Baptista et al., 2005). With noisy data, phase synchronization is defined in a statistical sense: two signals are phase synchronized if the difference between their phases is nearly constant over a selected time window, that is, it clusters around a single value (Pikovsky et al., 2001). Since phase difference is clearly defined in the unit circle, a measure of its variability is the circular variance (CV) of its distribution. This description means that if the series of phase differences over a time window are locked on the average, regardless of the specific mean value, the respective signals are considered to be "in synchrony" or phase-locked. The mean phase coherence statistic (MPC), described in Mormann et al. (2000), is simply 1-CV, a value defined between zero and one. In this scheme, maximal synchrony corresponds to $CV = 0$ or $MPC = 1$. Even very complex, apparently nonperiodic signals can synchronize, as mentioned in the introduction. In this regard, stochastic phase synchronization has been studied in several nervous systems, from crayfish (Bahar and Moss, 2003) to humans (Tass et al., 1998), as well as in computer simulations (Postnova et al., 2007; Galan et al., 2007).

A notable recent development that may be fruitful in the determination of the brain's functional connectivity has emerged from graph theory and the study of complex networks (Sporns et al., 2000; Strogatz, 2001; Stam, 2004; Gomez-Gardenes and Moreno, 2007a; Gomez-Gardenes et al., 2007b). This technique, intimately related to synchronization, offers the possibility of simplification, that is, of finding some fundamental elements that capture essential properties of the complex system under study. As illustrations of this drive toward simplification, some studies sought to convert the functional connectivity matrix derived from synchrony analysis to connected graphs, and from these to determine features of neural information processing (Stam, 2004). For instance, two different mechanisms to reach synchronization in complex networks (coupled oscillators models) have been identified using these methods: in one a large node of synchronized oscillators recruits others linked to it, and in another, small synchronized clusters initially appear and then recruit neighboring nodes (Gomez-Gardenes et al., 2007c). Hence, for those interested in mechanistic approaches, these analyses can be of particular value (even at these global levels of description there is room for some reductionism!). Some results from these studies relevant to brain information processing will be discussed in a section below.

The timescales at which synchronization occurs are also worth considering, because it is conceivable that there could be, for example, desynchronization in the spike firing of individual neurons (short timescales, high frequencies) and synchronization at slower timescales (low frequencies) of bursting activity and synaptic inputs. Some studies have addressed these and related considerations (Netoff and Schiff, 2002; Chavez et al., 2005), and the results indicated that there are transient periods of synchronization occurring at different timescales. In studies of synchronization in epileptiform activity, synchrony at higher frequencies was commonly observed between local neighborhoods during seizure onset (Chavez et al., 2005). A dynamic portrayal of synchronization, occurring at different frequencies as the brain activity approaches a seizure event, can be seen in Fig. 1 in Garcia Dominguez et al. (2005). Related to information processing, computer simulations of model neurons shown that the maximal capacity for information transfer occurs when neurons synchronize in the slow timescale (bursting in this case) and desynchronize in the fast scale, action potential firing (Baptista and Kurths, 2008), as discussed in more detail in the section on neural information processing below. Multifrequency locking (in addition to the 1:1 relation normally explored in most studies) has also been described in seizure activity (Perez Velazquez et al., 2007c), which suggests that, rather than synchronization through the adjustment of frequencies, forcing of one brain area by strong input from another area occurs, at least during seizures.

A question for the practitioner of synchrony analysis is which method to use amongst the assorted collection, which seems to be ever increasing with time. Comparisons between several methods revealed that a similar quantification of synchronization was obtained using distinct analytical methods (Lachaux et al., 1999; Quiroga et al., 2002). Hence, the choice will depend upon the consideration of the nature of the experimental measurements (what they represent, acquisition method, etc.) and what correlation is sought (neuronal spiking, synaptic potentials, or more abstract representations of the activity in a particular state space...). As Kreuz et al. (2007) recently advised: "the measure to be applied to a certain task can not be chosen according to a fixed criterion but rather pragmatically as the measure which most reliably yields plausible information in test applications... although certain

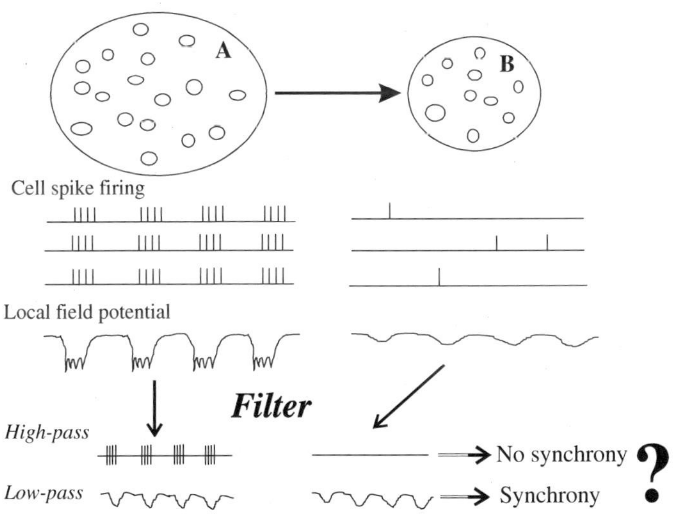

Fig. 1 Schematic representation of ambiguities arising from data processing in synchrony analysis

dynamical features... may render certain measures more suitable". In sum, while it is not a trivial matter, adequate methods exist to extract information from brain recordings that are appropriate for further analysis of synchronization and pattern formation in neuronal ensembles.

4 Expectations from Synchrony Analysis of Neural Signals

In the field of cognitive neuroscience, experiments are devoted to finding correlations between brain activity derived from the brain signals recorded and particular behavioral tasks, normally those that involve perception or motor actions. As already mentioned, coordination between brain regions is to be expected. Therefore, the end result may be a matter of the degree of synchrony. For those who wish to start this type of research, there is, at the least, one fundamental aspect that should be carefully considered and acknowledged to avoid false expectations. This point is that there must always be some type or another of coordinated activity in brain, nicely illustrated in the words of W. Klimesch (1999): "It is a matter of resolution, whether or not we may speak of synchronization or desynchronization. Even if the EEG desynchronizes, a large number of different networks may still show synchronous oscillations on a microscale level" (this comment is related to the synchrony at different timescales mentioned in the previous section); and in those of Trujillo et al. (2005): "These considerations suggest that neural synchrony should be present when any stimulus is perceived, although perhaps to a differential extent in magnitude or topography". Both comments stress the importance of the perspective that the investigator wishes to take in the determination of neuronal synchrony. If cognitive processes result in the transient formation and dissolution of cellular ensembles in disparate brain areas, then we can expect to find fluctuating patterns of synchronization, and thus even when the brain (the participant in a study) is idle, it is erroneous to think that there will be no synchrony at all. There will be synchrony, but perhaps not as marked as that found during the performance of particular cognitive tasks. Related to this, it should be mentioned that there is an extensive neuroimaging literature on the "default system" of the brain, a set of brain areas that is preferentially active when individuals are idle and not focused on the external environment (Buckner et al., 2008). This set of areas exhibits a slow waxing and waning of activity, as measured by functional magnetic resonance imaging (fMRI). Thus, even at rest, some brain areas are coordinating their activity in a significant manner. And, finally, always beware of the experimental setting to understand where your signals come from, since even concentrated mental activity produces gamma band activity recorded in scalp EEG... that is mostly unrelated to the neural tissue! Gamma frequency enhancement during thinking is mainly due to muscle activity under the EEG electrodes, which transmit both electroencephalographic and electromyographic signals (Whitham et al., 2008). Thus, because synchrony is present in "background" brain activity, we must rely on correlations between stimuli presentation and the magnitude of synchrony (Lachaux et al., 2000). Hence, it is all a matter of degree, perspective, or resolution.

We can in fact speak of several types of coordinated activity or synchronization. Figure 1 illustrates this point using very simple concepts. Cell network A is synaptically connected and sends input to network B. Assume the spike firing in network A is synchronous, as depicted in the figure, due to a strong synaptic input from other networks, but that the input from A to B is not enough to evoke consistent firing activity in B, thus resulting in a few unsynchronized spikes in several B cells. Depicted in the figure are the individual cell recordings (spikes) and the local field potentials (LFPs) in both areas. Because we know that local field potentials mostly represent the summation of synaptic activity, these recordings will look like those shown in the figure, with some small amplitude high-frequency activity in the LFP from A because we assumed these cells fire spikes synchronously and therefore there is a better chance to be detected in the LFP. To perform synchrony analysis, some filtering of the signals is normally used, so let us assume, for simplicity, that we subject the signals to a high-pass and a low-pass filter. The resultant traces are shown in the figure, and the comparison between both will reveal synchrony in the case of low-pass filtered signals, and no synchrony in the other case. What synchrony, then, if any, is present in this simple example? The answer depends on what we would like to consider: if we consider synchronization between the synaptic potentials received in both areas, then there is synchrony (phase locking to be more precise, because there will be a lag due to synaptic delay and axonal spike propagation) because both areas are receiving synaptic potentials of enough intensity to be detected as LFPs. If we consider spikes, then there is little synchrony. The moral of this very simple tale is that it is worth mentioning, in each particular study, what synchrony is most likely to be expected and analysed. For the majority of EEG or MEG recordings, it is reasonable to assume that correlation between synaptic potentials is what is measured. If individual spikes from several neurons are detected using tetrodes or other methods, then it is feasible to analyze spike correlations using joint peristimulus time histograms (PSTH), for example (Nicolelis, 1999).

5 Notes on the Physical Interpretation of the Computed Synchrony in Neural Systems

If we assume that processing information relies on the coordination of activity in disparate brain areas, then it is reasonable to apply methods to estimate synchrony, even though this "synchrony" may have little to do with the physicomathematical definitions mentioned in the first paragraphs of this chapter, and may convey no information about specific mechanisms that take place. Common criticisms of synchrony studies of brain activity stem from difficulties with the physiological interpretation of the computed synchrony in terms of the physicomathematical descriptions of the synchronization process. Consider, for instance, that a subcortical area (the thalamus, for example) is receiving sensory input and is relaying this information (basically as trains of action potentials) to two separate neocortical areas that, for the sake of this thought experiment, are not mutually connected. The final result from our synchrony analysis using recordings in these two cortical areas will reveal high synchronization in their activity because they receive a common strong input from the thalamus, even though it has nothing to do with their own corticocortical interactions. Nevertheless, one piece of information we seek is not whether these areas are synaptically interacting and adjusting their activity thus (which would represent the physical notion of the synchronization process as mentioned in the introduction), but rather whether *these areas are involved in processing that sensory stimulus*, which in fact they are because of the common drive from the thalamus that is relaying the sensory information. In sum, by studying the synchrony between neocortical areas we may not be sure of their functional, direct connectivity, but we will gain insight as to whether these cortical regions are participating in the processing of information. We can have a rough estimate of the possible degree of functional connectivity by considering the anatomical connections between specific cortical areas and the results of other neurophysiological experiments that may have been done in the study of synaptic interactions. But in the end, whether we care about this or whether we are content with just knowing, to some degree, if disparate areas are involved in processing particular stimuli, is a matter best left up to the science practitioner. Some investigators are clear in what they seek and want to show: take for example, Breakspear et al. (2004) "we use phase synchronization statistics as indices of information processing, regardless of the exact nature of their cause".

Experimental methodologies are also worthy of serious consideration in the interpretation of the analysis of synchronization. As mentioned above, most EEG, MEG, or local field potentials provide us information about synaptic potentials rather than action potential (spike) firing. In addition, it is well known that EEG recorded from the scalp suffers from the volume conduction predicament: coherence of scalp EEG signals was shown to be prominent even at distances of 20 cm (Nunez, 1995). A look at Fig. 1 in Nunez et al. (1997) reveals that more than 95% of the potential measured by one EEG electrode is generated by cortical sources within 6 cm of the electrode's centre. The Nunez et al. (1997) paper addresses in some detail many of these questions that are germane to synchrony (coherence) analysis using data obtained from scalp EEG recordings. The use of a reference electrode in EEG makes synchrony analysis interpretation even more complicated, as the referential montage will show spurious synchrony due to the common reference (Fein et al., 1988; Nunez et al., 1997; Guevara et al., 2005). MEG is not free from problems either: while there is no reference electrode problem and volume conduction is a lesser issue, summation of magnetic fields can occur at the MEG sensors, and depending on the amplitude of these fields, the derived synchrony will be, to some extent, not real, as depicted in Fig. 2 below (details in Garcia Dominguez et al., 2007). These considerations are detailed in the methodological sections below.

In cases where phase synchrony is evaluated, inspecting the differences in the angles of the phases may reveal some dubious results. For instance, if two distant brain areas are highly synchronized and have angle differences centered at 0, this suggests either a common input drive from other areas, as discussed above, or a

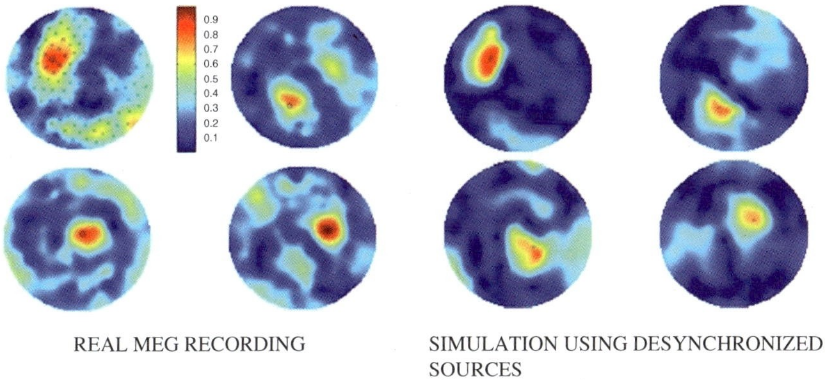

Fig. 2 Diagrammatic, color-coded mapping of synchronization values between the sensor marked with a *circle* (c-sensor) to each of the sensors marked with *crosses* (x-sensors). The *crosses* are shown only for the first graph (*upper left corner*) and omitted in the rest for clarity. Four different c-sensors are selected to show that the pattern of high synchronization with the neighboring channels is independent of the sensor selection. The space between the sensors has been filled with *color* by interpolation over adjacent synchrony values. The four patterns on the *left* are taken from real MEG data during an epileptic seizure in a patient, and the four on the *right* are obtained by a mathematical simulation of randomly positioned *unsynchronized sources*, where only the effect of field superposition has been taken into account. Reprinted, with permission, from Garcia Dominguez et al. (2007)

contamination with the common reference signal (if scalp EEG was used), or even a summation of magnetic fields detected by MEG gradiometers, in the case of very high amplitude signals in one area (Garcia Dominguez et al., 2007). Nevertheless, it is fair to note that zero-lag synchronization has been shown to occur in neuronal models if the dynamics of two delay-coupled oscillators are relayed through a third element connected to both (Fischer et al., 2006). Hence, while zero-lag synchrony could be suspicious, in the end it is often difficult to be certain what is going on. In the sections below, we explore a few extra steps in the analysis of phase synchronization that can be taken to address these potential problems.

6 The Significance of Considering Fluctuations in Coordinated Cellular Activity

We would like to bring attention to a point that, so far, remains seldom addressed in brain synchrony studies. The aforementioned notions about brain function imply that coordinated activity among brain networks *should be transient* for efficient information processing to take place. The dynamics of nervous systems, especially at higher levels, are determined not by long-lasting steady states but by continuous omnipresent fluctuating patterns of activity at almost all levels of nervous system function. Hence, while the majority of studies focus on measuring the amplitude (strength) of synchronization, examination of the fluctuations in synchronization is as fundamental as knowing the magnitude of the synchrony between two networks. Brain function seems to rely on the very transient establishment of synchronization and desynchronization patterns. Hans Flohr (1991) suggested that degrees of consciousness differ in the rate at which cell assemblies are generated, a sort of "production rate" of brain cell assemblies that determines the complexity and duration of the representations of sensory patterns. There are other proposals supporting functional roles of fluctuating transitions in synchronization (Engel et al., 2001; Varela et al., 2001; Lutz et al., 2002; Fries, 2005). This is clearly illustrated in the words of the late Varela and colleagues: "One needs to study neurons as members of large ensembles that are constantly disappearing and arising through their cooperative interactions and in which every neuron has multiple and changing responses in a context-dependent manner" (Varela et al., 1991). In coupled model systems, a variety of synchronous states have been noted in the transition between cellular firing patterns (Postnova et al., 2007). The comment of Nikolai Bernstein is still worth taking into consideration, in an era in which so many experiments study brains in relative isolation: "The classical physiology of the last hundred years is characterized by ... the study of the operation of the organism under quiescent inactive conditions ... under conditions of maximal isolation from the external world" (Bernstein, 1961). Brain activity cannot be isolated from the ongoing internal and external fluctuating environments (McIntosh, 2004). While in the past the recognition of variability, in general, has been uncommon to say the least (Fatt and Katz, 1950; Smith and Smith, 1965; Adey, 1972), the current situation is more favorable to a serious consideration of variability in neuroimaging and other neuroscientific experiments. This perspective change was brought about, in part, by a number of studies that stressed the importance of ongoing "background" activity (Arieli et al., 1996; Lutz et al., 2002; Kenet et al., 2003), as well as by the finding of the constructive nature of noise in signal processing in nervous systems. For instance, more efficient processing occurs for certain noise levels, which in signal processing is associated with the phenomenon of stochastic resonance, where noise improves the response of a nonlinear system (Moss et al., 2004). Hence, the classical view regarding noise as a general degrader of information processing in nervous tissue is shifting toward recognition of the importance of fluctuating and spontaneous neural activity. Fluctuations in synchrony were detected even during seizure activity (Perez Velazquez et al., 2007b), an instance where we could expect a robust synchronization rather than variability due to the accepted nature of a highly synchronous activity during seizures. In fact, spatiotemporal synchrony patterns were found to be variable (Aarabi et al., 2008) and unique to each patient (Schevon et al., 2007). Comments on stochastic approaches to synchronized activities can be found in Perez Velazquez (2006). In general, science practitioners take good care to eliminate variability by methods such as averaging, but the study of that variability, in and of itself, may reveal novel aspects of brain function.

While the microscopic sources of noise in neural systems are not completely defined (since these are multitude, ranging from fluctuating ion channel opening to synaptic variability), nevertheless, some fluctuations can be reasonably described at certain higher (macroscopic) levels of description. By comparison with fluid dynamics, it is conceived that phase differences (phase-locking values) undergo an analogous process to Brownian motion (Haken, 2002), and thus a phase diffusion coefficient has been defined and used in some studies to assess the correlation of fluctuations in phase synchrony with behavioral outcomes like perception (Perez Velazquez et al., 2007a). The phase diffusion coefficient represents the fluctuations in phase synchronization (Neiman et al., 1998) and is used as a measure to determine the spatiotemporal dynamics of the transient formation and dissolution of neuronal assemblies in cellular activity. Other methods to assess the dynamics of phase synchronization, based on estimating the time derivative of the phase differences to document the dynamics of synchronization and desynchronization, have been proposed (Breakspear et al., 2004). At the same time it should be noted that neuronal synchrony is affected by noise (Zorzano and Vazquez, 2003; Ward et al., 2006; McMillen and Kopell, 2003; Pakdaman and Mestivier, 2004). These matters are elaborated in Perez Velazquez et al. (2007b).

7 Neural Information Processing and Coordination of Activity

Nervous system operations are based on the transfer of information, which is normally achieved via action potentials (even though this may seem a simplistic concept of "information" in neural function, and indeed the concept could be widened to include other relevant features). Since information processing relies on the stable propagation of action potentials, this implies that some sort of coordinated (correlated) activity, call it synchronized or not, must exist in nervous systems, because most neurons need a relatively synchronous synaptic input from many other cells to depolarize enough to reach the threshold for spike generation (Alonso, 2006). Hence, the existence of synchronized activity seems to be fundamental for proper brain function. Suffice to recall the distributed nature of memory and other cognitive processes (Goldman-Rakic, 1988; Brown and Marsden, 2001; Fuster, 2003), where several brain areas are active at the same time and, presumably, coordinate their activities. Substantial empirical evidence indicates the importance of the serial triggering of activity in connected cell assemblies (Castelo-Branco et al., 1998; Kimpo et al., 2003; Harris, 2005). Consider again the network in Fig. 1, and the experimental recordings that can be either cell spike firing or local field potentials. The question arises as to how many cells in set B need to be active to pass information onto another network down the line. Recording simultaneously from many individual cells could help solve the query, but it is usually not feasible to record from many at the same time, as only a sample can be obtained even using tetrodes, for example. If we record field potentials, we are never sure whether or not individual cells are firing as the field potential represents mostly postsynaptic activity. Then, is it feasible to study the transfer of (neural) information using methods to characterize coordinated activity?

Computer simulations shed some light onto these questions. The impact of synchronization on the transmission of information has been studied in model networks, which indicated that the capacity of information transfer depends on whether a channel is self-excitable or not (Baptista and Kurths, 2008), an important observation for neural tissue as cells are considered self-excitable. While it is certainly conceivable that synchronous activity will improve reliability of information transfer (spikes), the study by Baptista and Kurths (2008) showed that, in particular cases of self-excitable information channels, *desynchronization* increases the robustness of transfer of larger amounts of information: when neurons synchronize in their slow timescale (bursting) and desynchronize in the fast timescale (spiking) is when the channel has the maximal capacity for information transmission. This is important because in nervous systems the constituents are dynamical systems themselves, and something that needs to be considered is the impact of the unit's internal dynamics on the synchronicity, which has been addressed in some studies (Hansel and Sompolinsky, 1996; Makarov et al., 2001; Postnova et al., 2007; Galan et al., 2007). These themes are obviously related to the comments on fluctuations in correlated activity in the previous section. From a more theoretical perspective, it can be stated that synchrony resides on a torus, the synchronization manifold, which is relatively stable to temporal fluctuations in neuronal activity, and thus the transmission of spikes occurs reliably (Diesmann et al., 1999). Other theoretical investigations have addressed the topological structure of the phase locking (Glass and Perez, 1982) studies that are not easy to derive from experimental recordings.

To understand the impact of coordinated activity on neural information processing it is important to study the recruitment of cells into patterns of activity, and the influence of the degree of synchrony present in the inputs on the recruitment of neurons, themes that have been studied using geometric approaches, which can be seen as a somewhat abstract level of description, but useful in many instances. Geometric conditions (revealed using one-dimensional maps of cellular activity in some state space) were found to determine whether specific temporal relations in the inputs to a network will lead to the network's recruitment into a localized activity pattern (Rubin and Bose, 2006).

To complicate matters, coupling between units in nervous systems is variable: synapses are "plastic", that is, synaptic activity alters synaptic strength. Could this influence coordinated activity? Apparently yes: enhancement of synchronization by activity-dependent synaptic coupling has been demonstrated using a mapping function that described the generality of the phenomenon (Zhigulin et al., 2003).

A conceptual framework that may be particularly fruitful for understanding the relation between patterns of coordinated activity and information processing is self-organized criticality. These ideas originate in proposals that living systems self-organize to operate near critical points (Langton, 1990; Kauffman, 1993). From a dynamical perspective, synchronization is one fundamental mechanism of self-organization in complex systems. Self-organized criticality in nervous system function is now being studied in models (Levina et al., 2007) and experimentally (Beggs and Plenz, 2004). There is support for the idea that neuronal systems operate close to dynamical bifurcation points, at the edge of stability, both in computer simulations and in theoretical analyses of experimental brain recordings (Makarov et al., 2001; Perez Velazquez and Wennberg, 2004; Perez Velazquez, 2005). Relations between some characteristics of nervous systems such as anatomy and the optimal organization for information processing have been sought using graph theory, particularly the "small-world" approach. In this regard, some studies have focused on the analysis of synchrony patterns using the random graph approach. Results based on MEG recordings indicated that, depending on the frequencies at which synchronization was evaluated, the brain synchrony patterns had characteristics typical of small-world networks, especially for low and high frequencies (Stam, 2004). This observation is of interest because the small-world network has been related to optimal information processing. The general significance of these results, for nervous systems, is that in these networks a small number of synapses will connect two neuronal networks, and the frequency dependence may indicate that the resulting synchronization processes may differ depending upon the frequency of the oscillation, as previously commented upon when the timescales of synchronization were discussed. Correlations in synaptic potentials, which depend on frequencies as well, establish temporal windows for the response of the receiving neurons, as detailed in Fries (2005). The spike firing patterns that result from the subtle interaction between subthreshold membrane potential oscillations and incoming synaptic inputs have been explored in vitro and in vivo (Hölscher et al., 1997; Volgushev et al., 1998). Considering that short re-entrant loops are associated with complex dynamics in networks (reviewed in Tononi et al., 1998a) and that long loops may facilitate global coherency (Sporns et al., 2000), an answer to the question posed above (about the need for functional integration and segregation for proper brain function) can start to be envisaged. In networks of model neurons using the Hodgkin–Huxley formalism, small-world topologies provided the system with *fast responses* associated with *coherent oscillations* (Lago Fernandez et al., 2000). This observation can be rephrased in the vocabulary of the concepts proposed by Tononi and Edelman (2000), as optimal integration (*fast response*) and segregation (because *coherent oscillations* can integrate sub-networks into a functional network in a principal frequency band and segregate others with different frequencies); hence small-world networks possess higher complexities than either randomly or completely regularly connected ensembles. There may be a critical level of connectivity in brain networks optimal for brain function (Beggs, 2008), suggestions that have originated mostly from computer models, with some empirical evidence as well (reviewed in Beggs, 2008). The relation between criticality and the studies on metastability and dynamical bifurcations in neural systems should not be underestimated (Perez Velazquez and Wennberg, 2004).

The following sections deal with technical considerations that are important when studying synchronization derived from neurophysiologic brain signals, specifically phase synchrony.

8 Some Methodological Considerations on the Assessment of Phase Synchronization

In previous sections, synchronization was defined from the point of view of signal processing as a measure of the degree of phase difference stability between two signals over a given time window. Any periodic motion or motion that repeats itself after a fixed period can be described by its phase. If we take a point on the oscillation as phase zero, any other point of the loop would have a phase equal to the fraction of time that has passed since departure from this zero-phase point relative to the period of the oscillation. This ratio is frequently expressed in radians where 2π is made equivalent to the period of the oscillation. Notice that the specific phase of an oscillation depends on the definition of this zero-phase point or the initial condition from which the movement has started or been recorded. However the synchronization measure does not depend on these ambiguities as it is only based on phase differences and not the phase itself. Regardless the starting points of two oscillations, they are considered to be synchronized if their phase differences remain constant during a period of time. In order to maintain a constant phase difference they must oscillate at the same angular speed, in other words, they have to move at the same instantaneous frequency.

Phase and phase differences can be defined as bounded in the interval $[0, 2\pi]$. Consequently a phase difference can be geometrically represented as a point on the unit circle and a sequence of the phase differences recorded during a period of time as a distribution of points on the unit circle. If these phase differences are relatively constant, this distribution must show a distinctive sharp single peak. On the other hand, under completely uncorrelated phase relationships, an even distribution of values is expected. These differences can be conveniently represented by a single parameter: the variance of the distribution or the circular variance (CV), since the distribution is defined in the unit circle. This parameter is maximal ($\text{CV} = 1$) when the distribution is even and minimal ($\text{CV} = 0$) when the distribution consists of a single value. Since maximal synchronization corresponds to minimal variance, it is more convenient to express synchronization as 1-CV (Mormann et al., 2000).

The formula that expresses this new parameter is given by

$$R = \left|\left\langle e^{(i\Delta\alpha)}\right\rangle\right|$$

where $\Delta\alpha$ is the series of phase differences, $< >$ the average, and $|\ |$ the absolute value of the complex vector $e^{(i\Delta\alpha)}$. This is known as the "mean phase coherence statistic" mentioned previously, which has also received other names in different works. This parameter is not the only one currently in use in neural synchrony research, however, in practice they are all equivalent.

As discussed in previous sections, neurophysiological recordings in EEG are far from being periodic. How then can the phase of the oscillation be defined objectively? To accomplish this, we should be guided by the specific question we would like to answer. Sometimes there is a distinct high amplitude frequency we may wish to explore. In this case, typically, the series are band-passed at the specific central frequency. In other situations, we may just explore a wide range of frequencies. Here also a band-pass must be applied at a number of frequencies inside the range. The importance of this operation is that band-passed signals, like purely sinusoidal waves, have a better defined instantaneous phase. Importantly, the filter must be one that does not produce phase distortions. Still, real band-passed signals are not ready to unambiguously produce phase values. Another trick is in place, a certain transformation to map each real time series into a complex one, basically embedding the unidimensional oscillation into a two-dimensional rotation. The phase of this rotation is unequivocally obtained as the angle of the complex movement around the origin. There are a number of ways that this two-dimensional representation can be achieved; the most popular is the Hilbert transform, a transformation that produces a complex series from a real one with identical power spectrum. The real part of this new complex number is the original time series and the imaginary part is simply a shifted version of this same series. Technical details can be found in many publications (Pikovsky et al., 2001; Anishchenko and Vadivasova, 2002).

These series of phase differences can now be subtracted between the two oscillators and evaluated in the previous equation for R.

Does the band-pass alter the value of synchronization? Of course it does, a band-passed signal produces a narrower set of instantaneous frequencies. This results, usually, in an increased value of synchronization compared to the original time series. The band-pass solves a problem and creates another. Without a band-pass, the instantaneous phases of most neurophysiological signals are not well defined (fluctuating back and forth incoherently). With the band-pass, the number of instantaneous frequencies narrows, increasing the likelihood of synchronization. However this last problem can be correctly addressed by testing the synchronization values obtained against those obtained from a set of surrogate signals. For example, a set of signals can be constructed with identical power spectrum as the originals (but otherwise uncorrelated) and then tested to see whether the synchronization values obtained with this set of signals, after the application of the same band-pass, are significantly different from those obtained with the originals. If the actual synchronization values tend to be significantly higher than those of the surrogates that should indicate the presence of a phase relationship that is not explained by chance. The band-pass effect is now eliminated since it has been applied in identical conditions to both data sets, original and surrogate.

Are we then ready to apply this algorithm to neurophysiological signals? Well, this is a complex question that is being currently debated. The problem can be better understood by noting the difference between the following two scenarios: do we want to know if two *signals* are synchronized or whether two *brain regions* are synchronized. The first question (synchrony between signals) can be immediately solved by applying the synchronization analysis just as has been presented. On the other hand, if we want to make physiological inferences about synchronization between brain areas, we must be concerned about the presence of a number of distortions of physical nature. Let us start by analyzing a simple case, the synchronization derived from scalp EEG recordings. It is always the case that the recording of each electrode of a scalp EEG is subtracted from the recording of a second electrode, or a combination of electrodes. Take the case where this second electrode (reference electrode) or combination of electrodes is the same for every other electrode. A natural question arises, is the synchronization measured from electrodes e and e' undisturbed by a common reference r? Or in a more clear fashion: does the following equality hold?

$$R(e, e') = R(e - r, e' - r)$$

where $R(x, y)$ means synchronization between signals recorded directly from electrodes x and y.

The answer is no. In fact, synchronization is greatly distorted to the extent that it is not possible to use the synchronization under this reference montage as an estimator of the actual one. The interested reader can find more details in Guevara et al. (2005). The limitations that a common reference electrode may impose when performing coherence analysis using EEG recordings have been addressed previously (Fein et al., 1988; Nunez et al., 1997; Zaveri et al., 2000), indicating that the activity recorded by the reference channel will significantly affect the coherence measurements. To overcome this problem, the synchronization must be estimated after an appropriate post-recording transformation capable of eliminating this referential scheme, provided it does not introduce a new distortion! The Laplacian derivation is one method (Nunez et al., 1997; Winter et al., 2007). Note that the distortion is produced by the amalgamation of two signals. We can still fairly ask: what is the synchronization of those traces we are observing? But that is not equivalent to asking the (physiological) question: what is the synchronization between the areas just below the two electrodes, since there is a third brain area (the one below the reference) distorting the original time series. Further comments on these queries can be found in Shaw et al. (1979), Nunez (1981), and Nunez et al. (1997).

A similar problem arises in a more general context. In MEG recordings, although not a referential montage, there is however one phenomenon that is equivalent. Each magnetic sensor records the magnetic field generated not only in the local neighborhood of the cortical structures just below, but also receives important contributions from more distant activity. The strength of the magnetic field decays approximately as the inverse of the square of the distance between the generator and the sensor. If the distance from a given sensor located very close to an area A is d and to another area A' is d', we can express the relative contributions or weights of these two brain areas to the specific signal recorded by the sensor as $(d/d')^2$. If this MEG sensor were to be more displaced from the head, then both distances increase in approximately the same magnitude and the relative contributions of both become more even. In general, the more distance between the sensors to the sources of activity the more integration from many sources occur at each place of measurement, since their specific weights to each sensor become more similar. This is the case of MEG where the recording sensors are more than 2 cm away from the cortical surface, producing substantial integration from a widespread region. Thus, two nearby sensors are likely to integrate signals from two sets of neural sources that largely overlap. As a consequence, their respective recordings are similar in structure, producing spurious high synchronization.

This phenomenon is demonstrated in Fig. 2. Note that we were able to reproduce an extremely similar pattern of synchronized activity to the one obtained from real MEG recordings, assuming only that the head is filled with a number of discrete *desynchronized* sources. The details of this and similar experiments can be found in Garcia Dominguez et al. (2007). Hence, by simulating signal superposition at the MEG sensors, an "epileptic pattern" derived from completely unsynchronized sources, indistinguishable from a real seizure pattern, is produced!

This phenomenon of source superposition is not exclusive to MEG recordings; it is present also in EEG in addition to the aforementioned EEG reference electrode problem. Other complications in scalp EEG recordings are associated with the phenomenon of "volume conduction" which includes effects due to passive charge movement as a consequence of the activity of neural generators (neurons and sets of neurons). This last effect has more weight in EEG recordings and tends to "blur" actual synchronization values.

9 Conclusions

Complexity arises not because of the great number of constituents in a system but from the manner in which these elements are interconnected and interacting. Hence, an understanding of the brain's apparent complexity hinges on the characterization of the interactions amongst brain elements. As Friston and colleagues proposed, a "better metric of neuronal interactions could be framed in terms of dynamic changes in correlations" (Friston et al., 1997). Several methodologies that can extract information about the dynamic establishment of coordinated activity in neural systems have been reviewed here. While these analytical tools provide us with many measures of coordinated activity, the results must always be carefully interpreted in a physiological sense, because there are different types of coordinated activity that may occur at different levels of description. These are all equally significant for brain function, as an activity pattern at one level (spike firing for instance) may result in different activity patterns at other levels (for example, synchronous oscillations).

Acknowledgments Our research is supported by the Natural Science and Engineering Research Council of Canada (NSERC), the Hospital for Sick Children Foundation, and the Bial Foundation.

References

Aarabi A, Wallois F, Grebe R (2008) Does spatiotemporal synchronization of EEG change prior to absence seizures? Brain Res 1188:207–221.

Adey WR (1972) Organization of brain tissue: is the brain a noisy processor? Int J Neurosci 3: 271–284.

Afraimovich VS, Verichev NN, Rabinovich MI (1986) Stochastic synchronization of oscillations in dissipative systems. Izv Vyssh Uchebn Zaved Radiofiz 29:795–803.

Alonso JM (2006) Neurons find strength through synchrony in the brain. Science 312:1604–1605.

Anishchenko VS, Astakhov VV, Neiman AB, Vadivasova TE, Schimansky-Geier L (2002) *Nonlinear Dynamics of Chaotic and Stochastic Systems,* Springer, Berlin.

Anishchenko VS, Vadivasova TE (2002) Synchronization of self-oscillations and noise-induced oscillations. J Commun Technol Electron 47:117–148.

Arieli A, Sterkin A, Grinvald A, Aertsen A (1996) Dynamics of ongoing activity: explanation of the large variability in evoked cortical responses. Science 273:1868–1871.

Bahar S, Moss F (2003) Stochastic phase synchronization in the crayfish mechanoreceptor photoreceptor system. Chaos 13:138–144.

Baptista MS, Pereira T, Sartorelli JC, Caldas IL, Kurths J (2005) Non-transitive maps in phase synchronization. Physica D 212:216–232.

Baptista MS, Kurths J (2008) Transmission of information in active networks. Phys Rev E 77 (2 pt 2):026205.

Beggs JM, Plenz D (2004) Neuronal avalanches are diverse and precise activity patterns that are stable for many hours in cortical slice cultures. J Neurosci 24:5216–5229.

Beggs JM (2008) The criticality hypothesis: how local cortical networks might optimise information processing. Phil Trans R Soc A 33:329–343.

Bernstein N (1961) Trends and problems in the study of physiology of activity. In Whiting HT (Ed) *Human Motor Actions: Bernstein Reassessed*, North Holland, Amsterdam, 1984.

Binkofski F, Buccino G, Posse S, Seitz RJ, Rizzolatti G, Freund H (1999) A fronto-parietal circuit for object manipulation in man: evidence from an fMRI study. Eur J Neurosci 11:3276–3286.

Breakspear M, Terry JR (2002) Nonlinear interdependence in neural systems: motivation, theory and relevance. Int J Neurosci 112:1263–1284.

Breakspear M, Williams LM, Stam CJ (2004) A novel method for the topographic analysis of neural activity reveals formation and dissolution of "dynamic cell assemblies". J Comp Neurosci 16:49–68.

Bressler SL, Kelso JAS (2001) Cortical coordination dynamics and cognition. Trends Cogn Sci 5:26–36.

Brown P, Marsden JF (2001) Cortical network resonance and motor activity in humans. Neuroscientist 7:518–527.

Buckner RL, Andrews-Hanna JR, Schacter DL (2008) The brain's default network. Anatomy, function and relevance to disease. Ann N Y Acad Sci 1124:1–38.

Castelo-Branco M, Neuenschwander S, Singer W (1998) Synchronization of visual responses between the cortex, lateral geniculate nucleus, and retina in the anesthetized cat. J Neurosci 18:6395–6410.

Chavez M, Adam C, Navarro V, Boccaletti S, Martinerie J (2005) On the intrinsic time scales involved in synchronization: a data-driven approach. Chaos 15:023904.

Damasio AR (1989) The brain binds entities and events by multiregional activation from convergence zones. Neural Comput 1:123–132.

Dehaene S, Changeux JP, Naccache L, Sackur J, Serhgent C (2006) Conscious, preconscious, and subliminal processing: a testable taxonomy. Trends Cogn Sci 10:204–211.

Diesmann M, Gewaltig M-O, Aertsen A (1999) Stable propagation of synchronous spiking in cortical neural networks. Nature 402:529–532.

Eckhorn R (2000) Cortical processing by fast synchronization: high frequency rhythmic and non-rhythmic signals in the visual cortex point to general principles of spatiotemporal coding. In Miller R (Ed) *Time and the Brain*, Harwood Academic Publishers, Amsterdam.

Engel AK, Fries P, Singer W (2001) Dynamic predictions: oscillations and synchrony in top-down processing. Nature Rev Neurosci 2:704–716.

Erb M, Aertsen A (1992) Dynamics of activity in biology-oriented neural network models: stability at low firing rates. In *Information Processing in the Cortex*, Springer, Berlin.

Fatt P, Katz B (1950) Some observations on biological noise. Nature 166:597–598.

Fein G, Raz J, Brown FF, Merrin EL (1988) Common reference coherence data are confounded by power and phase effects. Electroencephalogr Clin Neurophysiol 69:581–584.

Fischer I, Vicente R, Buldu JM, Peil M, Mirasso CR, Torrent MC, Garcia-Ojalvo J (2006) Zero-lag long-range synchronization via dynamical relaying. Phys Rev Lett 97:123902.

Flohr H (1991) Brain processes and phenomenal consciousness. A new and specific hypothesis. Theory & Psychology 1:245–262.

Fries P (2005) A mechanism for cognitive dynamics: neuronal communication through neuronal coherence. Trends Cogn Sci 9:474–480.

Friston KJ, Stephan KM, Frackowiak RSJ (1997) Transient phase-locking and dynamic correlations: are they the same thing? Hum Brain Mapp 5:48–57.

Friston KJ (2001) Brain function, nonlinear coupling, and neuronal transients. Neuroscientist 7:406–418.

Fujisaka H, Yamada T, Kinoshita G, Kono T (2005) Chaotic phase synchronization and phase diffusion. Physica D 205:41–47.

Fuster JM (2003) *Cortex and Mind*. Oxford University Press, Oxford.

Galán RF, Fourcaud-Trocmé N, Ermentrout GB, Urban NN (2006) Correlation-induced synchronization of oscillations in olfactory bulb neurons. J Neurosci 26:3646–3655.

Galan RF, Ermentrout GB, Urban NN (2007) Stochastic dynamics of uncoupled neural oscillators: Fokker-Planck studies with the finite element method. Phys Rev E 76:056110.

Garcia Dominguez L, Wennberg R, Gaetz W, Cheyne D, Snead OC III, Perez Velazquez JL (2005) Enhanced synchrony in epileptiform activity? Local versus distant phase synchronization in generalized seizures. J Neurosci 25:8077–8084.

Garcia Dominguez L, Wennberg R, Perez Velazquez JL, Guevara Erra R (2007) Enhanced measured synchronization of unsynchronized sources: inspecting the physiological significance of synchronization analysis of whole brain electrophysiological recordings. Int J Phys Sci 2: 305–317.

Glass L, Perez R (1982) Fine structure of phase locking. Phys Rev Lett 48:1772–1775.

Glass L (2001) Synchronization and rhythmic processes in physiology. Nature 410:277–284.

Goldman-Rakic P (1988) Topography of cognition: parallel distributed networks in primate association cortex. Ann Rev Neurosci 11:137–156.

Gomez-Gardenes J, Moreno Y (2007a) Synchronization of networks with variable local properties. Int J Bif Chaos 17:2501–2507.

Gomez-Gardenes J, Moreno Y, Arenas A (2007b) Synchronizability determined by coupling strengths and topology on complex networks. Phys Rev E 75:66106.

Gomez-Gardenes J, Moreno Y, Arenas A (2007c) Paths to synchronization in complex networks. Phys Rev Lett 98:34101.

Guevara R, Pérez Velazquez JL, Nenadovic V, Wennberg R, Senjanovic G, García Dominguez L (2005) Phase synchronization measurements using electroencephalographic recordings: what can we really say about neuronal synchrony? Neuroinformatics 3:301–313.

Haken H (2002) *Brain Dynamics*. Springer Verlag, Berlin.

Hansel D, Sompolinsky H (1996) Chaos and synchrony in a model of a hypercolumn in visual cortex. J Comp Neurosci 3:7–34.

Harris KD (2005). Neural signatures of cell assembly organization. Nature Rev Neurosci 6: 399–407.

Hölscher C, Anwyl R, Rowan MJ (1997) Stimulation on the positive phase of hippocampla theta rhythm induces long-term potentiation that can be depotentiated by stimulation on the negative phase in area CA1 in vivo. J Neurosci 17:6470–6477.

Hoppensteadt FC, Izhikevich EM (1997) *Weakly Connected Neural Networks*, Springer Verlag, Berlin.

Huygens C (1673) *Horologium Oscillatorium*, A.F. Muguet, Paris, France

Jirsa VK (2004) Connectivity and dynamics of neural information processing. Neuroinformatics 2:183–204.

John ER (2002) The neurophysics of consciousness. Brain Res Rev 39:1–28.

Kauffman SA (1993) *The Origins of Order: Self-Organization and Selection in Evolution*, Oxford University Press, Oxford.

Kelso JAS (1995) *Dynamic Patterns: The Self-Organization of Brain and Behaviour.* MIT Press, Cambridge, MA.

Kelso JAS, Engstrøm DA (2006) *The Complementary Nature.* MIT Press, Cambridge, MA.

Kenet T, Bibitchkov D, Tsodyks M, Grinvald A, Arieli A (2003) Spontaneously emerging cortical representations of visual attributes. Nature 425:954–956.

Kestler J, Kopelowitz E, Kanter I, Kinzel W (2008) Patterns of chaos synchronization. Phys Rev E 77:046209

Kimpo RR, Theunissen FE, Doupe AJ (2003) Propagation of correlated activity through multiple stages of a neural circuit. J Neurosci 23:5750–5761

Klimesch W (1999) EEG alpha and theta oscillations reflect cognitive and memory performance: a review and analysis. Brain Res Rev 29:169–195.

Kreuz T, Mormann F, Andrzejak RG, Kraskov A, Lehnertz K, Grassberger P (2007) Measuring synchronization in coupled model systems: a comparison of different approaches. Physica D 225:29–42.

Lachaux J-P, Rodriguez E, Martinerie J, Varela FJ (1999) Measuring phase synchrony in brain signals. Hum Brain Mapp 8:194–208.

Lachaux J-P, Rodriguez E, Le van Quyen M, Lutz A, Martinerie J, Varela FJ (2000) Studying single-trials of phase synchronous activity in the brain. Int J Bif Chaos 10:2429–2439.

Lago Fernandez LF, Huerta R, Corbacho F, Sigüenza JA (2000) Fast response and temporal coherent oscillations in small-world networks. Phys Rev Lett 84:2758–2761.

Langton CG (1990) Computation at the edge of chaos—phase transitions and emergent computation. Physica D 42:12–37.

Levina A, Herrmann JM, Geisel T (2007) Dynamical synapses causing self-organized criticality in neural networks. Nature Phys 3:857–860.

Li C-L (1959) Synchronization of unit activity in the cerebral cortex. Science 129:783–784.

Livanov MN (1977) *Spatial Organization of Cerebral Processes*, Wiley, New York.

Luria, AR (1966) *Higher Cortical Functions in Man*. Basic Books, New York.

Lutz A, Lachaux J-P, Martinerie J, Varela FJ (2002) Guiding the study of brain dynamics by using first-person data: synchrony patterns correlate with ongoing conscious states during a simple visual task. Proc Natl Acad Sci USA 99:1586–1591.

Makarov VA, Nekorkin VI, Velarde MG (2001) Spiking behavior in a noise-driven system combining oscillatory and excitatory properties. Phys Rev Lett 86:3431–3434.

McIntosh AR (2004) Contexts and catalysts. A resolution of the localization and integration of function in the brain. Neuroinformatics 2:175–181.

McMillen D, Kopell N (2003) Noise-stabilized long-distance synchronization in populations of model neurons. J Comp Neurosci 15:143–157.

Menendez de la Prida L, Sanchez-Andres JV (1999) Nonlinear transfer function encodes synchronization in a neural network from the mammalian brain. Phys Rev E 60:3239–3243.

Mormann F, Lehnertz K, David P, Elger CE (2000) Mean phase coherence as a measure for phase synchronization and its application to the EEG of epilepsy patients. Physica D 144: 358–369.

Moss F, Ward LM, Sannita WG (2004) Stochastic resonance and sensory information processing: a tutorial and review of application. Clin Neurophysiol 115:267–281.

Neiman A, Silchenko A, Anishchenko V, Schimansky-Geier L (1998) Stochastic resonance: noise-induced phase coherence. Phys Rev E 58:7118–7125.

Netoff TI, Schiff SJ (2002) Decreased neuronal synchronization during experimental seizures. J Neurosci 22:7297–7307.

Nicolelis, MAL (1999) *Methods for Neural Ensemble Recordings*, CRC Press, Boca Raton, FL.

Nunez PL (1981) A study of origins of the time dependencies of scalp EEG: ii—experimental support of theory. IEEE Trans Biomed Eng 28:281–288.

Nunez PL (1995) *Neocortical Dynamics and Human EEG Rhythms*, Oxford University Press, New York.

Nunez PL, Srinivasan R, Westdorp AF, Wijesinghe RS, Tucker DM, Silberstein RB, Cadusch PJ (1997) EEG coherency I: statistics, reference electrode, volume conduction, Laplacians, cortical imaging, and interpretation at multiple scales. Electroencephalogr Clin Neurophysiol 103:499–515.

Nunez PL (2000) Toward a quantitative description of large-scale neocortical dynamic function and EEG. Behav Brain Sci 23:371–437.

Pakdaman K, Mestivier D (2004) Noise induced synchronization in a neuronal oscillator. Physica D 192:123–137.

Pecora LM, Carroll TL, Heagy JF (1995) Statistics for mathematical properties of maps between time series embeddings. Phys Rev E 52:3420–3439.

Pecora LM, Carroll TL, Johnson GA, Mar DJ, Heagy JF (1997) Fundamentals of synchronization in chaotic systems, concepts, and applications. Chaos 7:520–543.

Perez Velazquez JL, Wennberg R (2004) Metastability of brain states and the many routes to seizures: numerous causes, same result. In Pandalai SG (Ed) *Recent Research Developments in Biophysics*, vol. 3, Transworld Research Network, Kerala.

Perez Velazquez JL (2005) Brain, behaviour and mathematics: are we using the right approaches? Physica D 212:161–182.

Perez Velazquez JL (2006) Brain research: a perspective from the coupled oscillators field. NeuroQuantology 4:155–165.

Perez Velazquez JL, Garcia Dominguez L, Guevara Erra R (2007a) Fluctuations in neuronal synchronization in brain activity correlate with the subjective experience of visual recognition. J Biol Phys 33:49–59.

Perez Velazquez JL, Garcia Dominguez L, Guevara Erra R, Wennberg R (2007b) The fluctuating brain: dynamics of neuronal activity. In Wang CW (Ed) *Nonlinear Phenomena Research Perspectives*, Nova Science Publishers, New York

Perez Velazquez JL, Garcia Dominguez L, Wennberg R (2007c) Complex phase synchronization in epileptic seizures: evidence for a devil's staircase. Phys Rev E 75:011922.

Pikovsky A, Rosenblum M, Osipov G, Kurths J (1997) Phase synchronization of chaotic oscillators by external driving. Physica D 104:219–238.

Pikovsky A, Rosenblum M, Kurths J (2001) *Synchronization: A Universal Concept in Nonlinear Sciences*, Cambridge University Press, Cambridge.

Popovych OV, Hauptmann C, Tass PA (2008) Impact of nonlinear feedback on synchronized oscillators. J Biol Phys 34, 367–379 DOI 10.1007/s10867-008-9068-1.

Postnova S, Voigt K, Braun HA (2007) Neural synchronization at tonic-to-bursting transitions. J Biol Phys 33:129–143.

Quiroga RQ, Kraskov A, Kreuz T, Grassberger P (2002) Performance of different synchronization measures in real data: a case study on electroencephalographic signals. Phys Rev E 65:041903.

Romano MC, Thiel M, Kurths J, von Bloh W (2004) Multivariate recurrence plots. Phys Lett A 330:214–223.

Romano MC, Thiel M, Kurths J, Kiss IZ, Hudson JL (2005) Detection of synchronization for non-phase coherent and non-stationary data. Europhys Lett 71:466–472.

Rubin J, Bose A (2006) The geometry of neuronal recruitment. Physica D 221:37–57.

Schevon CA, Cappell J, Emerson R, Isler J, Grieve P, Goodman R, McKhann G Jr, Weiner H, Doyle W, Kuzneiecky R, Devinsky R, Gilliam F (2007) Cortical abnormalities in epilepsy revealed by local EEG synchrony. NeuroImage 35:140–148.

Schiff SJ, So P, Chang T, Burke RE, Sauer T (1996) Detecting dynamical interdependence and generalized synchrony through mutual prediction in a neural ensemble. Phys Rev E 54: 6708–6724.

Schild D (1984) Coordination of neuronal signals as structures in state space. Int J Neurosci 22:283–297.

Shaw JC, Brooks S, Colter N, O'Connor KP (1979) A comparison for schizophrenic and neurotic patients using EEG power and coherence spectra. In Gruzelier JH and Flor-Henry P (Eds.) *Hemisphere Asymmetries of Function in Psychopathology*, Elsevier/North-Holland Biomedical, Amsterdam, pp. 257–284.

Singer W, Gray CM (1995) Visual feature integration and the temporal correlation hypothesis. Ann Rev Neurosci 18:555–586.

Singer W (2006) Phenomenal awareness and consciousness from a neurobiological perspective. NeuroQuantology 4:134–154.

Smith DR, Smith GK (1965) A statistical analysis of the continual activity of single cortical neurones in the cat unanaesthetized isolated forebrain. Biophys J 5:47–74.

Sporns O, Tononi G, Edelman GM (2000) Theoretical neuroanatomy: relating anatomical and functional connectivity in graphs and cortical connection matrices. Cereb Cortex 10:127–141.

Stam CJ (2004) Functional connectivity patterns of human magnetoencephalographic recordings: a 'small-world' network? Neurosci Lett 355:25–28.

Stratonovich RL (1963, 1967) *Selected Topics in the Theory of Random Noise*, vols. 1 & 2, Gordon & Breach, New York.

Strogatz SH (2001) Exploring complex networks. Nature 410:268–276.

So P, Barreto E, Josic K, Sander E, Schiff SJ (2002) Limits to the experimental detection of nonlinear synchrony. Phys Rev E 65:46225.

Swanson LW (2000) What is the brain? Trends Neurosci 23:519–527.

Tass P, Rosenblum M, Weule J, Kurths J, Pikovsky A, Volkmann J, Schnitzler A, Freund H-J (1998) Detection of n:m phase locking from noisy data: application to magnetoencephalography. Phys Rev Lett 81:3291–3294.

Taylor CP, Dudek FE (1984) Synchronization without active chemical synapses during hippocampal afterdischarges. J Neurophysiol 52:145–155.

Tononi G, Edelman G, Sporns O (1998a) Complexity and coherency: integrating information in the brain. Trends Cogn Sci 2:474–484.

Tononi G, McIntosh AR, Russel DP, Edelman GM (1998b) Functional clustering: identifying strongly interactive brain regions in neuroimaging data. Neuroimage 7:133–149.

Tononi G, Edelman GM (2000) Schizophrenia and the mechanisms of conscious integration. Brain Res Rev 31:391–400.

Trujillo LT, Peterson MA, Kaszniak AW, Allen JJB (2005) EEG phase synchrony differences across visual perception conditions may depend on recording and analysis methods. Clin Neurophysiol 116:172–189.

Varela F, Thompson E, Rosch E (1991) *The Embodied Mind*, MIT Press, Cambridge, MA.

Varela FJ, Lachaux J-P, Rodriguez E, Martinerie J (2001) The brainweb: phase synchronization and large-scale integration. Nature Rev Neurosci 2:229–239.

Volgushev M, Chistiakova M, Singer W (1998) Modification of discharge patterns of neocortical neurons by induced oscillations of the membrane potential. Neuroscience 83:15–25.

von der Malsburg C (1981) *The Correlation Theory of Brain Function*. Internal report, Max-Planck Institute, Göttingen, Germany.

Ward LM, Doesburg SM, Kitajo K, MacLean SE, Roggeveen AB (2006) Neural synchrony in stochastic resonance, attention, and consciousness. Can J Exp Psychol 60:319–326.

Whitham EM, Lewis T, Pope KJ, Fitzgibbon SP, Clark CR, Loveless S, DeLosAngeles D, Wallace AK, Broberg M, Willoughby JO (2008) Thinking activates EMG in scalp electrical recordings. Clin Neurophysiol 119:1166–1175.

Winfree AT (1980) *The Geometry of Biological Time*, Springer-Verlag, Berlin.

Winter WR, Nunez PL, Ding J, Srinivasan R (2007) Comparison of the effect of volume conduction on EEG coherence with the effect of field spread on MEG coherence. Statist Med 26: 3946–3957.

Zaveri HP, Duckrow RB, Spenser SS (2000) The effect of a scalp reference signal on coherence measurements of intracranial electroencephalograms. Clin Neurophysiol 111:1293–1299.

Zhigulin VP, Rabinovich MI, Huerta R, Abarbanel HDI (2003) Robustness and enhancement of neural synchronization by activity-dependent coupling. Phys Rev E 67:021901.

Zorzano MP, Vazquez L (2003) Emergence of synchronous oscillations in neural networks excited by noise. Physica D 179:105–114.

Synchronization Between Sources: Emerging Methods for Understanding Large-Scale Functional Networks in the Human Brain

Sam M. Doesburg and Lawrence M. Ward

Abstract This chapter summarizes currently available techniques for measuring synchronization between neural sources identified through EEG and MEG recordings. First the evidence for the involvement of neural synchronization in the implementation of cognitive processes is described. This involvement is mainly through the provision of high-quality communication between active brain regions, allowing integration of processing activities through the exchange of information and control signals. Second, we describe several useful techniques for obtaining phase information from time series of EEG and MEG records, and measuring phase locking or phase coherence using these methods. These include wavelet analysis and the analytic signal using the Hilbert transform for obtaining phase information, and phase-locking value and coherence for obtaining useful indices of synchronization. Finally, we summarize several available techniques for locating neural sources of EEG and MEG records and describe the use of the phase-locking measurements in ascertaining synchronization between sources located with these techniques. The techniques include those involving blind separation of sources, such as independent component analysis or principle component analysis, and those involving use of brain anatomy to constrain source locations, such as beamformer or LORETA. We also provide a few examples of published or forthcoming research that has used these approaches. All of the techniques described are available either in commercial software (such as BESA and MATLAB) or in freeware that runs in MATLAB (such as EEGLAB, Fieldtrip, Brainstorm). Some custom programming might be required (e.g., in MATLAB using the Signal Processing Toolbox) to implement some of the measurements.

S.M. Doesburg (✉)
Department of Psychology, University of British Columbia, Vancouver, BC, Canada
e-mail: sam@psych.ubc.ca

J.L. Perez Velazquez, R. Wennberg (eds.), *Coordinated Activity in the Brain*, Springer Series in Computational Neuroscience 2, DOI 10.1007/978-0-387-93797-7_2,

1 Introduction: Oscillatory Synchronization and Dynamic Functional Neural Assemblies

The brain oscillates and synchronization of these oscillations have been theoretically and empirically linked to the dynamic organization of communication in the nervous system. Task-dependent neural synchronization is a general phenomenon. It has been observed at various scales ranging from single unit recordings to large-scale cortical dynamics, and has been associated with diverse functions including motor activity, working memory, associative memory, attention, object recognition, awareness, and perceptual organization (e.g., Schnitzler and Gross, 2005a; Tallon-Baudry et al., 1998; Miltner et al., 1999; Rodriguez et al., 1999; Bhattacharya et al., 2001; Doesburg et al., 2008; Burgess and Ali, 2002; Gruber et al., 2002). Moreover, disturbance of oscillatory rhythms and decreased synchronization have been associated with a wide ranging set of disorders including schizophrenia, Parkinson's disease, essential tremor, autism, dyslexia, and Alzheimer's disease (Llinás et al., 2005; Uhlhass and Singer, 2006; Schnitzler and Gross, 2005a). Given this evidence of involvement with both normal and abnormal psychological processes, it is not surprising that the study of oscillatory rhythms and their synchronization in the brain is a subject of growing interest. This chapter briefly summarizes the history of the field beginning with work performed using implanted electrodes and recordings from the scalp. Various approaches to the estimation of the brain generators giving rise to scalp activity, as well as how synchronization between those sources can be calculated, also will be surveyed. Calculation of synchronization between neural sources is a nascent field populated by several competing methods. We will discuss the strengths, weaknesses, and achievements of several of these approaches. It is beyond the purview of this chapter to provide a complete guide to synchronization analysis of EEG and MEG data. Instead, the chapter provides a guide for the investigator interested in surveying the available options or the student wishing an introductory overview. The chapter assumes a working knowledge of oscillation, synchronization, and basic signal processing (e.g., Fourier analysis).

Neurons oscillate, cortical columns oscillate, and from such choruses emerge large-scale oscillatory population dynamics reflecting the activity of millions of cells. When two neurons (or two neural populations) oscillate synchronously, bursts of action potentials can be consistently exchanged during the depolarized phase of the target neuron's ongoing membrane potential fluctuations, thereby increasing fidelity of communication between these neurons (Fries, 2005). This mechanism has been proposed as a means by which the brain is able to selectively integrate relevant neural populations at each moment in order to construct dynamic functional brain networks that perform various cognitive tasks, and/or to bind the spatially distributed brain representations of features of percepts (Varela et al., 2001). The first empirical evidence for this mechanism was the observation that gamma-band oscillations recorded from cat primary visual cortex synchronize when columns of neurons respond to a common object (Gray et al., 1989). It was subsequently observed that the perception of coherent, integrated objects also involves gamma-band synchronization across cortical regions (long-distance synchronization). This phenomenon expresses itself in EEG scalp recordings as synchronization between electrodes associated both with recognition of a familiar object, and with the perception of a meaningful figure in an ambiguous stimulus (Engel et al., 1991b, a; Frein et al., 1994; Gruber et al., 2002; Rodriguez et al., 1999). It is clear that selective synchronization of relevant neural populations across long distances is a general principle for organizing communication in the brain as it is also associated with many cognitive processes (Ward, 2003; Jensen et al., 2007). To refer such phenomena to a Hebbian frame, imagine that recurrently co-activated sets of connected neurons form stable, distributed networks relevant for a particular representation or function, and that when this constellation is ignited and integrated into a large-scale assembly it embodies the perception or task performance at hand (i.e., Hebb, 1949). Alternatively stated, the complex functions expressed by the brain require that particular sets of neural populations cooperate to perform a given task or to bind the features of a percept. Given the manifold mental processes that characterize human mental life, and the universe of perceptions of which we are capable, it must be the case that functional connectivity in the brain can be quickly reorganized, emerging from its constituent elements (Başar, 2006). Selective synchronization of oscillations between relevant neural populations provides a solution to this problem, as synchronously oscillating neurons exchange information more effectively (Varela et al., 2001; Fries, 2005).

2 Methods for the Analysis of Oscillatory Synchronization

Several different approaches have been employed for the calculation of synchronization between oscillating signals. Most successfully employed among these are the phase-locking value (PLV) approach, in which the phases of oscillators are obtained using wavelet analysis or by calculating the analytic signal using the Hilbert transform, and phase cross-coherence analysis. Such methods aim to identify the phases of pairs of oscillators within a relatively narrow bandwidth of frequency, and to assess synchronization between those oscillators by quantifying the stability of the phase relationship between the two. The PLV and coherence techniques will be reviewed here, as they constitute the core methods for the study of synchronization between localized sources of brain activity. Alternative methods exist for the analysis of synchronization, such as mutual information, Shannon entropy, and synchronization likelihood. We shall focus our discussion here, however, on PLV and phase cross-coherence analysis, as these have been most successfully employed in EEG and MEG studies. An important decision lies in the choice of how one partners a particular form of synchronization analysis with any of the numerous approaches to source analysis, and successful marriages of such techniques will be discussed later. It should also be noted that synchronization between EEG and MEG signals draws upon a rich body of work addressing oscillatory synchronization as a general concept for physical systems, and the aspiring investigator would be well served to seek an understanding of such principles (e.g., Pikovski et al., 2001).

2.1 Wavelet Analysis: Application to Phase-Locking Analysis

The most common method to obtain the phase of an oscillator for EEG and MEG PLV analysis is wavelet analysis. Here, instead of decomposing a signal into cosine waves with different phase offsets, as is done in Fourier analysis, the signal is decomposed into various versions of a standard wavelet (a short section of a cosine wave). The wavelet coefficients, which are the output of this analysis, represent the similarity of a particular wavelet, usually the Morlet wavelet (which is the product of a sinusoidal wave with a Gaussian or normal probability density) to the signal at various times and in various relevant frequency bands. In most applications (e.g., Le Van Quyen et al., 2001) an EEG or MEG signal, $h(t)$, is filtered into small frequency ranges using a digital band-pass filter and then the wavelet coefficients, $W_h(\tau, f)$, which are complex numbers, are computed as a function of time, τ, and center frequency of each band, f, from

$$W_h(\tau, f) = \int_{-\infty}^{\infty} h(t)\, \Psi^*_{\tau,f}(u)\, du \tag{1}$$

where $\Psi^*_{\tau,f}(u)$ is the complex conjugate of the Morlet wavelet defined by

$$\Psi_{\tau,f}(u) = \sqrt{f}\; e^{i2\pi f(u-\tau)}\; e^{-\frac{(u-\tau)^2}{2\sigma^2}} \tag{2}$$

The complex conjugate of a complex number $z = x + iy$ is defined as $z^* = x - iy$. For a given time and frequency, $\Psi^*_{\tau,f}(u)$ is a function only of σ, the standard deviation of the Gaussian density function (proportional to the inverse of f), which determines how many cycles of the wavelet are to be used. The number of cycles, $nc = 6f\sigma$. The frequency resolution of the analysis, i.e., the frequency range for which the phase is measured, is determined by σ because the range of frequencies analyzed is about from $f - 4f/nc$ to $f + 4f/nc$. For 40 Hz, for example, the frequency range would be from 20 to 60 Hz for $nc = 8$. This is in spite of the fact that we may have filtered so that the signal being analyzed is narrow band and only ranges from, say, 38 to 42 Hz. This illustrates an important property of wavelet and other multiresolution analyses: at low frequencies, frequency resolution is good but time resolution is poor whereas at high frequencies time resolution is good but frequency resolution is poor. Note that the poorer the frequency resolution, the less meaningful the phase. It is a common practice (e.g., Delorme and Makeig, 2004; Le Van Quyen, et al., 2001) to use fewer cycles of the wavelet for lower frequencies (e.g., three cycles for 6 Hz) and more cycles for higher frequencies (e.g., eight cycles for 40 Hz) with number of cycles increasing roughly linearly with frequency between these limits.

The wavelet transform supplies both the amplitude of the envelope of the signal and the phase at each time point available (see Fig. 1). This is because the wavelet is passed along the signal from time point to time point, with the wavelet coefficient for each time point being proportional to the match between the signal and the

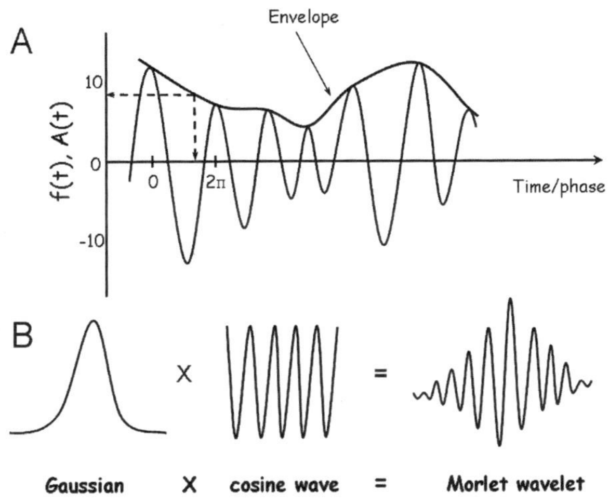

Fig. 1 (**A**) The amplitude envelope and phase of a sinusoidal signal. (**B**) A Morlet wavelet-like those used in wavelet analysis to obtain amplitude and phase of a sinusoidal signal like that in (**A**)

wavelet in the vicinity of that time point; this is closely related to the amplitude of the envelope of the signal at that time. The instantaneous phase at that time point, on the other hand, is the phase offset between the natural oscillation of the wavelet and the oscillation of the signal. Both of these are reflected in the wavelet coefficients. The difference between the phases of two signals, j and k, say one from each of two neural sources, can be computed from the wavelet coefficients for each time and frequency point from

$$e^{i(\phi_j(\tau,f)-\phi_k(\tau,f))} = \frac{W_j(\tau, f)\; W_k^*(\tau, f)}{\left|W_j(\tau, f)\; W_k(\tau, f)\right|} \tag{3}$$

Relative stability of this phase difference across time represents stochastic phase locking between the signals.

2.2 The Hilbert Transform: Application to Phase-Locking Analysis

Another effective technique available for obtaining the instantaneous phase of a signal is to use the Hilbert transform to obtain the analytic signal. The analytic signal, invented by Gabor (1946), is defined for a measured function of time, $h(t)$, as

$$\varsigma(t) = h(t) + i\,\tilde{h}(t) = A(t)e^{i\phi(t)} \tag{4}$$

where $\tilde{h}(t)$ means the Hilbert transform of $h(t)$,

$$\tilde{h}(t) = \frac{1}{\pi} P.V. \int_{-\infty}^{\infty} \frac{h(t)}{t - \tau} \, d\tau, \tag{5}$$

$i = \sqrt{-1}$, and P.V. indicates a special meaning of the improper integral (Pikovski et al., 2001). The Hilbert transform shifts a signal's phase by $\pi/2$. Thus, the real ($h(t)$) and imaginary ($i\tilde{h}(t)$) parts of the analytic signal are shifted by $\pi/2$.

In Eq. (4), the instantaneous amplitude is $A(t)$ and the instantaneous phase is $\phi(t)$. $A(t)$ is the amplitude of the envelope of the signal, and $\phi(t)$, the instantaneous phase, is the quantity that would be entered into computations involving measurement of synchronization, as when two sources are being compared (again, see Fig. 1). The Hilbert transform can be computed across relatively short epochs (say 1500 ms) or even across an entire EEG or MEG record (although available computer resources might make this infeasible). The latter approach should be viewed as preferable, given that the Hilbert transform produces distortions at the beginning and end of each analyzed data segment as it is based on an integral taken from $-\infty$ to $+\infty$. Such "edge effects" are particularly pronounced at lower frequencies. A practical solution for this problem is to apply the Hilbert transform to a time series that extends in both directions beyond the region of interest and to simply discard values near the beginning and end of each segment. Instantaneous phase is only meaningful for narrow band signals, so filtering must be done before the analytic signal is computed. This is usually accomplished, as for the wavelet analysis, by applying a digital band-pass filter to the recorded EEG or MEG signals. We have found that a filter having a pass-band of $f \pm 0.05f$ is effective (e.g., for $f = 10$ it is from 9.5 to 10.5 Hz).

2.3 Phase-Locking Value

Phase-locking value or PLV is a measure of synchronization that has been successfully employed in a variety of experiments and that we regard as the best method for the assessment of task-dependent EEG and MEG phase synchronizations. PLV measures the relative constancy of the difference of phases between two signals, here assumed to be from neural sources, as computed either by the wavelet transform or using the analytic signal, across the trials of an experiment. These trials are represented in the signals as epochs time-locked to a particular repetitive stimulus or response (Lachaux et al., 1999). PLV is defined as

$$PLV_{j,k,t} = N^{-1} \left| \sum_{N} e^{i[\phi_j(t) - \phi_k(t)]} \right| \tag{6}$$

where $\phi_j(t)$ and $\phi_k(t)$ are the phases of sources j and k at time point t for each of the N epochs considered (Lachaux et al., 1999). PLV ranges from a maximum of 1, when the phase differences are exactly the same across all N epochs, to a minimum of 0, when the phase differences vary randomly across the different epochs. In real data neither of these extreme values can be observed, but values close to 1 or 0 are often seen. PLV is the length of a resultant vector in the complex plane when each phase difference, $\phi_j(t) - \phi_k(t)$, is represented by a unit-length vector in the complex plane, and it is proportional to the standard deviation of the distribution of phase differences (see Fig. 2). Changes in PLV over time, in the absence of confounding factors such as volume conduction, reflect changes in the synchronization of the neural activity in a particular frequency band inferred from the sources.

If the recorded EEG or MEG signal is filtered into several frequency ranges, and PLV is then calculated for each sample point available for a particular pair of neural sources, then a time–frequency plot of PLV can be made. Typically baseline-corrected PLVs (written as PLV_z) are plotted rather than raw PLVs. This is done to remove the record of ongoing synchronization unrelated to task demands. Normalization of PLVs is accomplished by subtracting the mean baseline PLV from the PLV for every data point and dividing the difference by the standard deviation of baseline PLV. We have found that a useful baseline is a time period of several hundred milliseconds just before the presentation of the stimulus to which the epochs are time-locked.

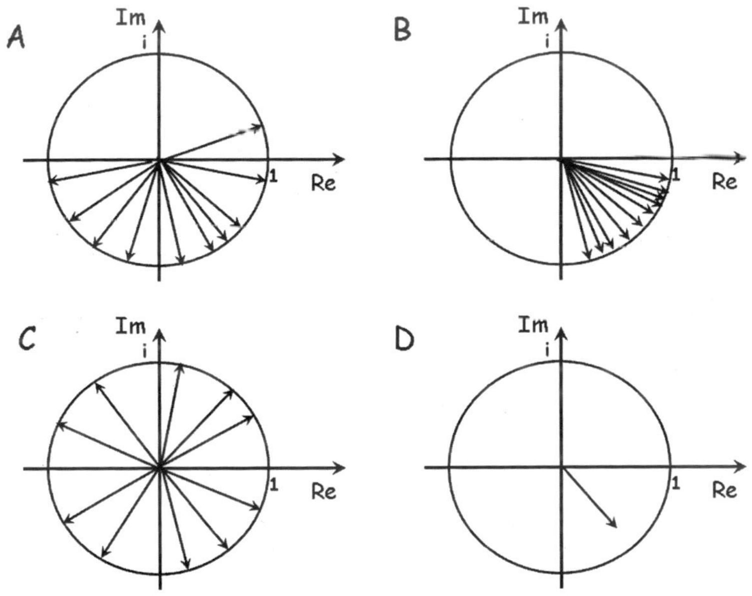

Fig. 2 (**A**) Unit vectors in the complex plane representing phase differences at a particular time–frequency point in the baseline period across 10 trials of an experiment. (**B**) Vectors representing phase differences that are more concentrated, and thus more phase locked, than during the baseline period, indicating an increase in synchronization. (**C**) Vectors representing a decrease in phase locking relative to the baseline period. (**D**) A vector representing a PLV value similar to that indicated by the fan of vectors in (**B**)

The study of phase synchronization in the human brain remains a nascent enterprise and consequently there is no developed statistical theory for the assessment of normalized PLVs, although progress in this direction has been made for phase concentration of single signals with respect to a repeated stimulus (e.g., Martínez-Montes et al., 2007). As we await the development of better techniques, surrogate statistics stand in as a common metric for ascertaining statistical reliability (Le Van Quyen et al., 2001). Surrogate PLV_z distributions are created for each frequency and time point available by randomly shuffling the epochs for one or both of the sensors and recomputing PLV_z for the scrambled data, a large number of times (in our work 200 times). As the temporal relationships between pairs of sources are random in the shuffled data, the surrogate distribution of PLV_z values represents a null distribution over the same actual data. If a PLV_z value exceeds either the 97.5th (or higher if a more conservative test is desired) or is less than the 2.5th (or lower) percentile of the surrogate distribution, then the PLV_z is considered to represent a significant increase (decrease) in synchronization at an alpha level of 0.05. Although our own analyses employ custom software, a similar PLV analysis has been implemented in commercially available EEG analysis software such as Brain Electrical Source Analysis (BESA 5.1, Megis Software, Germany). Moreover, the PLV analysis implemented in BESA has already been used to determine synchronization between neural sources (Sauseng et al., 2007).

2.4 *Phase Cross Coherence in the Assessment of Human Brain Synchronization*

Phase cross coherence should not be confused with linear coherence, which we do not recommend for estimating synchronization between EEG and MEG signals. Linear coherence (e.g., Delorme and Makeig, 2004) is defined in terms of wavelet coefficients as

$$LC_{1,2}(f,t) = \frac{\sum_{k=1}^{N} W_{1,k}(f,t) W_{2,k}^{*}(f,t)}{\sqrt{\sum_{k=1}^{N} \left|W_{1,k}(f,t)\right|^2} \sqrt{\sum_{k=1}^{N} \left|W_{2,k}(f,t)\right|^2}} \tag{7}$$

where $W_{1,k}(f,t)$ is the wavelet coefficient for signal 1 at frequency *f* and time *t* and $W_{2,k}^{*}(f,t)$ the complex conjugate of the corresponding wavelet coefficient for signal 2. The problem with linear coherence is that it confounds phase synchronization with correlated amplitude changes between the signals. It is not useful for EEG and MEG synchronization analysis because phase synchronization and local amplitude changes can reflect very different neural processes (i.e., Palva and Palva, 2007). Phase cross coherence between signals across trials, *N*, is a much more desirable measure as it does not suffer from this defect. It is defined in terms of wavelet coefficients as

$$CC_{1,2}(f,t) = \frac{1}{N}\sum_{k=1}^{N}\frac{W_{1,k}(f,t)W_{2,k}^{*}(f,t)}{\left|W_{1,k}(f,t)W_{2,k}(f,t)\right|} \tag{8}$$

(e.g., Delorme and Makeig, 2004). In Eq. (8) the denominator, which is the magnitude of the cross product of the wavelet coefficients, is responsible for canceling out the amplitude information and thus leaving only a function of the phase difference to be averaged across trials. This index too varies between 1 (for perfect phase locking) and 0 (for a random phase relationship). Phase coherence analysis can be implemented using freely downloadable EEG analysis toolboxes which run in MATLAB (The MathWorks, Inc.), such as EEGLAB (Delorme and Makeig, 2004).

3 Dynamic Brain Networks: Synchronization Between Sources

Although determination of synchronization between electrical or magnetic signals recorded from the human scalp has provided a vista of functional brain dynamics, such determinations inevitably suffer from an inability to claim any specifics about the putative brain regions involved in the generation of long-range synchronization effects. Determination of synchronization between scalp signals also suffers from contamination by volume conduction, which is the propagation of a signal from a single source to multiple recording sites, in this case giving rise to spurious synchronization (e.g., Doesburg et al., 2008; Lachaux et al., 1999). This problem is compounded by the certainty that any scalp-recorded signal consists of a superposition of signals from various neural sources (e.g., Domínguez et al., 2007; Menecke et al., 2005). Numerous advances have provided some means for the reduction of volume conduction and for the estimation of the magnitude of its effects (i.e., Doesburg et al., 2008; Nunez et al., 1997, 1999). Nevertheless, volume conduction and signal superposition remain concerns for studies reporting synchronization between electrodes or sensors placed on or near the scalp. Implanted electrodes, conversely, leave no doubt that their signals originate from near the sensor, but such studies are blind to any activity occurring away from the recording sites. This is a particularly pronounced deficit in studies using electrodes implanted in humans, as the placement of these sensors is typically performed according to the goals of medical procedures and not those of scientific investigation.

Elucidation of dynamic, synchronously oscillating networks in the brain requires the localization of the generators of task-relevant oscillatory brain activity, and subsequently, estimation of phase synchronization between those sources. This step is necessary if strong claims are to be made about the putative engagement of specific brain regions in large scale, synchronous, oscillatory brain networks. Moreover, such analyses make some progress toward solving the related problems of spurious synchronization induced by volume conduction and by signal superposition. In pursuit of this end, we will summarize various approaches to EEG/MEG source localization, survey their relative strengths and weaknesses, and discuss some studies that have successfully combined source localization with long-distance phase synchronization measures. We will also point to available software for the implementation of each of the source-localization techniques. The nascent merger of source localization with synchronization analysis signifies the crossing of an important threshold in the study of neural dynamics, and the perfection of such unions will be critical to further advances in the field.

3.1 Synchronization Between Sources Using "Blind" Source Separation

Source localization techniques used in EEG and MEG analysis fall broadly into two categories, those that are "blind" and rely solely on the identification of statistically independent components that account for some portion of the variance in the signal, and those that utilize some form of model of the human head, accounting for its dimensions and conductivities in an attempt to relate signals measured at the scalp to their neural origins. Blind source-separation techniques for EEG and MEG data most commonly employ either independent component analysis (ICA) or principal component analysis (PCA). ICA attempts to separate the superimposed recorded time series of potential, current, or magnetic field variations arising from different neural sources by reversing the superposition of such signals (Onton et al., 2006). This is accomplished for the recorded data matrix, X, consisting of n channels by t time points, by using a neural network to discover an "unmixing" matrix, W, that when multiplied by the data matrix produces a matrix, U, of maximally temporally independent components: $U = WX$. The criterion for independence is a non-parametric, informational one, i.e., the components of U share minimal mutual information with each other (e.g., Ungureanu et al., 2004). Thus, the various independent components also have minimal, higher-order statistical relationships with one another although they are *not* necessarily (or typically) orthogonal in the correlation sense. Scalp maps of the independent components that account for significant amounts of the variance in the recorded signal can be computed from the inverse of the unmixing matrix, W^{-1}, and these can be used to fit equivalent dipoles to characterize the anatomical location of the neural sources presumed to be responsible for those components. Equivalent dipole fitting can be accomplished using a range of software packages including BESA (BESA 5.1, Megis Software, Germany) and EEGLAB (dipfit 2 plugin, Delorme and Makeig, 2004). Often the location of the best-fitting dipole is anatomically implausible, i.e., outside the head or in white matter; such components are probably artifacts of some sort (e.g., muscle movements) and can be eliminated from consideration. When the location is anatomically plausible, and the fit is good enough, then the dipole can be taken to be a useful representation of a neural source and its relationships to other such sources can then be ascertained.

One problem that arises with the fitting of dipoles, unlike other source-localization methods discussed below, is that it does not result in a unique solution. Or rather, many different solutions with the same number of dipoles, varying only slightly in location, orientation, and strength, and having nearly the same residual variances, are possible. Moreover, by adding dipoles to the solution set, ever better solutions can be found. Thus, assumptions brought to bear by the researcher, such as whether or not bilateral activation is expected, must be used to discover the most useful solution. One approach that is relatively free of biases is to insist that each independent component be accounted for by a single, best-fitting dipole that has residual variance less than some criterion (often 5%) and is anatomically plausible or even predicted before the experiment. This ensures that researcher biases will not result in over-interpretation of the data.

ICA analysis has proven to be well suited to source localization in the study of dynamic networks, as many of the components identified by this method do represent the activities of single anatomical sources. Importantly, when the number of recorded data channels is large enough (31 or more), ICA can reveal multiple sources of this type (Onton and Makeig, 2006; Onton et al., 2006). Given this, it is unsurprising that ICA has been successfully combined with phase cross coherence analysis in high-density EEG data, such as in the study of oscillatory synchronization in pain networks (Drewes et al., 2006).

Synchronization between oscillatory sources, it should be noted, is itself a form of temporal dependency. The study of task-dependent phase synchronization, however, has found many synchronization effects to be relatively short lived, and transient synchronization effects are thus possible between components identified as independent over entire data segments. This is particularly true when ICA is applied to a time series of recordings before they have been divided into stimulus- or response-relevant epochs. Also, the maximal informational independence criterion in ICA implies only that a solution is found in which mutual information among all possible pairs of components is minimal, not zero. Moreover, mutual information is minimized for the total signal, which contains a mixture of amplitude and phase information, whereas the phase cross-coherence measure is unaffected by any correlations, or lack thereof, of the amplitude envelopes of the two sources. Thus phase cross-coherence can be observed between two sources even when variations in their activity over time have been identified as maximally temporally independent. Another practical question is whether ICA is to be applied to data that have already been subjected to narrow band-pass filtering or whether it is used to identify components of broadband signals and spectral information from these sources is determined thereafter. Identification of components within specific frequency ranges carries greater neural meaning. It is more likely, however, that task-relevant synchronization would be preserved between components showing maximal temporal independence across a wider frequency spectrum, given that functional synchronization is typically expressed in a narrow frequency range.

Principal component analysis is another method of blind separation that has been effectively employed for EEG source localization. It is a parametric method that creates a series of mutually orthogonal components that each explain as much of the remaining data variance as possible. It utilizes the linear correlations between pairs of signals as a starting point and creates successive components all of whose pairwise correlation coefficients are zero. Principal component analysis can be used to separate multichannel EEG or MEG data into temporally and spatially independent components that can often be associated with particular neural generators, as in ICA (Chapman and McCrary, 1995). Such neural sources could then be studied for synchronization in the same way as those derived from ICA. One drawback of PCA is that it attempts to lump together as much of the data variance as possible into each component, even though it may include several temporally independent sources. In contrast ICA attempts to split apart such sources. Thus, ICA is probably the preferred technique to discover temporally independent neural sources. In the context of oscillatory dynamics, however, it should be noted that PCA can also be used to decompose EEG data into its composite time–frequency components (Lagerlund et al., 2004; Bernat et al., 2005). PCA is a widely used tool in EEG analysis and is implemented in various software analysis suites, such as BESA and EEGLAB.

3.2 Synchronization Between Neural Sources Identified Using Anatomical Constraints

A second family of analysis techniques that has proven to be useful in the localization of EEG and MEG generators uses some form of head model to constrain solutions. Like "blind" separation techniques, this group of methods also uses spatial and temporal independence to assign activations to various neural locations. Such activations, however, are assigned directly in "brain space" by applying models of the human head that take into account various properties, such as the conductance of the skull, scalp, and brain.

A preeminent technique for the localization of oscillatory sources in cognition that has borne increasingly promising fruit is beamformer analysis. Beamformer analysis produces an estimate of the unique contribution of each voxel in a source space to activity measured at or near the scalp. It does this by creating a set of spatial filters that linearly pass contributions from a given voxel while reducing contributions from all other voxels within a designated time–frequency window (Van Veen et al., 1997). Beamformer and all similar approaches begin by assuming the equivalent dipole model of a neural generator for each voxel in a source space (brain). Here, however, additional constraints, such as minimizing the variance of the filter output while allowing the entire signal at the selected voxel to pass (Van Veen et al., 1997), are imposed that yield a unique solution, unlike unconstrained dipole fitting. Perhaps most seminal and prolific among this family is the "dynamic imaging of coherent sources" (DICS) approach, which has been principally employed in the study of functional networks in the motor system (Gross et al., 2001; Schnitzler and Gross, 2005b). The DICS method was designed specifically for the study of coherence across multiple localized sources of oscillatory activity and its efficacy has been robustly demonstrated in imaging task-dependent activation in the motor system (i.e., Pollok et al., 2005; Timmermann et al., 2004). Although much published work using DICS has been performed by the progenitors of the technique, it has now been implemented in software analysis packages such as Fieldtrip, which runs as a toolbox in MATLAB.

Other variants of beamformer analysis suitable for the determination of multiple sources can be found in commercially available software such as the BESA suite (BESA 5.1; Megis Software, Germany). Such software applications have proven to be effective in revealing the sources and timing of oscillations relevant to cognitive processes such as endogenous attention control. For example, Green and McDonald (2008) used BESA beamformer to uncover a series of activations in theta-band-filtered EEG signals as human subjects performed a visual attention orienting task. The order and timing of activations (early visual cortex, parietal lobe, frontal cortex, parietal lobe, and early visual cortex again) corresponded to the expected sequence of activations involved in orienting visual attention to a specific region of visual space under endogenous control. Recently we combined theta-band beamformer sources identified from EEG data by BESA with PLV analysis to reveal task-dependent synchronization effects between sources in the context of both auditory and visual endogenous attention orienting. In our analysis of visual attention orienting, parietal and occipital sources identified by Green and McDonald (2008) were synchronized in the alpha band, continuously from about 300 ms after cue onset until the target appeared, on the side contralateral to the visual field toward which orienting was taking place (Doesburg et al., 2008, unpublished). In our analysis of auditory attention orienting, synchronization between activated brain areas occurred mainly when activity in one area was declining while activity in another was increasing, indicating that information resulting from processing in the earlier area was being passed to the next area in the sequence (Green et al., 2008, unpublished). BESA beamformer is also effective at identifying neural sources within relatively short time windows, allowing for the mapping of sequentially activated sets of cortical regions embodying organized stages of processing underlying a complex process (Green and McDonald, 2008). Moreover, when PLVs are computed between generators identified in this manner, complementary long-range phase synchronization effects are revealed, suggesting that this method may prove fruitful in illuminating the evolution of functional network dynamics over relatively short periods of time (Green et al., 2008, unpublished).

Another productive manifestation of beamformer analysis can be found in synthetic aperture magnetometry (SAM) developed for MEG analysis (Vrba and Robinson, 2001). Event-related SAM (erSAM) has been used to uncover sources of neuro-magnetic activity associated with cognitive processing (Herdman et al., 2007; see Hillebrand and Barnes, 2005 for a review). Studies employing a combination of the beamformer technique (using both BESA algorithms and SAM) with PLV analysis have proven to be effective for elucidation of large-scale oscillatory synchronous brain networks. An example of this can be found in the analysis of phase synchronization between reconstructed sources of oscillatory activation. Convergent results have been found using BESA beamformer algorithms (EEG) and SAM analysis (MEG) in conjunction with the PLV technique (analytic signal approach). In both cases, when attention is endogenously deployed to one visual hemifield, increased synchronization in the alpha band is observed between reconstructed occipital and parietal sources in the contralateral hemisphere (Doesburg

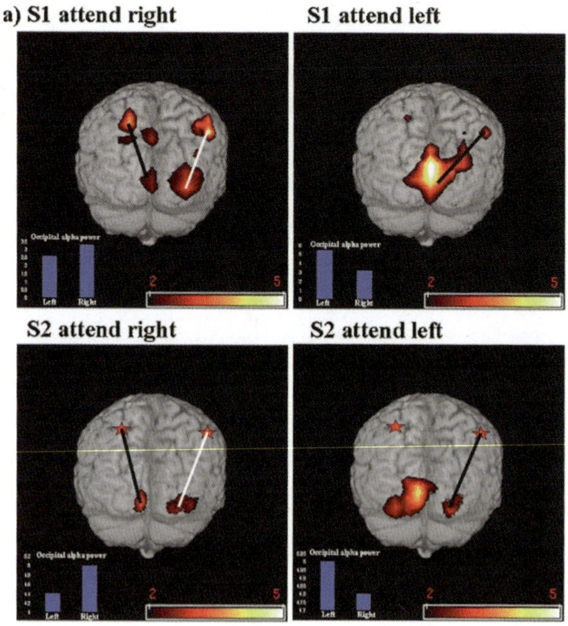

Fig. 3 Example of synchronization between SAM beamformer sources. *Black* (*white*) lines represent synchronization (desynchronization). *Reddish* areas represent alpha sources and *blue bars* represent local alpha power in early visual cortex for each hemisphere

et al., 2007, 2008, unpublished). Figure 3 shows SAM beamformer sources from the MEG experiment projected to the cortical surface, along with an indication of which sources are synchronized or desynchronized at the moment depicted (800 ms after an orienting cue was presented; black (white) lines indicate synchronization increases (decreases)). Interestingly, this lateralization of long-distance synchronization is inversely related to local occipital alpha power measurements in the same frequency range (Fig. 3, blue bar graphs), illustrating the complexity of the relationship between oscillatory synchronization and cognitive processing. More importantly, such results highlight the importance of using analysis techniques that separate phase synchronization from correlated local amplitude changes.

Perhaps the most elegant technique for the determination of locally coherent oscillatory sources and the extraction of epoched signals for phase-locking analysis is one in which minimum norm current solutions are employed in an iterative sequence to determine intertrial coherence and coherence between adjacent voxels for source reconstruction (David and Garnero, 2002; David et al., 2002, 2003). To date, however, this technique has only been used in the processing of brain responses entrained by flickering stimuli and has not been employed for the estimation of endogenous oscillatory activity (e.g., Cosmelli et al., 2004). Unfortunately, no widely accessible software exists for the performance of this state-of-the-art analysis. These methods, however, are well documented in the literature (above) and await reimplementation by an enterprising investigator.

Autoregressive techniques also have been employed for the estimation of multiple sources of task-relevant oscillatory activation within a particular frequency band. Sources identified using this method can be successfully combined with PLV analysis, as has been eloquently demonstrated in the study of gamma oscillations in the human object recognition network (Supp et al., 2007). The reliability of effects gained using this method has been demonstrated through replication (Gruber et al., 2008). Moreover, these source-localization effects and corresponding phase synchronization effects were obtained for high-frequency gamma-band signals. Localization of such rhythms is difficult as this frequency range contributes little to overall spectral power, but has been profusely evidenced as important to information processing in the cerebral cortex (Kaiser and Lutzenberger, 2005). Another advantage of this approach is that it has proven to be compatible with available methods for the calculation of Granger causality measures that can be used to index the causal direction of information flow between sources (Baccala and Sameshima, 2001). This is attractive because phase synchronization in the brain, which has been theoretically related to increased causal interaction, can thus be directly compared with just such a measure. The open source software used in all stages of this analysis, BioSig (Version 1.95), is freely available online at http://biosig.sourceforge.net/

Low resolution brain electromagnetic tomography (LORETA) (Pascual-Marqui et al., 1994) is another functional imaging method based on electrophysiological and neuroanatomical constraints. LORETA and its variants have been employed by many studies seeking to analyze ERP components as well as spectral components of EEG activation (see Pascual-Marqui et al., 2002 for a review). In view of this track record, LORETA also promises to be a useful method for the localization of neural generators in the study of long-distance neural synchronization.

4 Summary and Conclusion

The study of dynamic functional networks in the brain will require techniques to localize sources of activation and to determine synchronization between such sources. From our current point of departure there are many viable paths toward this goal. Which road is best will depend largely on the questions being addressed, and careful consideration must be given to which methods of synchronization analysis are combined with particular means of source imaging—not all methods may be compatible. Because these two streams of understanding have evolved, until relatively recently, as separate enterprises, the choosing of an appropriate pairing will be important. As the terrain here is varied, it is beyond the reach of this chapter to provide a walkthrough of each method, and providing one method would not adequately service those interested in various researches into this subject. We have accordingly provided an overview and primer to the various approaches thus far used to study synchronization between sources of brain activity.

Acknowledgments Preparation of this chapter was supported by a grant from the Natural Sciences and Engineering Research Council (NSERC) of Canada.

References

Baccala LA, Sameshima K (2001) Partial directed coherence: a new concept in neural structure determination. Biol Cybern 84:463–474
Başar E (2006) The theory of the whole-brain-work. Int J Psychophysiol 60:133–138
Bernat EM, Williams WJ, Gehring WJ (2005) Decomposing ERP time–frequency energy using PCA. Clin Neurophysiol 116:1314–1334
Bhattacharya J, Petsche H, Feldmann U, Rescher B (2001) EEG gamma-band phase synchronization between posterior and frontal cortex during mental rotation in humans. Neurosci Lett 311:29–32
Burgess AP, Ali L (2002) Functional connectivity of gamma EEG activity is modulated at low frequency during conscious recollection. Int J Psychophysiol 46:91–100
Chapman RM, McCrary JW (1995) ERP component identification and measurement by principal components analysis. Brain Cognit 27:288–310
Cosmelli D, David O, Lachaux JP, Martinerie J, Garnero L, Renault B, Varela F (2004) Waves of consciousness: ongoing cortical patterns during binocular rivalry. NeuroImage 23:128–140
David O, Cosmelli D, Hasboun D, Garnero L (2003) A multitrial analysis for revealing significant corticocortical networks in magnetoencephalography and electroencephalography. NeuroImage 20:186–201
David O, Garnero L (2002) Time-coherent expansion of MEG/EEG cortical sources. NeuroImage 17:1277–1289
David O, Garnero L, Cosmelli D, Varela FJ (2002) Estimation of neural dynamics from MEG/EEG cortical current density maps: application to the reconstruction of large-scale cortical synchrony. IEEE Trans Biomed Eng 49:975–987
Delorme A, Makeig S (2004) EEGLAB: an open source toolbox for analysis of single-trial EEG dynamics. J Neurosci Methods 134:9–21
Doesburg SM, Herdman AT, Ward LM (2007) MEG reveals synchronous neural network for visuospatial attention. Poster presented at the 2007 Meeting of the Cognitive Neuroscience Society in New York City
Doesburg SM, Roggeveen AB, Kitajo K, Ward LM (2008) Large-scale gamma-band phase synchronization and selective attention. Cereb Cortex 18:386–396 Epub 2007 Jun 7
Domínguez LG, Wennberg R, Perez Velázquez JL, Erra RG (2007) Enhanced measured synchronization of unsynchronized sources: inspecting the physiological significance of synchronization analysis of whole brain electrophysiological recordings. Int J Phys Sci 2:305–317
Drewes AM, Sami SA, Dimcevski G, Nielsen KD, Funch-Jensen P, Valeriani M, Arendt-Nielsen L (2006) Cerebral processing of painful oesophageal stimulation: a study based on independent component analysis of the EEG. Gut 55:619–629
Engel AK, König P, Kreiter AK, Singer W (1991a) Interhemispheric synchronization of oscillatory neuronal responses in cat visual cortex. Science 252:1177–1179
Engel AK, König P, Singer W (1991b) Direct physiological evidence for scene segmentation by temporal coding. Proc Natl Acad Sci USA 88:9136–9140
Frein A, Eckhorn R, Bauer R, Woelbern T, Kehr H (1994) Stimulus-specific fast oscillations at zero phase between visual areas V1 and V2 of awake monkey. Neuroreport 5:2273–2277
Fries P (2005) A mechanism for cognitive dynamics: neuronal communication through neuronal coherence. Trends Cogn Sci 9:474–479
Gabor D (1946) Theory of communication. J IEE (London) 93:429–457
Gray CM, König P, Engel AK, Singer W (1989) Oscillatory responses in cat visual cortex exhibit inter-columnar synchronization which reflects global stimulus properties. Nature 338:334–337
Green JJ, McDonald JJ (2008) Electrical neuroimaging reveals timing of attentional control activity in human brain. PLoS Biol 6:e81
Gross J, Kujala J, Hämäläinen M, Timmermann L, Schnitzler A, Salmelin R (2001) Dynamic imaging of coherent sources: studying neural interactions in the human brain. Proc Natl Acad Sci USA 98:694–699

Gruber T, Maess B, Trujillo-Barreto NJ, Müller MM (2008) Sources of synchronized induced gamma-band responses during a simple object recognition task: a replication study in human MEG. Brain Res 1196:74–84

Gruber T, Müller MM, Keil A (2002) Modulation of induced gamma band responses in a perceptual learning task in the human EEG. J Cogn Neurosci 14:732–744

Hebb DO (1949) The Organization of Behavior. Lawrence Earlbaum Associates, New York

Herdman AT, Pang AW, Ressel V, Gaetz W, Cheyne D (2007) Task-related modulation of early cortical responses during language production: an event related synthetic aperture magnetometry study. Cereb Cortex 17:2536–2543

Hillebrand A, Barnes GR (2005) Beamformer analysis on MEG data. Int Rev Neurobiol 68: 149–171

Jensen O, Kaiser J, Lachaux JP (2007) Human gamma-frequency oscillations associated with attention and memory. Trends Neurosci 30:317–324 Epub 2007 May 17

Kaiser J, Lutzenberger W (2005) Human gamma-band activity: a window to cognitive processing. Neuroreport 16:207–211

Lachaux JP, Rodriguez E, Martinerie J, Varela FJ (1999) Measuring phase synchrony in brain signals. Hum Brain Mapp 8:94–208

Lagerlund TD, Sharbrough FW, Busacker NE (2004) Use of principal component analysis in the frequency domain for mapping electroencephalographic activities: comparison with phase-encoded Fourier spectral analysis. Brain Topogr 17:73–84

Le Van Quyen M, Foucher J, Lachaux J-P, Rodriguez E, Lutz A, Martinerie JM, Varela FJ (2001) Comparison of Hilbert transform and wavelet methods for the analysis of neural synchrony. J Neurosci Methods 111:83–98

Llinás R, Urbano FJ, Leznik E, Ramírez RR, van Marle HJF (2005) Rhythmic and dysrhythmic thalamocortical dynamics: GABA systems and the edge effect Trends Neurosci 28:325–333

Martínez-Montes E, Cuspineda-Bravo ER, El-Deredy W, Sánchez-Bornot JM, Lage-Castellanos A, Valdés-Sosa PA (2007) Exploring event-related brain dynamics with tests on complex-valued time-frequency representations. Statist Med doi: 10.1002/sim. 3132

Menecke FC, Ziehe A, Kurths J, Müller K-R (2005) Measuring phase synchronization of superimposed signals. Phys Rev Lett 94:084102 (1–4)

Miltner WH, Braun C, Arnold M, Witte H, Taub E (1999) Coherence of gamma-band EEG activity as basis for associative learning. Nature 397:434–436

Nunez PL, Srinivasan R, Wetrdorp AF, Wijesinghe RS, Tucker DM, Silberstein RB, Cadusch PJ (1997) EEG coherency I: statistics, reference electrode, volume conduction, Laplacians, cortical imaging and interpretation at multiple scales. EEG Clin Neurophysiol 103:499–515

Nuncz PL, Silberstein RB, Shi Z, Carpenter MR, Srinivasan R, Tucker DM, Doran SM, Cadusch PJ, Wijesinghe RS (1999) EEG coherency II: experimental comparisons of multiple measures. Clin Neurophysiol 110:469–486

Onton J, Makeig S (2006) Information-based modeling of event related brain dynamics. Prog Brain Res 25:199–120

Onton J, Westerfield M, Townsend J, Makeig S (2006) Imaging human EEG dynamics using independent component analysis. Neurosci Biobehav Rev 30:808–822

Palva S, Palva JM (2007) New vistas for alpha-frequency band oscillations. Trends Neurosci 30:150–158

Pascual-Marqui RD, Esslen M, Kochi K, Lehmann D (2002) Functional imaging with low resolution brain electromagnetic tomography (LORETA): a review. Methods Find Exp Clin Pharmacol 24C:91–95

Pascual-Marqui RD, Michel CM, Lehmann D (1994) Low resolution electromagnetic tomography: a new method for localizing electrical activity in the brain. Int J Psychophysiol 18:49–65

Pikovski A, Rosenblum M, Kurths J (2001) Synchronization: A Universal Concept in Nonlinear Sciences. Cambridge University Press, Cambridge

Pollok B, Gross J, Müller K, Aschersleben G, Schnitzler A (2005) The cerebral oscillatory network associated with auditorily paced finger movements. NeuroImage 24:646–655

Rodriguez E, George N, Lachaux JP, Martinirie J, Renault B, Varela FJ (1999) Perception's shadow: long-distance synchronization of human brain activity. Nature 397:430–433

Sauseng P, Hoppe J, Klimesch W, Gerloff C, Hummel FC (2007) Dissociation of sustained attention from central executive functions: local activity and interregional connectivity in the theta range. Eur J Neurosci 25:578–593

Schnitzler A, Gross J (2005a) Normal and pathological oscillatory communication in the brain. Nat Rev Neurosci 6:285–296

Schnitzler A, Gross J (2005b) Functional connectivity analysis in magnetoencephalography. Int Rev Neurobiol 68:173–195

Supp GG, Schlogl A, Trujillo-Barreto N, Muller M, Gruber T (2007) Directed cortical information flow during human object recognition: analyzing induced gamma-band responses in brain's source space. PLoS One 2,(8):e684 (1–11)

Tallon-Baudry C, Bertrand O, Peronet F, Pernier J (1998) Induced γ-band activity during the delay of a visual short-term memory task in humans. J Neurosci 18:4244–4254

Timmermann M, Gross J, Butz M, Kircheis G, Haussinger D, Schnitzler A (2004) Pathological oscillatory coupling within the human motor system in different tremor syndromes as revealed by magnetoencephalography. Neurol Clin Neurophysiol 26:1538–4098

Uhlhass PJ, Singer W (2006) Neural synchrony in brain disorders: relevance for cognitive dysfunctions and pathophysiology. Neuron 52:55–68

Ungureanu M, Bigan C, Strungaru R, Lazarescu V (2004) Independent component analysis in biomedical signal processing. Meas Sci Rev 4:1–8

Van Veen BD, van Drongelen W, Yuchtman M, Suzuki A (1997) Localization of brain electrical activity via linearly constrained minimum variance spatial filtering. IEEE Trans Biomed Eng 44:867–880

Varela F, Lachaux JP, Rodriguez E, Martinerie J (2001) The brainweb: phase synchronization and large-scale integration. Nat Rev Neurosci 2:229–239

Vrba J, Robinson SE (2001) Signal processing in magnetoencephalography. Methods 25:249–271

Ward LM (2003) Synchronous neural oscillations and cognitive processes. Trends Cogn Sci 17:553–559

Approaches to the Detection of Direct Directed Interactions in Neuronal Networks

Ariane Schad, Jakob Nawrath, Michael Jachan, Kathrin Henschel, Linda Spindeler, Jens Timmer, and Björn Schelter

Abstact In this chapter, we address the challenge of detecting interactions among neuronal processes by means of bivariate and multivariate linear analysis techniques. For linear systems, both undirected and directed measures exist. Coherence is a commonly used undirected bivariate measure to detect the interaction between two nodes of a network, while multivariate measures like the partial coherence distinguish direct and indirect connections. The partial directed coherence additionally features the direction influences between nodes. We introduce the theoretical framework of these analysis techniques, discuss their estimation, and present their application to simulated and real-world data.

1 Introduction

In the last years, the research on large complex systems and networks has gained increasing interest in various fields of science like physics, biology, and neuroscience. The emerging interest is further advanced by the rapid development of novel and innovative measurement devices and high-performance computers for providing, storing, and processing huge amounts of signals. A major goal is the determination of the connection topology of the constituent parts and the understanding of the characteristic behavior of the complex system. The challenge thereby is the inference of the interactions from measured time series without prior knowledge about the system. This topic has traditionally been addressed in the time domain as well as in the frequency domain by bivariate methods, like cross-correlation and coherence analysis. However, if more than two processes are to be analyzed, one has to face the problem that rather complex interaction structures may arise between the processes. For instance, consider two processes of a multivariate system which do not interact directly, but are both affected by a third process. Then, bivariate measures reveal an interrelation

A. Schad (✉)
FDM, Freiburg Center for Data Analysis and Modeling, University of Freiburg, Eckerstr. 1, 79104 Freiburg, Germany
e-mail: schad@fdm.uni-freiburg.de

J.L. Perez Velazquez, R. Wennberg (eds.), *Coordinated Activity in the Brain*, Springer Series in Computational Neuroscience 2, DOI 10.1007/978-0-387-93797-7_3,

between the two processes although it is an indirect interrelationship mediated by a common influence. Hence, bivariate methods are often not sufficient to adequately describe the interaction structure. Multivariate methods are needed, like the partial coherence, which allow to distinguish between direct and indirect influences.

Moreover, especially for the investigation of brain networks, uncovering the direction of interactions enables deeper insights into the mechanisms underlying such networks. In order to measure the direction of influences, directed multivariate measures are used, like the directed transfer function or the partial directed coherence. However, comparisons of both methods have shown that in contrast to the partial directed coherence, the directed transfer function is not able to differentiate direct and indirect influences (Winterhalder et al., 2005).

The chapter is separated into four sections. First, we give a short introduction to the framework of linear multivariate time series analysis in the frequency domain. Based on the spectral matrix of multivariate processes, we review coherence and partial coherence for the detection of undirected interactions. We describe their estimation and provide measures to judge their statistical significance. Subsequently, we address the challenge of the detection of directed interactions. The notion of Granger-causality is introduced based on modeling multivariate time series by vector autoregressive processes. We specify the partial directed coherence as well as its estimation and statistical evaluation. In the second section, we elaborate on the application of the presented methods to simulated linear and nonlinear systems. We compare the different methods and illustrate their specific abilities and limitations. Particular emphasis is laid on the investigation of multivariate systems with unobserved components. In the third section, we consider an application of the partial directed coherence to signals originating from Parkinsonian tremor and demonstrate the ability of this measure to infer pathophysiological relationships in tremor. Finally, the limitations and perspectives of the presented methods are pointed out and discussed.

2 Measuring Interactions: Theory and Methods

In this section, we present methods of multivariate linear time series analysis for the investigation of interactions in complex systems. We review theoretical and practical aspects of undirected nonparametric methods like coherence and partial coherence (Sect. 2.1) and introduce the concept of partial directed coherence based on the notion of multivariate autoregressive modeling (Sect. 2.2).

2.1 Undirected Interactions: Linear Spectral Analysis

Fundamental for the definition of various undirected measures in the frequency domain, like coherence and partial coherence, is the spectral representation of multivariate linear processes in the framework of linear spectral analysis.

2.1.1 The Spectral Matrix

Consider a complex system whose dynamical and statistical properties can be formally represented by a multivariate stochastic process $\mathbf{X} = (X_1, \ldots, X_M)^T$ composed of M processes X_j. Henceforth it is assumed that the processes under investigation are real-valued, time continuous, of zero-mean, and stationary, i.e., the characteristics of the underlying system are time-invariant. Then, one can represent the process X_j in the frequency domain by its Fourier transform

$$f_{X_j}(\omega) = \frac{1}{\sqrt{2\pi}} \int_{-\infty}^{\infty} X_j(t)\, e^{-i\omega t} \,. \tag{1}$$

The frequency domain representation of a process is useful especially for the investigation of oscillatory contents in the corresponding signal. Such contents can be readily identified in the frequency domain using the spectrum. For a multivariate process, the spectral properties at a certain frequency ω are described by the spectral matrix (Priestley, 1989)

$$\mathbf{S}(\omega) = \begin{pmatrix} S_{X_1X_1}(\omega) & S_{X_1X_2}(\omega) & \ldots & S_{X_1X_M}(\omega) \\ S_{X_2X_1}(\omega) & S_{X_2X_2}(\omega) & \ldots & \vdots \\ \vdots & \ldots & \ddots & \vdots \\ S_{X_MX_1}(\omega) & \ldots & \ldots & S_{X_MX_M}(\omega) \end{pmatrix} . \tag{2}$$

The entries of $\mathbf{S}(\omega)$ are defined by

$$S_{X_iX_j}(\omega) =< f_{X_i}(\omega)\, f^*_{X_j}(\omega) > , \tag{3}$$

where $(\cdot)^*$ denotes complex conjugate and $< \cdot >$ the expected value. Thus, $S_{X_iX_j}(\omega)$ represents the covariance between the Fourier transforms $f_{X_i}(\omega)$ and $f_{X_j}(\omega)$ of X_i and X_j.

The diagonal elements $S_{X_jX_j}(\omega)$ of the spectral matrix correspond to the auto-spectra of the processes X_j. The auto-spectrum quantifies the variance of a process in dependence on frequency. Thus, pronounced large peaks in the auto-spectrum correspond to predominant oscillatory contents in a signal. For simplicity we will denote $S_{X_jX_j}(\omega)$ in the following as $S_{X_j}(\omega)$.

The off-diagonal elements $S_{X_iX_j}(\omega)$ of the spectral matrix are of special interest in multivariate analysis. They define the bivariate cross-spectrum $CS_{X_iX_j}(\omega) = S_{X_iX_j}(\omega)$ of two processes X_i and X_j. As the cross-spectrum measures the covariance between two processes in dependence on frequency, it contains information about linear interrelationships between them.

2.1.2 Coherence

In contrast to the auto-spectrum, the cross-spectrum is complex-valued. In order to measure a linear relationship between two processes, it is therefore meaningful to calculate the absolute value of the cross-spectrum $|CS_{X_i X_j}(\omega)|$. Normalization of the absolute cross-spectrum by the auto-spectra of X_i and X_j yields the real-valued coherence spectrum

$$Coh_{X_i X_j}(\omega) = \frac{|CS_{X_i X_j}(\omega)|}{\sqrt{S_{X_i}(\omega)\, S_{X_j}(\omega)}}, \qquad Coh_{X_i X_j}(\omega) \in [0, 1]. \tag{4}$$

The coherence spectrum is a powerful measure for the investigation of linear relationships between two processes at a given frequency. Due to the normalization, the value of the coherence is restricted between 0 and 1 and may be interpreted as the correlation coefficient between the spectral components of X_i and X_j at frequency ω. Thus, $Coh_{X_i X_j}(\omega) = 1$ indicates the presence of a linear relationship between X_i and X_j at a certain frequency, while $Coh_{X_i X_j}(\omega) = 0$ indicates its absence.

2.1.3 Partial Coherence

An essential challenge that may occur, when bivariate cross-spectral analysis is applied to signals of actually multivariate complex systems, is the detection of indirect correlations. The cross-spectrum and the coherence spectrum are purely bivariate measures and consequently they fail to correctly distinguish direct from indirect linear relationships in multivariate systems. To illustrate this, we consider a system consisting of three components, (X_1, X_2, X_3), where X_3 linearly influences X_1 and X_2. A pairwise bivariate coherence analysis of the processes X_1, X_2, and X_3 would lead to significant coherence between all components, including X_1 and X_2. But the correlation between X_1 and X_2 in this system is actually a pseudo correlation or indirect correlation mediated by X_3.

In order to decide whether correlations between any two components X_i and X_j of a multivariate process $(X_1, \ldots, X_M)$ are direct or indirect, the bivariate coherence analysis can be extended to the partial coherence analysis. The basic idea is to subtract those linear influences on both processes under investigation which originate from the remaining processes $\mathbf{X}_Z$. Thereby, $\mathbf{X}_Z$ consists of all the other processes of the multivariate system, except for X_i and X_j, i.e., $\mathbf{X}_Z = \mathbf{X}_{\{k=1,\ldots,M;\, k \neq i,j\}}$. This idea leads to the definition of the partial cross-spectrum which is defined as the cross-spectrum between X_i and X_j without the linear influence of $\mathbf{X}_Z$,

$$PCS_{X_i X_j|\mathbf{X}_Z}(\omega) = CS_{X_i X_j}(\omega) - CS_{X_i \mathbf{X}_Z}(\omega)\, S_{\mathbf{X}_Z}^{-1}(\omega)\, CS_{\mathbf{X}_Z X_j}(\omega)\,. \tag{5}$$

For detecting directly interacting processes in multivariate systems, the partial coherence spectrum is defined as in the case of the coherence spectrum by normalization of the respective partial cross-spectra with the corresponding partial spectra, i.e.,

$$PCoh_{X_i X_j|\mathbf{X}_Z}(\omega) = \frac{|PCS_{X_i X_j|\mathbf{X}_Z}(\omega)|}{\sqrt{PCS_{X_i|\mathbf{X}_Z}(\omega)\, PCS_{X_j|\mathbf{X}_Z}(\omega)}} . \tag{6}$$

In the case that a linear relationship between X_i and X_j can be explained completely by $\mathbf{X}_Z$, one obtains $PCoh_{X_i X_j|\mathbf{X}_Z}(\omega) = 0$, whereas a direct linear relationship between two components is indicated by a partial coherence of one.

2.1.4 Nonparametric Estimation of the Spectral Matrix

Suppose that observations $\mathbf{x}(t_1), \mathbf{x}(t_2), \ldots, \mathbf{x}(t_N)$ of a multivariate process $\mathbf{X}$ are measured at N equally spaced discrete time samples $t_1, \ldots, t_N$. From such a multivariate time series of finite length, we aim to estimate the coherence spectra and partial coherence spectra, respectively. For this purpose, the spectral matrix has to be estimated. That means according to Eq. (2) we have to estimate the auto-spectra and cross-spectra which define the spectral matrix entries.

There exist several methods to estimate the auto-spectrum of a stationary process from measured time series $x(t_1), \ldots, x(t_N)$ (Brillinger, 1981; Brockwell and Davis, 1998; Priestley, 1989). A commonly used nonparametric spectral estimation method is given by *smoothing the periodogram*. This approach is based on the discrete Fourier transform

$$f_X(\omega) = \frac{1}{\sqrt{2\pi N}} \sum_{n=1}^{N} h(t_n)\, x(t_n)\, \mathrm{e}^{-\mathrm{i}\omega t_n} \text{ , with } -\pi \leq \omega \leq \pi . \tag{7}$$

In this definition, the time series $x(t_n)$ is "tapered", i.e., modified by multiplication with a taper window $h(t_n)$. The taper window smoothes the transition from nonzero to zero values at both ends of the finite time series. A prevalent taper window is for example a triangular window. The Fourier transform of a time series of length N is commonly evaluated at the Fourier frequencies $\omega_k = 2\pi \frac{k}{N}$ with $k = -N/2, \ldots, -1, 0, 1, \ldots, N/2-1$. Based on $f_X(\omega_k)$ the periodogram is defined as

$$Per_X(\omega_k) = |f_X(\omega_k)|^2 . \tag{8}$$

Except for harmonic periodic signals, the periodogram is an extremely poor representation of the spectrum of a process as it follows a scaled χ^2_2 distribution, resulting in a highly fluctuating estimate (cf. Fig. 1). The variance of the periodogram ordinates depends on the values of the spectrum of the process, i.e., $var(Per_X(\omega_k)) = 2S_X(\omega_k)$. Moreover, the variance is independent of the number of observations N. Consequently, increasing N would not decrease the variance of $Per_X(\omega_k)$, but it would instead increase the resolution in the frequency domain.

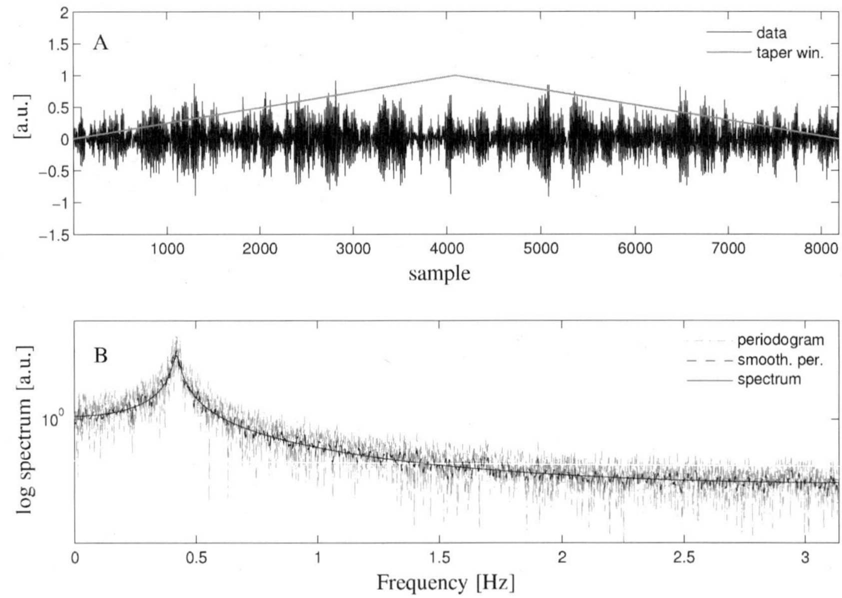

Fig. 1 (**A**) Time series of an AR[2] process and triangular taper window. (**B**) Raw periodogram and smoothed periodogram calculated from the AR time series and the analytical spectrum of the process

According to Eq. (3), the expected value of $Per_X(\omega_k)$ has to be estimated to obtain the spectrum. One can show that the spectrum of a mixing process, i.e., a stationary stochastic or chaotic process, is a smooth function of frequency. In order to estimate the spectrum, the periodogram is smoothed with a normalized smoothing window w_l of width $2m + 1$, leading to

$$\widehat{S}_X(\omega_k) = \sum_{l=-m}^{m} w_l \, Per_X(\omega_{k+l}), \text{ with } \sum_{l=-m}^{m} w_l = 1 . \tag{9}$$

This smoothing procedure is equivalent to a weighted averaging of the periodogram. For increasing N and m, i.e., $N \to \infty$ and $m \to \infty$ such that $m/N \to 0$, the smoothed periodogram provides a consistent estimator of the spectrum. Different choices of smoothing windows are possible, e.g., a triangular window (Priestley, 1989).

Example 1 Nonparametric estimation of the spectrum The nonparametric estimation of a spectrum is illustrated in Fig. 1 using a time series from an autoregressive process of second order, denoted as AR[2]. An AR[2] process is defined by $X(t) = a_1 X(t-1) + a_2 X(t-2) + \varepsilon(t)$ (Priestley, 1989). In this example, the parameters were set to $a_1 = 1.9718$ and $a_2 = -0.9964$. The term $\varepsilon(t) \sim N(0, \sigma^2)$ represents a Gaussian white noise process $N(0, \sigma^2)$ with mean 0 and standard deviation $\sigma = 0.5$. This AR process can be assigned a physical meaning, where it represents a damped oscillator driven at each instant t by a stochastic input $\varepsilon(t)$. A particularity of this process is that its spectrum can be calculated analytically, leading to

$$S(\omega) = \frac{1}{2\pi} \frac{\sigma^2}{|1 - a_1 \mathrm{e}^{-\mathrm{i}\omega} - a_2 \mathrm{e}^{-2\mathrm{i}\omega}|^2}.$$

In Fig. 1, the analytical spectrum of the AR process is compared to its highly fluctuating raw periodogram and the estimated spectrum obtained by smoothing the periodogram.

2.1.5 Estimation of Coherence

The cross-spectrum of two time series $x_i(t_n)$ and $x_j(t_n)$ of equal length can be estimated in the same way as the auto-spectrum by smoothing the cross-periodogram, i.e.,

$$\widehat{CS}_{X_i X_j}(\omega_k) = \sum_{l=-m}^{m} w_l \, f_{X_i}(\omega_{k+l}) \, f^*_{X_j}(\omega_{k+l}) \,. \tag{10}$$

By replacing the cross- and auto-spectra in Eq. (4) with the respective sample spectra one obtain the estimate of the coherence spectrum

$$\widehat{Coh}_{X_i X_j}(\omega_k) = \frac{|\widehat{CS}_{X_i X_j}(\omega_k)|}{\sqrt{\widehat{S}_{X_i}(\omega_k) \, \widehat{S}_{X_j}(\omega_k)}} \,. \tag{11}$$

2.1.6 Estimation of Partial Coherence

The partial cross-spectrum can be estimated from Eq. (5) by substitution of the respective spectra with their sample estimates, such that

$$\widehat{PCS}_{X_i X_j | \mathbf{X}_Z}(\omega_k) = \widehat{CS}_{X_i X_j}(\omega_k) - \widehat{CS}_{X_i \mathbf{X}_Z}(\omega_k) \, \widehat{S}^{-1}_{\mathbf{X}_Z}(\omega_k) \, \widehat{CS}_{\mathbf{X}_Z X_j}(\omega) \,. \tag{12}$$

The estimated partial coherence spectrum is obtained from Eq. (6) as

$$\widehat{PCoh}_{X_i X_j | \mathbf{X}_Z}(\omega_k) = \frac{\left|\widehat{PCS}_{X_i X_j | \mathbf{X}_Z}(\omega_k)\right|}{\sqrt{\widehat{PCS}_{X_i | \mathbf{X}_Z}(\omega_k) \, \widehat{PCS}_{X_j | \mathbf{X}_Z}(\omega_k)}} \,. \tag{13}$$

A numerically more efficient procedure to estimate the partial coherence is given by inversion and rescaling of the spectral matrix $\widehat{\mathbf{S}}(\omega_k)$ (cf. Dahlhaus, 2000).

2.1.7 Testing for Significance

For the evaluation of the significance of the results obtained by coherence or partial coherence analysis of a time series, statistical criteria are necessary. From the statistical properties of the estimators, critical values can be deduced in order to test whether or not an obtained value of these measures is significant. Based on the assumption or null hypothesis that there is no linear interrelationship between the two processes X_i and X_j, a critical value

$$c_{Coh} = \sqrt{1 - \alpha^{\frac{2}{\nu-2}}} \tag{14}$$

can be derived for the coherence estimator $\widehat{Coh}_{X_i,X_j}(\omega_k)$ at each frequency ω_k. The parameter α denotes the statistical significance level and ν is the equivalent number of degrees of freedom of the used spectral estimator (Bloomfield, 1976; Hannan, 1970).

With a slight modification of Eq. (14) one further obtains the critical value for the partial coherence (Timmer et al., 2000) as

$$c_{PCoh} = \sqrt{1 - \alpha^{\frac{2}{\nu-2L-2}}}\,, \tag{15}$$

where L denotes the dimension of the remaining process $\mathbf{X}_Z$. Note that both relations describe pointwise critical values at single frequencies. However, the values itself are independent of the frequency.

By means of the critical values we are able to discuss the meaning of a coherence or a partial coherence estimate at a given frequency ω_k as follows: if $\widehat{Coh}(\omega_k) > c_{Coh}$ for a significance level $\alpha = 0.05$, then the null hypothesis of no linear relationship, direct or indirect, between two processes at ω_k can be rejected with probability $(1-\alpha) \times 100\% = 95\%$. Correspondingly, one can deduce from a significant partial coherence $\widehat{PCoh}(\omega_k) > c_{PCoh}$ that a direct linear relationship at ω_k is likely.

2.2 *Directed Interactions: Linear Autoregressive Models*

The above defined partial coherence is able to reveal linear relationships and to disentangle direct and indirect connections in multivariate systems. However, especially for neuronal systems one is moreover interested in revealing the directions of interactions, which correspond to causal relationships between the respective subsystems. This cannot be inferred by use of the partial coherence.

2.2.1 Granger-Causality and Vector Autoregressive Processes

A prerequisite for the investigation of direction of interactions between the components of a multivariate system is the operational specification of the term “causality”. Following the common concept that causes precede their effects,

Granger introduced a notion of causality for multivariate dynamical systems based on terms of predictability (Granger, 1969). In this notion, a process X_j Granger-causes a process X_i, denoted as $X_j \to X_i$, if the incorporation of the knowledge of the past of process X_j significantly improves the prediction of the process X_i compared to a prediction solely based on the knowledge of the past of process X_i and other observable processes.

In the case of linear relationships, the concept of Granger-causality can be formulated in terms of multivariate vector autoregressive processes. Univariate autoregressive processes (cf. Example 1 for an AR[2] process) can be generalized to M-dimensional multivariate vector autoregressive processes VAR[p] of order p by (Lütkepohl, 1993)

$$\mathbf{X}(t) = \sum_{n=1}^{p} \mathbf{a}_n \, \mathbf{X}(t-n) + \boldsymbol{\varepsilon}(t) , \tag{16}$$

with p coefficient matrices $\mathbf{a}_n$, $n = 1, \ldots, p$, each of dimension $M \times M$. The term $\boldsymbol{\varepsilon}(t)$ represents a M-dimensional Gaussian white noise process with covariance matrix $\boldsymbol{\Sigma}$, i.e., $\boldsymbol{\varepsilon}(t) \sim N(\mathbf{0}, \boldsymbol{\Sigma})$. The off-diagonal matrix elements $a_{ij,n}$, $i, j = 1, \ldots, M$ of the linear VAR model contain the information about Granger-causal interactions between the components of the multivariate process. For instance, $a_{ij,n} \neq 0$ describes a Granger-causal influence of X_j at time lag n on the present value of X_i, i.e., $X_j \to X_i$.

Example 2 Graphical model of a three-dimensional VAR[2]-process Interdependencies and directed influences among components of a multivariate complex system are commonly visualized by graphical models (Dahlhaus and Eichler, 2003; Eichler, 2005, 2006). They are composed of nodes and directed edges, which represent the subsystems and the direction of influences between them. In Fig. 2, a graph is presented belonging to the three-dimensional VAR[2]-process

$$\mathbf{X}(t) = \begin{pmatrix} -0.4 & 0 & 0 \\ -0.3 & 0.5 & 0 \\ 0.2 & 0 & -0.9 \end{pmatrix} \mathbf{X}(t-1) + \begin{pmatrix} 0.6 & 0 & 0 \\ 0.1 & -0.4 & 0 \\ -0.4 & 0 & 0.6 \end{pmatrix} \mathbf{X}(t-2) + \begin{pmatrix} \varepsilon_1(t) \\ \varepsilon_2(t) \\ \varepsilon_3(t) \end{pmatrix} . \tag{17}$$

The non-zero off-diagonal elements of the coefficient matrices, i.e., $a_{21,1}$, $a_{21,2}$ and $a_{31,1}$, $a_{31,2}$, describe directed Granger-causal influences of X_1 on X_2 and of X_1 on X_3 which are indicated in the graph by directed edges (cf. Fig. 2).

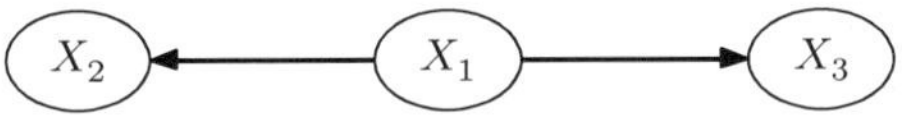

Fig. 2 Directed graph of a three-dimensional VAR[2]-process. Processes are displayed by nodes. Directed edges indicate directed interdependences between the processes. Process X_1 influences process X_2 and process X_3

2.2.2 Partial Directed Coherence

In various multivariate systems the information flow between subsystems is limited to distinct frequency bands. For such systems it is reasonable to transfer the concept of Granger-causality to the frequency domain. Granger-causal relationships can be detected in the frequency domain by means of the partial directed coherence (PDC) which is based on modeling the investigated multivariate system by vector autoregressive processes (Baccalá and Sameshima, 2001). Therefore, the PDC is regarded as a parametric approach. Based on the difference between the M-dimensional identity matrix $\mathbf{I}$ and the Fourier transform of the autoregressive coefficients

$$\mathbf{A}(\omega) = \mathbf{I} - \sum_{n=1}^{p} \mathbf{a}_n \, \mathrm{e}^{-\mathrm{i}\omega n} , \tag{18}$$

the PDC between processes X_j and X_i is defined as

$$\Pi_{i \leftarrow j}(\omega) = \frac{|A_{ij}(\omega)|}{\sqrt{\sum_k |A_{kj}(\omega)|^2}} , \qquad i \neq j . \tag{19}$$

The PDC is normalized to [0, 1]. In contrast to coherence and partial coherence it is an asymmetric measure, which enables to detect the direction of influences. A direct Granger-causal influence from process X_j to X_i is inferred by a non-zero value, i.e., $\Pi_{i \leftarrow j}(\omega) > 0$. Since VAR processes are by construction linear, strictly speaking, linear Granger-causality is inferred by the PDC.

2.2.3 Estimation of Partial Directed Coherence

The PDC is estimated according to Eqs. (19) and (18). The coefficient matrices $\widehat{\mathbf{a}}_n$ are estimated by fitting a VAR model of order p to the multivariate observations under investigation. The VAR model fit can be performed by different standard methods, e.g., via multichannel Yule-Walker VAR estimation or by a least squares fit (Brockwell and Davis, 1998; Lütkepohl, 1993). The model order p is usually unknown. Likelihood ratio tests and statistical model selection criteria can be used to estimate p, e.g., Bayesian Information Criterion (BIC) or Akaike's Information Criterion (AIC), which optimizes the tradeoff between minimizing the squared residual error of the VAR model fit and model order p (Akaike, 1974; Lehmann, 1997; Lütkepohl, 1993).

2.2.4 Testing for Significance

As for the coherence and partial coherence, the statistical properties of the PDC have to be considered in order to evaluate a PDC estimate at a given frequency and to detect directed interrelations from observations. The statistical significance of a nonzero value $\widehat{\Pi}_{i \leftarrow j}(\omega)$ can be investigated by means of a critical value c_{PDC}. Based on the assumption that there are no Granger-causal influences between two processes X_j and X_i, i.e., $|A_{ij}(\omega)| = 0$ for $i \neq j$, this value can be derived from theoretical considerations for each frequency ω at a given α-significance level by (Schelter et al., 2006b)

$$c_{PDC}(\omega) = \left(\frac{\widehat{C}_{ij}(\omega)\, \chi^2_{1,1-\alpha}}{N \sum_k |\widehat{A}_{kj}(\omega)|^2} \right)^{1/2} . \tag{20}$$

The term $\chi^2_{1,1-\alpha}$ denotes the (1–α)-quantile of the χ^2 distribution with one degree of freedom and $\widehat{C}_{ij}(\omega)$ is an estimate of the expression

$$C_{ij}(\omega) = \mathbf{\Sigma}_{ii} \left\{ \sum_{k,l=1}^{p} \mathbf{R}^{-1}_{jj}(k,l) \left[\cos(k\omega)\cos(l\omega) + \sin(k\omega)\sin(l\omega)\right] \right\} , \tag{21}$$

where $\mathbf{R}^{-1}_{jj}(k,l)$ denotes the entries of the inverse of the covariance matrix $\mathbf{R}$ of the VAR process (Lütkepohl, 1993). Estimates of the covariance matrix $\widehat{\mathbf{R}}$ and the variance matrix $\widehat{\mathbf{\Sigma}}$ of the driving Gaussian noise of the VAR process can be obtained from the VAR model fit.

The critical value $c_{PDC}(\omega)$ is a pointwise test statistic. In contrast to the critical values for the coherence and partial coherence it depends on frequency ω, cf. Eq. (20). The choice of a meaningful frequency for an investigation of directed interrelations by means of the PDC depends on the application and the spectral properties of the investigated multivariate system itself. Usually the frequencies of pronounced spectral peaks are of interest.

The variability of the PDC and the critical value depends on the number of investigated data points N. According to Eq. (20), increasing N decreases $c_{PDC}(\omega)$ and thus the performance of testing for significant directed connectivities is improved. Typically, the critical value $c_{PDC}(\omega)$ increases with increasing model order p. Therefore, even if the investigated system is strongly overparameterized, the above described critical value for the PDC prevents false positive conclusion about the directed influences (Schelter et al., 2006b).

3 Simulations: What Can Be Inferred from Multivariate Analysis

In this section, we illustrate the abilities as well as the limitations of the presented methods by simulating an autoregressive system (Sect. 3.1) and a nonlinear stochastic system (Sect. 3.3). We elucidate pitfalls together with false positive conclusions about the underlying interaction structures. Particular emphasis is laid on the difficulty of unobserved processes (Sect. 3.2). The following results, coherence spectra, partial coherence spectra, and partial directed coherence spectra were estimated using the FDMa-software package accompanying this book. VAR model fits for the PDC analysis were computed using the ARfit software package (Neumaier and Schneider, 2001; Schneider and Neumaier, 2001).

3.1 Comparison of Coherence, Partial Coherence, and Partial Directed Coherence

The coherence, partial coherence, and PDC analysis feature different abilities and are subject to certain restrictions. Some practical aspects of their application on time series and a comparison of these methods is demonstrated in the following. For this purpose, the methods were applied to a time series of the following four-dimensional VAR[3] process

$$\begin{aligned} X_1(t) &= 0.8X_1(t-1) - 0.7X_1(t-2) + 0.4X_3(t-1) + \varepsilon_1(t) \\ X_2(t) &= -0.5X_2(t-3) - 0.4X_1(t-2) + \varepsilon_2(t) \\ X_3(t) &= 0.6X_2(t-1) + \varepsilon_3(t) \\ X_4(t) &= -0.5X_3(t-3) + \varepsilon_4(t) \end{aligned} \tag{22}$$

with Gaussian white noise $\varepsilon_i \sim N(0,1)$, $i = 1, ..., 4$. In this system, there are linear influences from $X_1 \rightarrow X_2$, $X_2 \rightarrow X_3$, $X_3 \rightarrow X_4$, and $X_3 \rightarrow X_1$. For the estimation of the PDC a VAR model order $p = 10$ was chosen, which actually overparameterize the true model.

The coherence spectra $\widehat{Coh}_{ij}(\omega)$ and the partial coherence spectra $\widehat{PCoh}_{ij}(\omega)$ estimated from the VAR time series are shown in Fig. 3A. The auto-spectra $\widehat{S}_{jj}(\omega)$ estimated by smoothing the periodogram are arranged in subfigures on the diagonal of Fig. 3A. The PDC is depicted in Fig. 3B. Moreover, the auto-spectra obtained from the VAR model fit are shown on the diagonal of Fig. 3B.

In both figures, the estimated critical values of the corresponding measures for a significance level $\alpha = 0.05$ are represented by gray lines. A pronounced peak can be observed in the auto-spectra at frequency $\omega_0 = 2\pi \times 156\,\text{Hz}$, which is marked by a vertical dashed line in both figures. This frequency was selected for an evaluation of the different measures in order to deduce significant connectivities between the components of the system. Significant values, i.e., values above the critical values at the considered frequency, are interpreted as the presence of an interrelationship between the corresponding components. Not significant values at ω_0 are interpreted as the absence of a connection. As described in Sects. 2.1.7 and 2.2.4 false positive conclusions are expected in accordance to the chosen significance level with a probability of 5%. Significant results of coherence, partial coherence, and partial directed coherence are summarized by edges in graphs depicted in Fig. 4.

For instance, in Fig. 3A the coherence as well as the partial coherence between X_1 and X_2, i.e., $\widehat{Coh}_{12}$ and $\widehat{PCoh}_{12}$, are both significant at ω_0. Thus, we conclude a direct relationship between both components of the VAR process. However, for the components X_2 and X_4, we obtain that $\widehat{Coh}_{24}(\omega_0)$ is significant while $\widehat{PCoh}_{24}(\omega_0)$

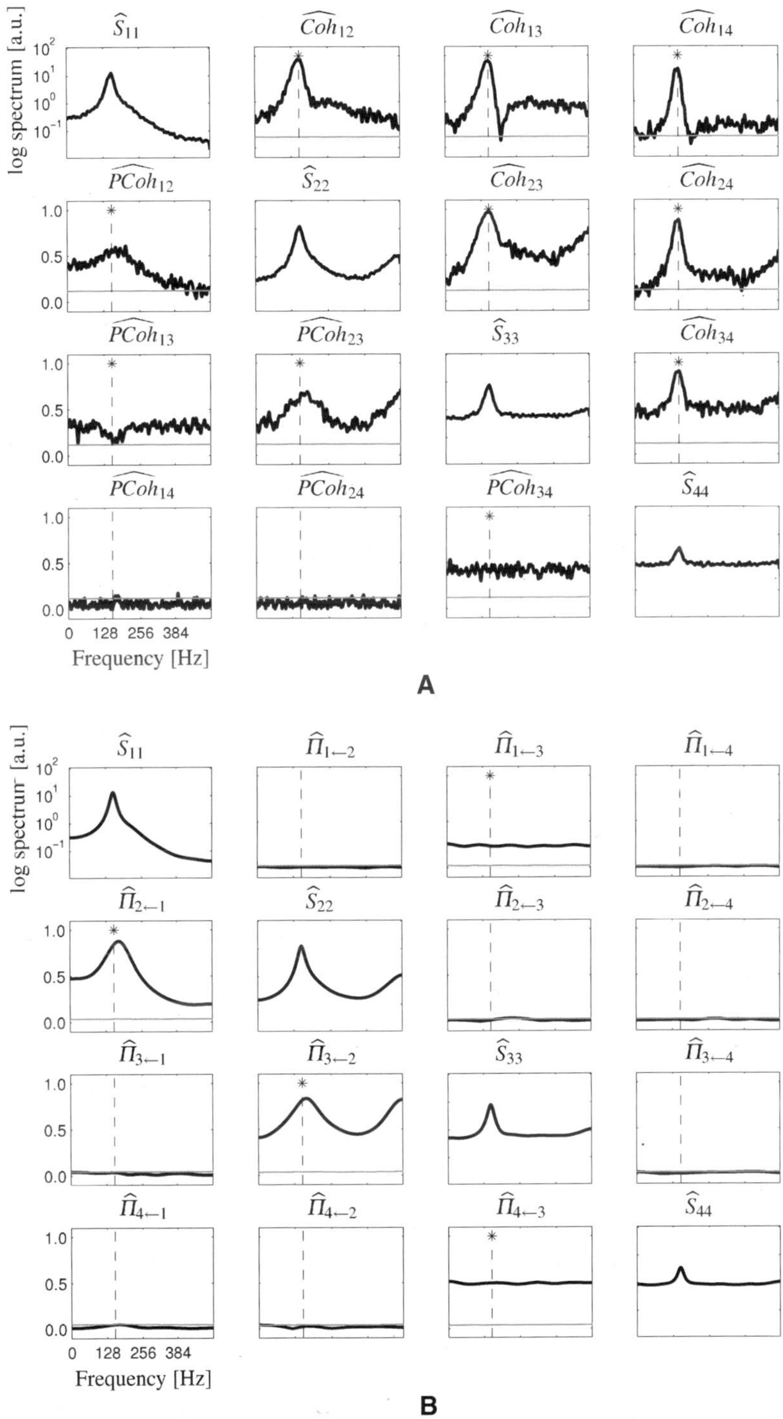

Fig. 3 (continued)

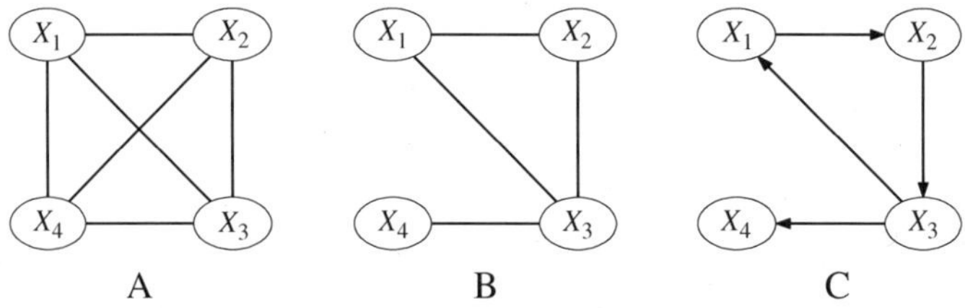

Fig. 4 Comparison of graphs representing the estimated interaction structure of a four-dimensional VAR[3] process obtained from different multivariate analysis methods. (**A**) Coherence: all edges are present denoting the presence of significant coherence between each pairwise combination of processes, (**B**) Partial coherence: edges are missing due to indirect interrelations, and (**C**). Partial directed coherence: the direction of influences between directly interrelated components are indicated by directed edges in the graph

is not significant. Hence, there is no direct relationship between these two components. From the bivariate coherence analysis, we obtain significant coherence for all pairwise combinations of processes leading to a fully connected graph (Fig. 4A). The graph deduced from the partial coherence spectra (Fig. 4B) is in agreement with the simulated influences of the VAR[3] process, as edges between processes X_1 and X_4 as well as between processes X_2 and X_4 are missing.

The direction of influences of the simulated VAR[3] systems can be concluded from the partial directed coherence analysis. For instance, in Fig. 3B we observe a significant value of $\widehat{\Pi}_{2\leftarrow 1}(\omega_0)$, while $\widehat{\Pi}_{1\leftarrow 2}(\omega_0)$ is not significant. Thus, we conclude that the influence between the processes X_1 and X_2 is unidirectional, since X_1 is influencing X_2 whereas there is no direct influence of X_2 on X_1. Unidirectional influences obtained from the PDC are summarized in the graph by directed edges (Fig. 4C). Although the model order for the PDC analysis was oversized, all direct interrelationships as well as their directions are discovered correctly by the PDC for the simulated VAR[3] process.

This example illustrates the inability of the bivariate coherence to distinguish between direct and indirect influences in multivariate systems. Whereas both methods, the partial coherence and partial directed coherence, depict the coupling scheme of direct influences correctly. Additionally, the partial directed coherence contains information about the direction of influences.

Fig. 3 Spectra of a four-dimensional VAR[3] process. (**A**) *top right triangular*: Coherence spectra; *bottom left triangular*: Partial coherence spectra. (**B**) Partial directed coherence spectra. Corresponding critical values are depicted by *gray lines*. Auto-spectra are shown on the *diagonal*. The frequency $\omega_0 = 2\pi \times 156\,Hz$ is marked by a *vertical dashed line*. For each measure, significant values at ω_0 are highlighted by *an asterik*

3.2 Unobserved Processes

The inference of the coupling scheme from a PDC analysis of multivariate observations can lead to unexpected results if there are unobserved processes influencing the system under investigation. Due to lack of relevant information from the unobserved components only restricted conclusions about Granger-causal influences and therefore the underlying coupling scheme can be drawn.

To illustrate the influence of unobserved processes on a PDC analysis we consider the four-dimensional VAR[3] process in Eq. (22). Now suppose that only the three-dimensional subsystem $\mathbf{X}_{\{1,3,4\}}$ is observed, i.e., X_2 is unobserved. For this subsystem one can derive (not shown) from Eq. (22) the following new VAR process

$$\begin{aligned} X_1(t) &= 0.8X_1(t-1) - 0.7X_1(t-2) + 0.4X_3(t-1) + \varepsilon_1(t) \\ X_3(t) &= -0.24X_1(t-3) + \tilde{\varepsilon}_3(t) \\ X_4(t) &= -0.5X_3(t-3) + \varepsilon_4(t) \end{aligned} \tag{23}$$

where $\tilde{\varepsilon}_3(t) = \varepsilon_3(t) - 0.3X_2(t-4) + 0.6\,\varepsilon_2(t-1)$. Following this equations, the process X_1 is now Granger-causal for X_3, i.e., $X_1 \to X_3$, since the process X_2 which mediated this link is not observed. The corresponding graph obtained by a partial directed coherence analysis of the subsystem $\mathbf{X}_{\{1,3,4\}}$ is shown in Fig. 5A. It correctly reflects the coupling scheme derived for the subsystem. In contrast to the graph obtained from the PDC analysis of the complete system (cf. Fig. 4C), we now obtain a bidirectional influence between X_1 and X_3, i.e., X_1 is influencing X_3 and vice versa.

A quite different graph is obtained if we consider the three-dimensional subsystem $\mathbf{X}_{\{1,2,4\}}$ where X_3 is unobserved. The corresponding VAR process of this subsystem is given as

$$\begin{aligned} X_1(t) &= 0.8X_1(t-1) - 0.7X_1(t-2) + 0.24X_2(t-2) + \tilde{\varepsilon}_1(t) \\ X_2(t) &= -0.5X_2(t-3) - 0.4X_1(t-2) + \varepsilon_2(t) \\ X_4(t) &= a\,X_1(t-2) - b\,X_1(t-3) - c\,X_1(t-4) + d\,X_2(t-4) + \tilde{\varepsilon}_4(t) \end{aligned} \tag{24}$$

with $\tilde{\varepsilon}_1(t) = \varepsilon_1(t) + 0.4\,\varepsilon_3(t-1)$ and $\tilde{\varepsilon}_4(t) = \varepsilon_4(t) - \alpha\,\varepsilon_1(t-2) + \beta\,\varepsilon_3(t-3)$ and nonzero constants $a, b, c,$ and d. This process exhibits Granger-causal

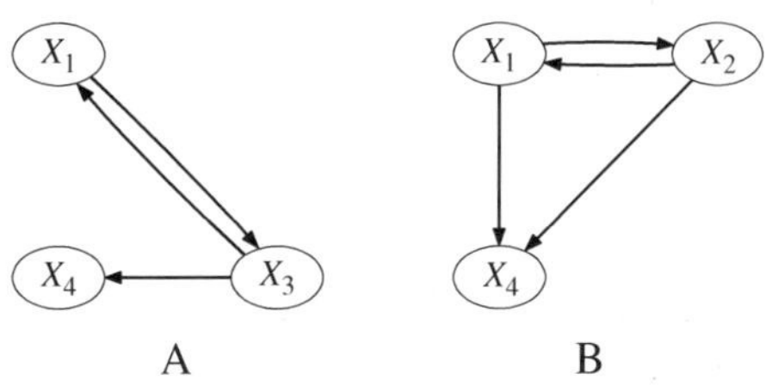

Fig. 5 Directed graphs obtained from PDC analysis of a four-dimensional VAR[3] process with unobserved components. (**A**) X_2 is unobserved, (**B**) X_3 is unobserved

dependencies from $X_2 \rightarrow X_1$ and $X_2 \rightarrow X_4$ which are mediated by X_3. Intriguingly, there is a further Granger-causal influence, namely $X_4 \rightarrow X_1$. The same coupling scheme is obtained by the partial directed coherence analysis of $\mathbf{X}_{\{1,2,4\}}$ shown in the graph in Fig. 5B. A formerly not present directed influence of X_1 on X_4 has emerged by just restricting the analysis to the particular subsystem. Intuitively this result has not been expected from the graph in Fig. 4C. This influence results from the different time lags at which the components affect each other. The influence of X_3 on X_1 has a smaller time lag than the influence of X_3 on X_4 and therefore X_4 at time t is affected by X_1 at time $t-2$.

These examples demonstrate that in the case of unobserved processes coupling schemes deduced from VAR modeling and PDC analysis can lead to unexpected results. Indirectly linked processes mediated by an unobserved process can become directly linked. Moreover, causally independent processes can become causally dependent and curious directions of influences can emerge. This should be borne in mind whenever a real-world system is analyzed. Therefore, whenever additional information about the investigated system is available it should be taken into account in order to test the reliability of the obtained graphs and for example to prevent unphysiological conclusion.

3.3 Application to Nonlinear Stochastic Systems

Although the PDC is based on linear VAR processes and thus, strictly regarded, restricted to applications to linear systems, the PDC is feasible to detect directed interrelations also in nonlinear stochastic systems. To demonstrate the performance of the PDC for nonlinear systems we simulated four coupled stochastic Van der Pol oscillators given by

$$\begin{aligned}
\ddot{x}_1 &= \mu(1-x_1^2)\dot{x}_1 - \omega_1^2 x_1 + \sigma\eta_1 + \varepsilon_{12}(x_2 - x_1)\\
\ddot{x}_2 &= \mu(1-x_2^2)\dot{x}_2 - \omega_2^2 x_2 + \sigma\eta_2 + \varepsilon_{24}(x_4 - x_2) + \varepsilon_{21}(x_1 - x_2)\\
\ddot{x}_3 &= \mu(1-x_3^2)\dot{x}_3 - \omega_3^2 x_3 + \sigma\eta_3 + \varepsilon_{31}(x_1 - x_3) + \varepsilon_{34}(x_4 - x_3)\\
\ddot{x}_4 &= \mu(1-x_4^2)\dot{x}_4 - \omega_4^2 x_4 + \sigma\eta_4 + \varepsilon_{42}(x_2 - x_4).
\end{aligned}$$

The parameters were chosen as $\omega_1 = 1.5$, $\omega_2 = 1.48$, $\omega_3 = 1.53$, and $\omega_4 = 1.44$. The interaction structure between the oscillators is determined by the coupling parameters ε_{ij} which were set to zero except for $\varepsilon_{12} = \varepsilon_{21} = \varepsilon_{31} = \varepsilon_{34} = \varepsilon_{24} = \varepsilon_{42} = 0.2$. Each oscillator was exposed to Gaussian distributed white noise η_i with standard deviation $\sigma = 1.5$.

For the estimation of the PDC a VAR model of order $p = 50$ was fitted to a time series of the system. Due to the nonlinearity of the system, a large model order is needed for an adequate VAR model fit in order to capture nonlinear effects in the data. The estimated PDC and critical values for a 5%-significance level as well as the auto-spectra are shown in Fig. 6A. A pronounced peak can be found in the

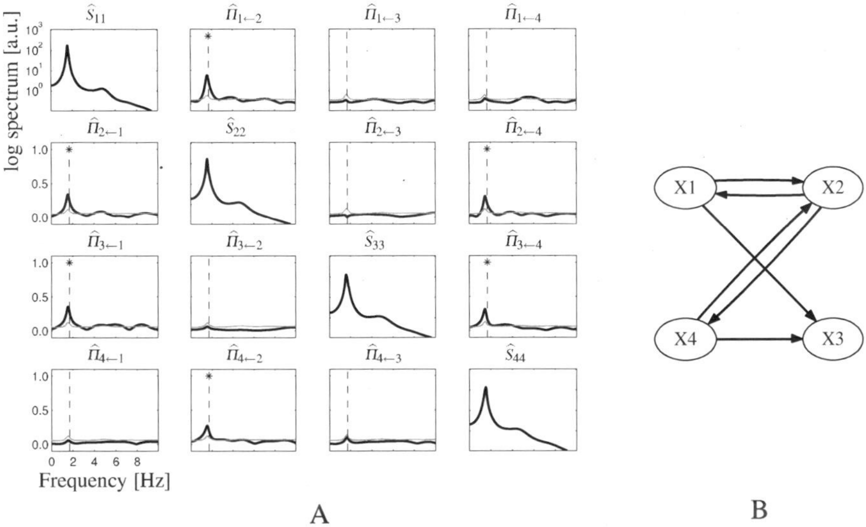

Fig. 6 (**A**) PDC and spectra of four coupled Van der Pol oscillators. The model order was set to $p = 50$. (**B**). Interaction graph deduced from significant PDC at the peak frequency at 1.7 Hz

auto-spectra. The coupling scheme deduced from the estimated PDC at the peak frequency is depicted in the graph in Fig. 6B. It is in accordance to the simulated coupling scheme.

Although the coupling scheme in this example is reproduced correctly by the PDC, the detection of coupling directions in a nonlinear stochastic system of coupled oscillators is a nontrivial task. For example, the coupling strengths of the oscillators can be essential for the applicability of the PDC (Winterhalder et al., 2007). As a general rule we strongly recommend to perform tailored simulation studies in advance of the analysis of nonlinear systems.

The presented example illustrates that partial directed coherence is a sensitive analysis technique. Although partial directed coherence was originally designed for linear systems, it is capable of correctly reproducing the network structure also for systems it has not been developed for. But partial directed coherence is not specific. That means, conclusions from a certain result to the underlying dynamics are not possible. One cannot tell that the underlying system consisted of coupled oscillators using the partial directed coherence analysis. This should also be kept in mind when analyzing data where the underlying dynamic is unknown. Extensions of techniques showing high sensitivity but poor specificity that enables increasing the latter have recently been discussed (Schelter et al., 2007; Winterhalder et al., 2006).

4 Application to Real-World Data

In this section, we present an application of the partial directed coherence to electroencephalographic (EEG) and electromyographic (EMG) signals of a patient suffering from Parkinsonian tremor to gather insights into the underlying mechanisms of this symptom of Parkinson's disease.

4.1 Application to Tremor Data

Tremor is an involuntary rhythmical movement predominantly of the upper limbs and occurs, for example, as a symptom of Parkinson's disease. The pathophysiology of the human tremor is still under debate. The partial directed coherence allows for the investigation of the relationship between trembling muscles and the brain activity in order to answer the question whether there is a tremor-correlated activity in the sensorimotor cortex.

We investigated the tremor activity of the left wrist flexor of a Parkinsonian patient using an EMG electrode. The EMG signal and simultaneously recorded EEG signals of the left and right sensorimotor cortices of this patient were analyzed by means of the partial directed coherence. The estimated PDC for this three-dimensional data set is depicted in Fig. 7A. The order of the VAR model was set to $p = 50$ to capture influences between the processes with a time lag up to the duration of 2.5 periods of the tremor process. The trembling frequency of this patient is about 5 Hz, which is typical for Parkinsonian patients. The trembling frequency and its higher harmonics can be identified by maxima in the estimated EMG spectrum $\widehat{S}_{33}(\omega)$. The EEG spectra of the right and left hemispheres, $\widehat{S}_{11}(\omega)$ and $\widehat{S}_{22}(\omega)$, contain activity over a broad range of frequencies.

An evaluation of the PDC at the frequency of the trembling muscle leads to a significant bidirectional Granger-causal influence between the right EEG and the

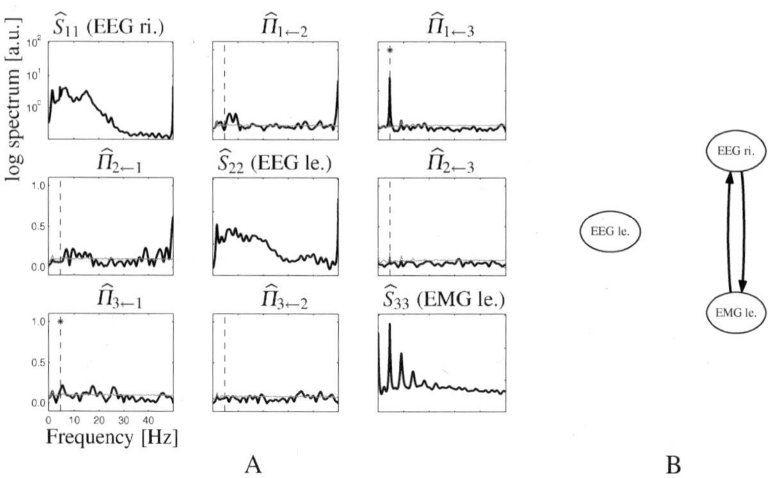

Fig. 7 Application of the PDC to tremor data, consisting of EEG recordings of the *right* and *left hemispheres* and an EMG recording of the *left* flexor muscle. (**A**). Estimated PDC and auto-spectra for chosen model order $p = 50$. (**B**). Graph obtained from significant PDC at frequency 4.7 Hz

EMG of the flexor of the left wrist, which is summarized in the graph in Fig. 7B. An ipsilateral coupling, i.e., between left hemisphere and left wrist, is not found. A further significant bidirectional coupling can be deduced from PDC between the left and the right hemispheres (c.f. Fig. 7A), but at frequencies higher than the tremor frequency and thus it is not depicted in the graph.

One can further observe that at the trembling frequency the value of the PDC for the influence of the muscle on the brain, $\widehat{\Pi}_{1\leftarrow 3}$, is larger than $\widehat{\Pi}_{3\leftarrow 1}$ for the opposite direction. But this should not be interpreted as an evidence for a stronger influence of the muscle on the brain, since the comparison of the magnitude of PDC values is not meaningful due the normalization of the PDC. The difference in the magnitude is ascribed to the higher signal-to-noise ratio of the EMG compared to the EEG. A low signal-to-noise ratio leads to an underestimation of the VAR parameters and therefore to a lower PDC.

In summary, by applying not only a technique measuring the interaction between the recorded processes but also the direction thereof, we could show not only that the sensorimotor cortex receives feedback from the muscles but also that the cortex plays an active role in tremor generation. We also emphasize that the interaction between the muscle and the sensorimotor cortex is contralateral, i.e., in our case between the left muscle and the right EEG, which is expected from physiological knowledge. The results related to the direction of information flow need to be verified by a tremor study including more patients and different types of tremor.

5 Discussion and Conclusion

In this chapter, we discussed theoretical and practical aspects for the detection of undirected and directed interactions within multivariate systems by means of coherence, partial coherence, and partial directed coherence. Since these measures are formulated in the frequency domain, they are especially meaningful for the investigation of systems in which pronounced spectral peaks or distinct frequency bands are of interest.

Particularly in the field of neuroscience, bivariate coherence analysis is widely used. But as demonstrated in simulations, this approach cannot distinguish between direct and indirect interrelations in multivariate systems. Indirect interrelations, however, are likely to occur especially in large systems due to causally mediated influences or common influences on actually mutually independent components. In order to discriminate between direct and indirect influences, multivariate measures like the partial coherence, which take the entire system into account, are necessary. The advantage of partial coherence compared to coherence when investigating multivariate systems was demonstrated in simulations.

Besides the determination of direct interrelationships, the investigation of the direction of influences is of current interest in many fields of research. Coherence and partial coherence, however, are both undirected measures and thus cannot reveal such information. Based on the notion of Granger-causality and vector autoregressive modeling of multivariate time series directed measures, like the partial directed coherence, are defined. By simulating VAR processes and nonlinear stochastic systems, we demonstrated the ability of the PDC to detect and to reveal direct influences and their direction correctly as long as all processes of a system are observed. In the case of unobserved components, we have shown that a PDC analysis can result in deceptive coupling schemes, for example, indirect influences mediated by unobserved processes emerge as direct ones. Hence, conclusions from graphs obtained by a PDC analysis are restricted to the investigated system itself. That means, for example, that the evidence for a coupling from significant PDC is no evidence for a real direct coupling between the investigated processes as long as influences of unobserved processes cannot be excluded. This has to be kept in mind when interpreting a coupling scheme deduced from a PDC analysis of real-world signals. However, a multivariate approach like the PDC is still favorable compared to bivariate approaches.

Since the PDC is a parametric approach, the adequate representation of the system under investigation by a VAR[p] model is crucial. Hence, the choice of model order p is important. In practice, the order is usually unknown. A comparison of the spectra of the non-parametric and parametric approaches for different p can yield a hint of the appropriate model order. Besides, likelihood ratio tests and statistical model selection criteria can be employed to select p (Akaike, 1974; Lehmann, 1997; Lütkepohl, 1993). If the model order p is chosen too small, especially smaller than the time lags at which the individual system components affect each other, it is not possible to conclude the underlying Granger-causal influences correctly. Otherwise, increasing the order p increases the variability of the PDC and thus its significance level, which prevents erroneous conclusions about probably not existent directed influences. Therefore, if the system under investigation is unknown it is preferable to oversize the order p of the VAR[p] model instead of using a small model order.

The choice of large model orders is furthermore advisable if nonlinear stochastic systems are investigated. By definition, VAR models and thus the PDC are designed for linear interrelations. Consequently, a misspecification of the model order is inevitable for nonlinear systems. Nevertheless, we demonstrated by means of simulated coupled stochastic Van der Pol oscillators that the correct coupling scheme can be revealed from partial directed coherence even for such systems, as long as the model order is chosen sufficiently large. This observation is in agreement to numerous simulation studies (Schelter et al., 2006b, c; Winterhalder et al., 2005, 2007). We note that various nonlinear methods for the investigation of nonlinear systems can be found in literature. However, they are mainly based on bivariate measures and thus again, cannot distinguish between direct and indirect influences. That means, linear multivariate approaches are still favorable compared to nonlinear bivariate approaches in order to adequately describe the coupling scheme of multivariate systems. Multivariate nonlinear methods are difficult to design and subject of recent research, e.g., the partial phase synchronization (Schelter et al., 2006a) which aims to investigate phase dynamics in synchronizing coupled self-sustained oscillators.

The presented methods are based on the assumption of investigating stationary processes. In practice, however, it may happen that one is confronted with nonstationary signals. For instance, long-term EEG recordings are strictly regarded not stationary due to different activity patterns and circadian rhythms which are passed over several hours. Especially the representation of a nonstationary system by a stationary VAR model is not appropriate. In this case, it is recommended to analyze signals blockwise over time by choosing a blocklength such that effects of nonstationarity are negligible. Such a time-resolved analysis allows to track transitions in the interdependence structure of nonstationary systems which may be of particular interest.

In summary, we demonstrated the importance of multivariate methods in order to investigate interaction structures in dynamical systems. We elaborated on possible pitfalls and emphasized the advantages of the partial directed coherence over coherence and partial coherence for the investigation of direct directed interrelationships. In an application to EMG and EEG signals of tremor, we illustrated the power of the partial directed coherence to reveal tremor-correlated cortical activity in the Parkinson's disease and to facilitate investigations of neurological diseases.

Acknowledgments Special thanks goes to Bernhard Hellwig, Florian Amtage, and Professor Carl Hermann Lücking who provided us not only with the tremor data but also with knowledge about the neurophysiology of Parkinsonian tremor. This work was supported by the German Science Foundation (Ti315/2-1) and by the German Federal Ministry of Education and Research (BMBF grant 01GQ0420).

References

H. Akaike. A new look at the statistical model identification. *IEEE Trans. Automat. Contr.*, AC-19:716–723, 1974.

L. A. Baccalá and K. Sameshima. Partial directed coherence: a new concept in neural structure determination. *Biol. Cybern.*, 84:463–474, 2001.

P. Bloomfield. *Fourier Analysis of Time Series: An Introduction*. John Wiley & Sons, New York, 1976.

D. R. Brillinger. *Time Series: Data Analysis and Theory*. Holden-Day, San Francisco, 1981.

P. J. Brockwell and R. A. Davis. *Time Series: Theory and Methods*. Springer, New York, 1998.

R. Dahlhaus. Graphical interaction models for multivariate time series. *Metrika*, 51:157–172, 2000.

R. Dahlhaus and M. Eichler. Causality and graphical models for time series. In P. Green, N. Hjort, and S. Richardson, editors, *Highly Structured Stochastic Systems*, pp. 115–137. Oxford University Press, Oxford, 2003.

M. Eichler. A graphical approach for evaluating effective connectivity in neural systems. *Phil. Trans. R. Soc. B*, 360:953–967, 2005.

M. Eichler. Graphical modeling of dynamic relationships in multivariate time series. In B. Schelter, M. Winterhalder, and J. Timmer, editors, *Handbook of Time Series Analysis*, Chapter 14, pp. 335–372. Wiley-Vch, Weinheim, 2006.

J. Granger. Investigating causal relations by econometric models and cross-spectral methods. *Econometrica*, 37:424–438, 1969.

E. J. Hannan. *Multiple Time Series*. Handbook of Statistics. Wiley, New York, 1970.

E. L. Lehmann. *Testing Statistical Hypotheses*. Springer, New York, 1997.

H. Lütkepohl. *Introduction to Multiple Time Series Analysis*. Springer, Berlin 1993.

A. Neumaier and T. Schneider. Estimation of parameters and eigenmodes of multivariate autoregressive models. *ACM Trans. Math. Softw.*, 27:27–57, 2001.

M. B. Priestley. *Spectral Analysis and Time Series*. Academic Press, London, 1989.
B. Schelter, M. Winterhalder, R. Dahlhaus, J. Kurths, and J. Timmer. Partial phase synchronization for multivariate synchronizing systems. *Phys. Rev. Lett.*, 96:208103, 2006a.
B. Schelter, M. Winterhalder, M. Eichler, M. Peifer, B. Hellwig, B. Guschlbauer, C. H. Lücking, R. Dahlhaus, and J. Timmer. Testing for directed influences among neural signals using partial directed coherence. *J. Neurosci. Methods*, 152:210–219, April 2006b.
B. Schelter, M. Winterhalder, B. Hellwig, B. Guschlbauer, C. H. Lücking, and J. Timmer. Direct or indirect? Graphical models for neural oscillators. *J. Physiol. Paris*, 99:37–46, January 2006c.
B. Schelter, M. Winterhalder, J. Kurths, and J. Timmer. Phase synchronization and coherence analysis: Sensitivity and specificity. *Int. J. Bif. Chaos*, 17:3551–3556, 2007.
T. Schneider and A. Neumaier. Algorithm 808: ARfit – A Matlab package for the estimation of parameters and eigenmodes of multivariate autoregressive models. *ACM Trans. Math. Softw.*, 27:58–65, 2001.
J. Timmer, M. Lauk, S. Häußler, V. Radt, B. Köster, B. Hellwig, B. Guschlbauer, C. H. Lücking, M. Eichler, and G. Deuschl. Cross-spectral analysis of tremor time series. *Int. J. Bif. Chaos*, 10:2595–2610, 2000.
M. Winterhalder, B. Schelter, W. Hesse, K. Schwab, L. Leistritz, D. Klan, R. Bauer, J. Timmer, and H. Witte. Comparison of linear signal processing techniques to infer directed interactions in multivariate neural systems. *Sig. Proc.*, 85:2137–2160, 2005.
M. Winterhalder, B. Schelter, J. Kurths, A. Schulze-Bonhage, and J. Timmer. Sensitivity and specificity of coherence and phase synchronization analysis. *Phys. Lett. A*, 356:26–34, 2006.
M. Winterhalder, B. Schelter, and J. Timmer. Detecting coupling directions in multivariate oscillatory systems. *Int. J. Bif. Chaos*, 17:3735–3739, 2007.

Supplemental electronic material can be found at
http://www.springer.com/978-0-387-93796-0

The Phase Oscillator Approximation in Neuroscience: An Analytical Framework to Study Coherent Activity in Neural Networks

Roberto F. Galán

Abstract One of the major challenges in neuroscience today is to understand how coherent activity emerges in neural networks from the interactions between their neurons, as well as its role in normal and pathological brain function. Probably the two most relevant examples of coherent activity in the brain are synchronization and phase-locking. The phase oscillator approximation used in physics and the mathematical models based thereon provide a powerful tool to investigate the biophysical mechanisms underlying these phenomena. In particular, the experimental or numerical determination of the phase–response curve, which derives from the intrinsic properties of a real or a simulated neuron, allows one to greatly simplify the analytical treatment of their dynamics, as well as their computational modeling. In this way, it is possible to predict the existence and stability of phase-locked states and of synchronized assemblies in real and simulated neural networks with electrical or chemical synapses. The phase–response curve also determines the stability of repetitive neuronal firing thereby affecting spike-time reliability and noise-induced synchronization. Moreover, for neurons that fire repetitively, the phase–response curve is directly related to the waveform of the input that most likely precedes an action potential, i.e., to the spike-triggered average, or optimal stimulus. Here, we discuss these phenomena with special emphasis on the implications for neural function.

1 Introduction

Neural activity frequently organizes in assemblies of neurons firing synchronized action potentials. This synchronous, usually periodic activity generates the electrical waves registered in electroencephalograms (EEG) that have been reported in a variety of species. Since the frequency and amplitude of these oscillations depend

R.F. Galán (✉)
Department of Neurosciences, School of Medicine, Case Western Reserve University, 10900 Euclid Avenue, Cleveland, OH 44106, USA
e-mail: rfgalan@case.edu

J.L. Perez Velazquez, R. Wennberg (eds.), *Coordinated Activity in the Brain*, Springer Series in Computational Neuroscience 2, DOI 10.1007/978-0-387-93797-7_4,

on the state of awareness and on the brain area from which they are recorded, it has been proposed that neural synchrony is used to encode information and facilitate the communication between the different parts of the brain involved in processing such information (Singer and Gray, 1995; Engel and Singer, 2001). Whereas this hypothesis still remains controversial (Shadlen and Movshon, 1999), in the last few years it has been demonstrated that the alteration of certain brain rhythms correlates with different pathologies, suggesting a functional role for neural oscillations and synchrony. For example, certain types of epilepsy reflect an excess of neuronal synchrony, whereas a defect of synchrony in the gamma frequency band (20–80 Hz) in prefrontal areas correlates with short-term memory loss, schizophrenia, and autism (Uhlhaas and Singer, 2006). The biophysical mechanisms underlying this alteration of brain waves, and hence, of neural synchrony are still under study. At the same time, the general problem of neural synchronization has been intensively investigated in the context of computational neuroscience. The gamut of models ranges from highly detailed, describing synapses, ion channel kinetics and their distribution along the neuronal morphology, to highly reduced models that keep only the essential features of the neural dynamics leading to synchrony. Interestingly, the former type can usually be reduced to the latter by applying adequate mathematical techniques, which will be briefly overviewed here, namely, the phase oscillator approximation and the phase–response curve. An obvious advantage of having simple models that preserve relevant dynamical features of more complex ones is that computer simulations become faster and more efficient. Another advantage is that the analytical treatment of these models is easier and some important quantities can be exactly calculated. These simple models also allow for a more intuitive interpretation of the mechanisms underlying neuronal synchronization. In this chapter, we focus on these simplified models emerging from the phase oscillator approximation in neuroscience. In Sect. 2 we introduce this approach and in Sect. 3 we will discuss several applications. Whereas the mathematical details can be found in the references provided, here we make use of a rather intuitive description.

2 Mathematical Description of Biological Rhythms

Rhythms and synchrony are ubiquitous phenomena in biology. In contrast to the systems investigated in physics and chemistry, biological systems typically consist of several hierarchical levels of interacting parts, like the case considered here of neuronal networks. Quite frequently these systems organize in coherent patterns, like synchronized assemblies. This implies that their large number of degrees of freedom can be reduced, in certain circumstances, to a few order parameters, thereby reducing the dimensionality of the dynamics. In the case of systems displaying periodic activity in time, the dynamics can be reduced to as simple as a single parameter: the phase. Obviously, this dynamic is trivial: it is topologically equivalent to the motion of a particle with constant speed on a circumference. However, as already appreciated by Winfree over three decades ago, periodic systems can be investigated more deeply in a semi-quantitative way by measuring the effect of perturbations on their dynamics, i.e., their phase-resetting properties (Winfree, 2001). The rationale behind this approach is that a stable limit cycle underlies the periodic dynamics and thus, by perturbing the system, it will slip off the limit cycle and will come back to it typically at another point along the cycle, i.e., at another phase. This phase change provides us with information about the neighborhood (basin of attraction) of the stable limit cycle. If the perturbations are strong enough, it is possible for the system to explore a larger neighborhood and even interrupt the periodic motion by escaping from the basin of attraction of the limit cycle. Winfree applied these ideas to investigate a variety of biological systems displaying periodic activity including circadian rhythms, the respiratory cycle, and the heart beat (see Winfree, 2001 and references therein).

2.1 Time, Phase, and Phase Resetting

Entering the territory of definitions, we should start by saying that the definition of phase is not unique: several definitions are possible and they are not mathematically equivalent. Therefore, it is necessary to be consistent with the definition chosen. In general terms, the phase is just a monotonic mapping of the state of the system along the limit cycle onto the line of the real numbers, $\mathbb{R}$. Typical definitions of phase are different because they describe the limit cycle in different ways. For example, the limit cycle can be represented in an embedding of the dynamics using delay coordinates (Oprisan and Canavier, 2002). Similarly, it can be represented on the complex plane with the measured signal on the abscissa and its Hilbert transform on the ordinate (Schafer et al., 1999). The common feature of these representations is that they are topologically equivalent to a ring. This suggests that we can define a phase directly along a circumference. Following (Winfree, 2001), the phase can be defined as the *time*, t elapsed between two reference marks of the periodic dynamics:

$$\varphi = 2\pi \times (t \bmod T),$$

where T is the duration of the cycle. As mentioned above, this has trivial dynamics: it represents a particle moving along the circumference with constant angular velocity. The interesting aspects appear when one focuses on the response of the system to perturbations. For this reason one needs a more general definition that accounts for the modulation of the instantaneous frequency as a result of those perturbations:

$$\varphi(t) = 2\pi \frac{t - t_k}{t_{k+1} - t_k}, \qquad t_k \leq t < t_{k+1}, \tag{1}$$

where the t_k, $k = 1,2,3\ldots$ indicate the points in time chosen to be the beginning of successive cycles, for example, in the case of neurons, t_k are the times with largest spike amplitude in a spike train. This definition of phase allows us to compare the expected phase of the system in the absence of a perturbation, $\varphi_{\text{old}} = 2\pi(t - t_k)/T$ with the actual phase as a result of the perturbation, φ_{new} as in Eq. (1). Winfree referred to the plot of φ_{new} vs. φ_{old} as the phase-resetting curve. Alternatively, one can plot $\varphi_{\text{new}} - \varphi_{\text{old}}$ vs. φ_{old}, which tells us how much the phase of the oscillator is changed (reset) by a perturbation applied at φ_{old} and therefore can be thought of as the phase–response curve of the oscillator. Nonetheless, the terms "phase resetting" and "phase response" are frequently used indistinctly.

Strictly speaking, Eq. (1) is only valid for weak or brief perturbations of a stable limit cycle with very strong attraction, because in those cases the system will recover quickly from the perturbation. However, the concept of phase resetting can also be applied to strong or long stimuli. To this end, the old phase is taken to be the phase at which the stimulus starts and the new phase is defined as the *asymptotic phase*, i.e., as the phase that the oscillator has when it eventually comes back to the limit cycle after the perturbation. If the perturbation is too strong or too long it may pull the system out of the basin of attraction of the limit cycle, leaving the system in an aperiodic, phase-less state.

The existence of this strong form of resetting has been thoroughly reported by Winfree and other authors (Winfree, 2001) in circadian, respiratory, and cardiac rhythms. These studies can explain the existence of fatal perturbations causing arrhythmias or even terminating the heart beat. At the same time, this approach can also provide a semi-quantitative explanation for the benefits of deep-brain stimulation to interrupt pathological synchronization in the basal ganglia of patients with Parkinson's disease and in other brain areas during epileptic seizures.

In contrast to the case of strong resetting, which cannot be treated mathematically in the general case, the case of weak resetting allows for detailed mathematical studies in many situations of great relevance for neuroscience. In the rest of this chapter, we will focus on weak resetting.

2.2 Phase Oscillator Approximation and the Kuramoto Model

Let us start with a general dynamical system of the form

$$\frac{d\vec{X}}{dt} = \vec{F}(\vec{X}),$$

which possesses a stable limit cycle, $\vec{X}_0$. We now consider a small, perturbing force $\vec{p}(t)$:

$$\frac{d\vec{X}}{dt} = \vec{F}(\vec{X}) + \vec{p}(t). \tag{2}$$

Following Kuramoto (Kuramoto, 1984), we define the scalar function $\varphi(\vec{X})$ and take its time derivative applying the chain rule

$$\frac{d\varphi}{dt} = \frac{d\varphi}{d\vec{X}} \cdot \frac{d\vec{X}}{dt} = \frac{d\varphi}{d\vec{X}} \cdot \left(\vec{F}(\vec{X}) + \vec{p}\,(t)\right) = \frac{d\varphi}{d\vec{X}} \cdot \vec{F}(\vec{X}) + \frac{d\varphi}{d\vec{X}} \cdot \vec{p}\,(t). \quad (3)$$

So far, the scalar $\varphi(\vec{X})$ is a phase arbitrarily defined. We now make our choice unique by imposing

$$\left.\frac{d\varphi}{d\vec{X}}\right|_{\vec{X}_0} \cdot \vec{F}(\vec{X}) = \omega, \quad (4)$$

where ω is a constant representing the angular frequency of the model, i.e., we are mapping the unperturbed limit cycle onto a ring. We then define the phase-dependent sensitivity (Kuramoto, 1984), a.k.a. phase-resetting or *phase–response curve* as

$$\vec{Z}(\varphi) = \left.\frac{d\varphi}{d\vec{X}}\right|_{\vec{X}_0}. \quad (5)$$

One can now define what weak perturbation means mathematically. Indeed, the perturbing force $\vec{p}(t)$ can be considered as small, if it satisfies

$$\omega \gg \vec{Z}(\varphi) \cdot \vec{p}\,(t),$$

which implies that the frequency of the oscillation is only slightly modified by the perturbation. This in turn implies that the oscillator remains very close to the limit cycle despite the perturbation, so that phase and asymptotic phase are equivalent and a one-to-one mapping of time. Therefore, we can replace $\vec{p}(t)$ with $\vec{p}(\varphi)$. Then, taking into account (4) and (5) Eq. (3) becomes

$$\frac{d\varphi}{dt} = \omega + \vec{Z}(\varphi) \cdot \vec{p}\,(\varphi). \quad (6)$$

Equation (6) is the celebrated Kuramoto model of a phase oscillator (Kuramoto, 1984), which for neurons is schematically represented in Fig. 1.

In the case of neurons, the perturbing force $\vec{p}(\varphi)$ typically affects only one degree of freedom of the dynamical system (2). For example, if one thinks of the Hodgkin–Huxley model (Hodgkin and Huxley, 1952), a synaptic input or an injected current pulse will only affect the equation of the membrane potential (V), leaving unaffected the equations of the ionic gates (h, m, and n). This means that only one component of $\vec{p}(\varphi)$ contributes to the dot product $\vec{Z}(\varphi) \cdot \vec{p}(\varphi)$. Consequently, Eq. (6) reduces to a simpler form

$$\frac{d\varphi}{dt} = \omega + Z(\varphi)p(\varphi). \quad (7)$$

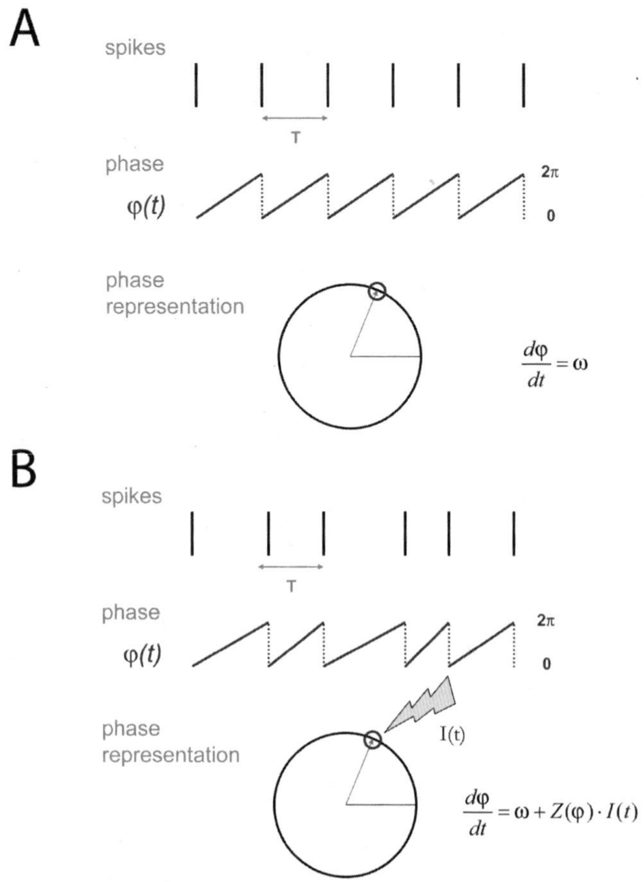

Fig. 1 Phase oscillator approximation. (**A**) An unperturbed, regularly spiking neuron, its associated phase, and its geometrical representation as a particle moving on a circumference with constant angular velocity. (**B**) The spiking neuron is weakly perturbed and as a result the angular velocity is modulated in time in a phase-dependent manner. $Z(\varphi)$ is the phase response and $I(t)$ is the perturbing current

There are three remarkable features of Eq. (7). The first one is that it is the perturbing stimulus itself, $p(\varphi)$ that enters the equation and not a projection or any other transformation of it. The second feature is that the effect of the perturbation is simply a product of the stimulus with the phase–response curve, and not something more complicated, like a convolution. The third feature is that it is a scalar, first order differential equation with no delays, i.e., the dynamics of the general model (2), say for example, the Hodgkin–Huxley model, has been greatly simplified to model (7). The price that one pays for this is the determination of the phase–response curve, $Z(\varphi)$ of model (2).

2.3 Estimation of the Phase–Response Curve

Several approaches have been proposed to estimating the phase–response curve of neurons. A popular approach consists of measuring changes in the interspike interval as a result of a perturbation (e.g., a current pulse) at each phase (Rinzel and Ermentrout, 1998). In particular, if T is the unperturbed period and $T'(\varphi)$ is the period perturbed at phase φ, the phase–response curve $\Delta(\varphi)$ is calculated as

$$\Delta(\varphi) = \frac{T - T'(\varphi)}{T}. \tag{8}$$

It is worth noting the relationship between this definition and the phase–response curve $Z(\varphi)$ of the Kuramoto model, when the perturbation is weak. From (8) we obtain

$$T'(\varphi) = T - T\Delta(\varphi). \tag{9}$$

Taking the inverse of each side in (9) and multiplying by 2π we get

$$\frac{2\pi}{T'(\varphi)} = \frac{2\pi}{T - T\Delta(\varphi)} \approx \frac{2\pi}{T} + \frac{2\pi}{T}\Delta(\varphi), \tag{10}$$

where we have used the following Taylor expansion assuming that the perturbation is weak, i.e., $T \gg T\Delta(\varphi)$:

$$\frac{1}{a+x} \approx \frac{1}{a} - \frac{x}{a^2}.$$

Comparing (7) with (10) and recalling the phase definition in (1) we finally obtain

$$\omega = \frac{2\pi}{T}, \quad Z(\varphi) = \omega\Delta(\varphi),$$

i.e., both phase–response curves are equivalent and identical up to a constant factor.

An alternative method to estimate the phase–response curve consists of calculating the phase via (1) in response to a sequence of pulses and fitting the experimental data to the Kuramoto model (Galán et al., 2005). In particular, the phase–response curve, $Z(\phi)$ is expanded as a Fourier series:

$$\frac{d\varphi}{dt} = \omega + \sum_{n=1}^{N}[a_n \cos(n\varphi) + b_n \sin(n\varphi)]p(\varphi). \tag{11}$$

The Fourier coefficients a_n and b_n are the only unknowns in (11), since the phase and its derivative are measured experimentally and $p(\varphi)$ is the applied stimulus (current pulses). Then, a_n and b_n can be easily calculated with a linear regression of the experimental data to Eq. (11). A typical choice for the number of Fourier components in (11) is $N=2$ or 3 to avoid overfitting. This method has been applied to estimate the phase response of mitral cells in the olfactory system of mice, from which their synchronization properties can be studied (Galán et al., 2005; Galán et al., 2006). Along this line, more recent studies have applied a similar approach to inhibitory (Tateno and Robinson, 2007; Mancilla et al., 2007) and excitatory (Tsubo et al., 2007) cortical neurons.

The previous methods can be applied to both real and simulated neurons. In the case of simulated neurons, like the Hodgkin–Huxley model, the phase–response curve can be directly calculated from the equations, usually involving numerical techniques. This was already appreciated by Kuramoto who described a way of performing these calculations (Kuramoto, 1984). Similar and perhaps easier calculations are also provided in Brown et al. (2004) which also compiles the methods from Hoppensteadt and Izhikevich (1997) and Ermentrout (1996). The latter is implemented in the XPPAUT software package (Ermentrout, 2002).

In general, the phase–response curves of real neurons and realistic neural models have simple forms like those in Fig. 2 which can be divided into two qualitative types. Type I are strictly positive curves, which implies that a small pulse delivered at any phase always advances the next spike. Type II are partially positive and partially negative, i.e., depending on the phase at which the pulse is delivered, the next spike will be advanced or delayed. Type I possess an integrator behavior, i.e.,

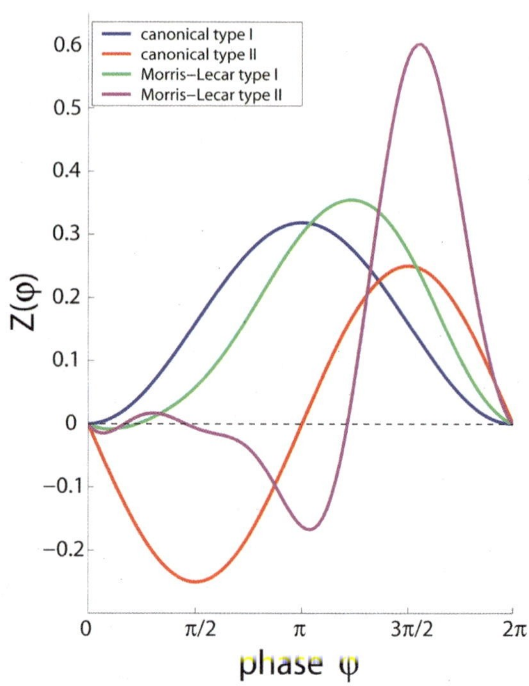

Fig. 2 Some examples of phase–response curves. The curves from a modified Morris–Lecar model (Rinzel and Ermentrout, 1998) in two different regimes resemble the phase–response curves of real neurons (see references in text)

the neuron integrates the input until the next spike is generated. In two-dimensional models of single-neuron dynamics these neurons come to fire from a quiescent state through a saddle–node bifurcation (Ermentrout, 1996). Type II neurons on the other hand do not just integrate inputs, since a pulse can delay the next spike. In real neurons, this type of behavior emerges, for example, when a small current pulse accelerates the activation of a potassium current, which in turn tends to repolarize the cell. As a result, the net effect of the positive perturbation is to slightly hyperpolarize the cell delaying the next spike. In two-dimensional models of single-neuron dynamics, type II neurons come to fire from a quiescent state through a Hopf bifurcation (Ermentrout, 1996), which endows the neurons with a voltage-dependent subthreshold resonance. Therefore they are frequently referred to as resonators. Some examples of phase–response curves are shown in Fig. 2.

3 Applications of the Phase Oscillator Approximation in Neuroscience

The purpose of introducing phase oscillator models is to simplify the quantitative analysis of interacting oscillatory systems. In neuroscience, the oscillators are neurons (or large neural masses, as in EEG studies) that fire roughly regularly either permanently or transiently, and which excite or inhibit other neurons in the network. The mathematical analysis based on the phase–response curve permits us to understand the features of neural excitability and synaptic interactions that lead to synchrony and phase-locking, which are thought to underlie information processing in the brain. Moreover, the phase–response curve is closely related to the optimal stimulus of the neuron and therefore, it also affects spike-time reliability. Thus, the knowledge of the phase–response curve allows us to quantify several aspects of the neural dynamics that are relevant for encoding and processing information in neural networks, as summarized in Fig. 3.

3.1 Dynamics of Oscillator Networks

We now need to modify model (7) in order to take into account interactions between two periodic systems (Kuramoto, 1984; Hoppensteadt and Izhikevich, 1997)

$$\begin{cases} \dfrac{d\vec{X}_1}{dt} = \vec{F}_1(\vec{X}_1) + \vec{V}_1(\vec{X}_1, \vec{X}_2) \\ \dfrac{d\vec{X}_2}{dt} = \vec{F}_2(\vec{X}_2) + \vec{V}_2(\vec{X}_2, \vec{X}_1) \end{cases}$$

where the terms $\vec{V}_1(\vec{X}_1, \vec{X}_2)$ and $\vec{V}_2(\vec{X}_2, \vec{X}_1)$ account for the weak, reciprocal coupling between the systems. In analogy to Sect. 2.2, the perturbing forces are now the mutual coupling terms, $\vec{P}_i = \vec{V}_i(\vec{X}_i, \vec{X}_j)$, so that one obtains

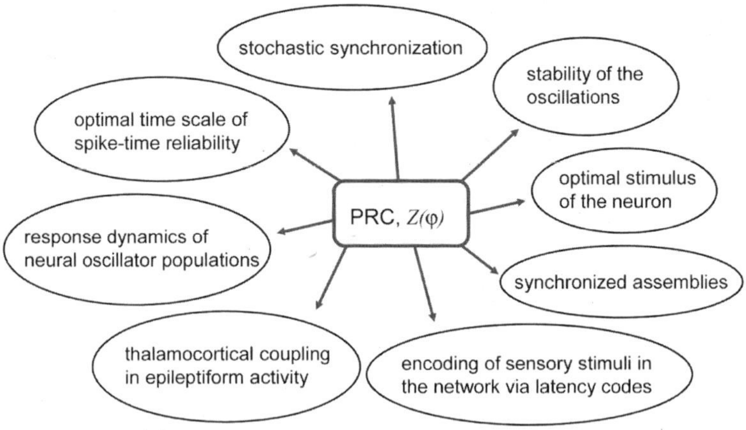

Fig. 3 Relevant information that can be extracted from the phase–response curve (PRC) in applications to neuroscience. See text for details

$$\begin{cases} \dfrac{d\varphi_1}{dt} = \omega_1 + \vec{Z}_1(\varphi_1).\vec{V}_1(\varphi_1, \varphi_2) \\ \dfrac{d\varphi_2}{dt} = \omega_2 + \vec{Z}_2(\varphi_2).\vec{V}_2(\varphi_2, \varphi_1) \end{cases} \tag{12}$$

Equations (12) can be simplified even more to study mutual synchronization. To this end, we first note that the intrinsic frequencies must be sufficiently close, i.e., $\omega = \omega_1 \approx \omega_2$. We also note that $\vec{V}_1(\varphi_1, \varphi_2)$ and $\vec{V}_2(\varphi_1, \varphi_2)$ are periodic in ϕ_1 and ϕ_2 with period $T = 2\pi/\omega$. Then, in order to get rid of the phase drift of the uncoupled dynamics, we make the change of variables, $\theta_i = \varphi_i - \omega t$ and average (12) over the natural period, encountering the following integrals

$$H_i(\theta_i, \theta_j) \equiv \frac{1}{T}\int_0^T \vec{Z}(\omega t + \theta_i)\cdot \vec{V}_i(\omega t + \theta_i, \omega t + \theta_j)dt \equiv \frac{1}{T}\int_0^T \vec{G}_i(\omega t, \theta_i, \theta_j)dt. \tag{13}$$

To simplify these integrals we expand the integrands in Fourier components

$$\frac{1}{T}\int_0^T \sum_{p,q=-\infty}^{\infty} \vec{g}_{pq} e^{ip(\omega t+\theta_i)+iq(\omega t+\theta_j)}dt = \frac{1}{T}\sum_{p,q=-\infty}^{\infty} \vec{g}_{pq} e^{ip\theta_i+iq\theta_j}\int_0^T e^{i(p+q)\omega t}dt,$$

where $\vec{g}_{pq}$ are the Fourier coefficients (do not confuse i as imaginary unit with i as oscillator index). The integral on the right hand yields

$$\frac{1}{T}\int_0^T e^{i(p+q)\omega t}dt = \begin{cases} 1 & \text{if } p = -q, \\ 0 & \text{otherwise} \end{cases}$$

which implies that integrals (13) are functions of the phase difference only, i.e., $H_i(\theta_i, \theta_j) = H_i(\theta_i - \theta_j)$. Finally, we rewrite (12) as $d\theta_i/dt = H_i(\theta_i - \theta_j)$, which with respect to the original phase variables becomes

$$\begin{cases} \dfrac{d\varphi_1}{dt} = \omega + H_1(\varphi_1 - \varphi_2) \\ \dfrac{d\varphi_2}{dt} = \omega + H_2(\varphi_2 - \varphi_2). \end{cases} \tag{14}$$

The functions, H_i, are called interaction functions of the coupled oscillators. In many cases of interest in neuroscience, the coupling is fast compared to the duration of the period. For example, fast synaptic transmission in glutamatergic and GABAergic neurons occurs within a few milliseconds (2–5 ms), whereas the period of typical beta activity (˜10 Hz) is around 100 ms. Thus, the interaction can be regarded as an instantaneous pulse, i.e., a delta function. In addition, in neural models like the Hodgkin–Huxley model, the synaptic coupling only affects the equation of the membrane potential, so only one term survives in the dot product $\vec{Z}_i(\varphi_i) \cdot \vec{V}_i(\varphi_i, \varphi_j)$. Together, these observations imply that in the case of fast synaptic transmission the interaction function in (13) can be approximated by the phase–response curve of the oscillators, i.e., $H_i(\varphi_i - \varphi_j) \approx Z_i(\varphi_i - \varphi_j)$.

We can now study a system of two-coupled phase oscillators with a sinusoidal interaction function that resembles the phase response of a type II neuron, i.e., $Z(\varphi) \approx -\sin(\varphi)$:

$$\begin{cases} \dfrac{d\varphi_1}{dt} = \omega_1 - J_{21}\sin(\varphi_2 - \varphi_1) \\ \dfrac{d\varphi_2}{dt} = \omega_2 - J_{12}\sin(\varphi_1 - \varphi_2) \end{cases} \tag{15}$$

where J_{21} is the synaptic strength of oscillator 2 onto oscillator 1, and J_{12} is the synaptic strength of oscillator 1 onto oscillator 2. Using the following change of variables

$$\begin{aligned} \psi &= \varphi_2 - \varphi_1 \\ \Omega &= \omega_2 - \omega_1 \\ J &= J_{12} + J_{21}, \end{aligned}$$

system (15) reduces to

$$\frac{d\psi}{dt} = \Omega - J_{12}\sin(-\psi) + J_{21}\sin(\psi) = \Omega + J\sin(\psi). \tag{16}$$

The fixed points ψ^* satisfy Eq. (14) with $d\psi/dt = 0$, which yields

$$\psi^* = \arcsin\left(-\frac{\Omega}{J}\right).$$

The linear stability of the fixed points is given by

$$\frac{d}{d\psi}\left(\frac{d\psi}{dt}\right)_{\psi^*}\begin{cases} < 0 \rightarrow \psi^* & \text{is stable} \\ \geq 0 \rightarrow \psi^* & \text{is unstable} \end{cases}$$

which applied to (16) yields

$$\frac{d}{d\psi}\left(\frac{d\psi}{dt}\right) = J\frac{d}{d\psi}\sin(\psi) = J\cos(\psi).$$

If the oscillators have the same frequency, then $\Omega = 0$ and the fixed points are $\psi^* = 0, \pi$. For inhibitory coupling, $J<0$ the former is stable and the latter unstable, which implies that the oscillators will perfectly synchronize. On the other hand, for excitatory coupling, $J>0$ the former is unstable and the latter is stable, which implies that the oscillators will phase-lock with a relative lag of π, i.e., in anti-phase. In other words, pairs of type II neurons tend to synchronize through inhibitory synapses.

The previous model of coupled oscillators easily explains the role of shunting inhibition in synchronizing interneurons of the hippocampus recently reported in Vida et al. (2006). These interneurons are connected with each other through GABAergic synpases and their resting potential lies between the reversal potential of chloride ions and the spiking threshold. This implies that GABAergic inputs are excitatory close to the resting potential and inhibitory close to and above the spiking threshold. Thus, right after a spike, during the refractory period the phase response with respect to GABAergic pulses is positive and becomes negative as one approaches the next spike. As a result, this system is equivalent to (15) with $J<0$, which means that these interneurons perfectly synchronize, as shown with simulations in Vida et al. (2006).

A similar analysis can be performed for neurons whose phase–response curves have been estimated experimentally and with known synaptic interactions. Several authors have studied synchronization capabilities with this method in various neural systems: hybrid networks of real and simulated neurons with dynamic clamp (Netoff et al., 2005; Oprisan et al., 2004), mitral cells of the olfactory system (Galán et al., 2005), inhibitory cortical neurons (Tateno and Robinson, 2007), also coupled with gap junctions (Mancilla et al., 2007), pyramidal cells of the neocortex (Tsubo et al., 2007), central pattern generators (Canavier et al., 1997; Oprisan et al., 2004), simple circuits in mollusks (Preyer and Butera, 2005), thalamo-cortical coupling in EEG signals during epileptic seizures (Pérez Velázquez et al., 2007), etc.

3.2 Predicting the Emergence of Synchronized Assemblies

System (15) can be easily generalized for a population of oscillators with arbitrary interaction functions H_i and synaptic strengths J_{ji}. The model for N oscillators then becomes

$$\frac{d\varphi_i}{dt} = \omega_i + \sum_{j\neq i}^{N} J_{ji}\cdot H_i(\varphi_j - \varphi_i). \tag{17}$$

This model is completely general and can already be used to simulate the network behavior. In many cases, systems of this type generate synchronized neural assemblies or perfect synchrony. To study this phenomenon analytically (Galán et al., 2006), we make some simplifications. We consider identical frequencies and interaction functions, global, homogeneous coupling, and a large number of neurons, $N \to \infty$. In this case, one can take the continuum limit by defining an oscillator density, ρ

$$\rho(\theta) = \frac{1}{N} \sum_{i=1}^{N} \delta(\theta - \varphi_i).$$

So that Eq. (17) becomes

$$\frac{d\varphi}{dt} = \omega + J \int_0^{2\pi} H(\varphi - \theta)\rho(\theta)d\theta.$$

In addition, the oscillator density satisfies a continuity equation, i.e., the number of oscillators moving along the circumference is preserved

$$\frac{d\rho}{dt} = -\frac{\partial}{\partial \varphi}\left(\rho \frac{d\varphi}{dt}\right) = -\frac{\partial}{\partial \varphi}\left[\rho \left(\omega + J \int_0^{2\pi} H(\varphi - \theta)\rho(\theta)d\theta\right)\right]. \tag{18}$$

Now we study the stability of the asynchronous state, i.e., we check whether the phases of the oscillators can remain uniformly distributed along the circumference, with density $\rho = 1/(2\pi)$. To this end, we linearize equation (18) around the state of uniform density. Let y be a small deviation from the asynchronous state. Then, y follows the equation

$$\frac{dy}{dt} = -\omega \frac{dy}{d\varphi} - J \frac{dy}{d\varphi} \int_0^{2\pi} H(\varphi - \theta)y(\theta)d\theta + \text{nonlinear terms}, \tag{19}$$

whose linear part admits solutions of the form $\exp(\lambda_n t + in\varphi)$. Upon substitution of these solutions in (19) one obtains the eigenvalues

$$\lambda_n = J\pi n(b_n - i\omega a_n),$$

where a_n and b_n are the nth cosine and sine coefficients of the Fourier expansion of the interaction function H, respectively. Note that the real part of the nth eigenvalue is positive if J and b_n are both positive or both negative, making the nth mode linearly unstable. If one of the modes is linearly unstable, the asynchronous state cannot last, which implies that neurons will cluster in synchronized assemblies. If only one mode, say the kth mode, is linearly unstable, the number of clusters of the steady state is determined by the order of the mode, k as well as its symmetry, i.e., the distance or phase lag between clusters, which will be $2\pi/k$. If more than one mode is linearly stable, the steady state of the system will consist of a nonlinear mixture of the modes including higher order terms in (19), which in general precludes an analytical prediction of the number of clusters and their symmetry. An example of this type of behavior is shown in Fig. 4 for neurons with the phase–response curve estimated from mitral cells and two different types of synaptic interactions.

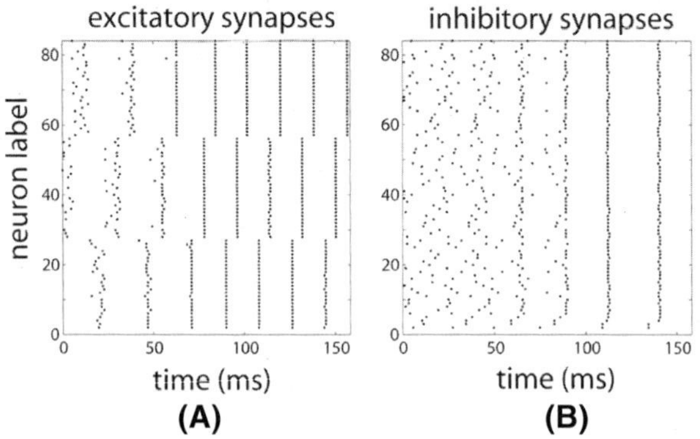

Fig. 4 Emergence of synchronized neural assemblies in a network of identical neurons and homogeneous connectivity. (**A**) The network with only excitatory synapses splits into three evenly interleaved synchronized assemblies. (**B**) The same network but with only inhibitory synapses now splits into two assemblies, one of which is barely populated

Despite its simplicity, this model offers relevant insights on the emergence of synchronized assemblies and brain rhythms. First, we note that clusters can form spontaneously through a symmetry-breaking process that is fully determined by the shape of the interaction function. Note that the synaptic connections are equal and constant in the model. Thus, synaptic plasticity is not necessary to form synchronized neural assemblies. Second, because different clusters are phase lagged with respect to each other, an extracellular recording of the population activity will display several frequency bands. For example, consider the case of a network of identical neurons oscillating at 20 Hz that breaks up into two synchronized clusters separated by π. A local field potential, or EEG of the neural population will reflect the intrinsic oscillation of 20 Hz (high beta band) and also an apparent global oscillation of 40 Hz (gamma band). In reality, all neurons oscillate at 20 Hz, but due to the phase lag between clusters, the *average activity* of the population fluctuates at 40 Hz. This may explain why EEG recordings sometimes, especially during epileptic seizures, display very high frequency oscillations (up to 300 Hz), which seems to be far beyond the physiological limit of neuronal firing rates (Foffani et al., 2007). In effect, this population frequency may emerge, e.g., from three interleaved clusters of neurons firing each at 100 Hz.

3.3 Encoding Models

In the previous paragraph we have focused on networks with homogeneous connectivity, with all synapses being either excitatory or inhibitory. Despite the simplicity of such architectures the network can self-organize in synchronized assemblies. We now consider a more complex type of behavior that emerges from oscillator networks with both excitatory and inhibitory synapses. Quite often, these networks rapidly converge to states in which the neurons are phase-locked preserving their relative phase lags. This phenomenon is reminiscent of the concept of *frustration* in solid state physics and spin glass theory. In effect, in lattices of spins with ferromagnetic and anti-ferromagnetic interactions, the spins try to arrange themselves obeying the command of their neighbors. If a neighbor exerts a ferromagnetic force, the spin will try to align parallel to that neighbor, whereas if the neighbor exerts an anti-ferromagnetic force, the spin will try to align anti-parallel to the neighbor. However, if a given spin receives mixed, ferromagnetic, and anti-ferromagnetic signals from different neighbors, it will be unable to satisfy both and frustration occurs. In this case, the spins of the lattice arrange in complex mixtures of differently oriented spins trying to minimize the magnetic energy. In the case of neural networks, if a given neuron receives mixed signals, i.e., excitation and inhibition from different neighbors at the same time, the neurons will try to find frustrated states where relative phase changes are minimal, i.e., phase-locked states, as shown in Fig. 4.

The state to which a network with excitatory and inhibitory synapses converges typically depends on the initial conditions. This allows for encoding capabilities of networks of oscillators. First, we note that the initial conditions of model (16) with mixed J_{ji} represent the phases at which the neurons start firing, i.e., the delay of the neural responses or response latency. Interestingly, the mapping of sensory stimuli into neural latencies has been reported or hypothesized in the visual (Kupper et al., 2005), auditory (Hopfield and Brody, 2001) and the olfactory system (Spors et al., 2006; Brody and Hopfield, 2003). This form of sensory mapping is usually referred to as a *latency code* (Fig. 5).

The next question to solve is the decoding mechanism. How can these phase-locked states be decoded by a downstream neural network? As in many other models of pattern recognition, a downstream neuron can decode and classify the patterns of the network by a mechanism of coincidence detection that supports delays. In other words, a "mother cell" will respond to a given pattern of the neural network if the transmission delays between the neurons and the mother cell match the phase lags. In that case all inputs to the mother cell will arrive at the same time pushing the "mother cell" beyond firing threshold. A similar decoding mechanism that does not require delays can be accomplished with a perceptron, analogue to the model of olfactory coding and decoding in insects proposed in Galán et al. (2004). In this case, the synaptic strengths of the neurons onto the mother cell can be tuned, e.g., through Hebbian learning to make the downstream mother cell respond to a specific input pattern.

3.4 Stability of Oscillations and Liapunov Exponent

The phase–response curve describes the phase-dependent sensitivity of a limit cycle to small perturbations of whatever nature. Therefore, one expects that the robustness

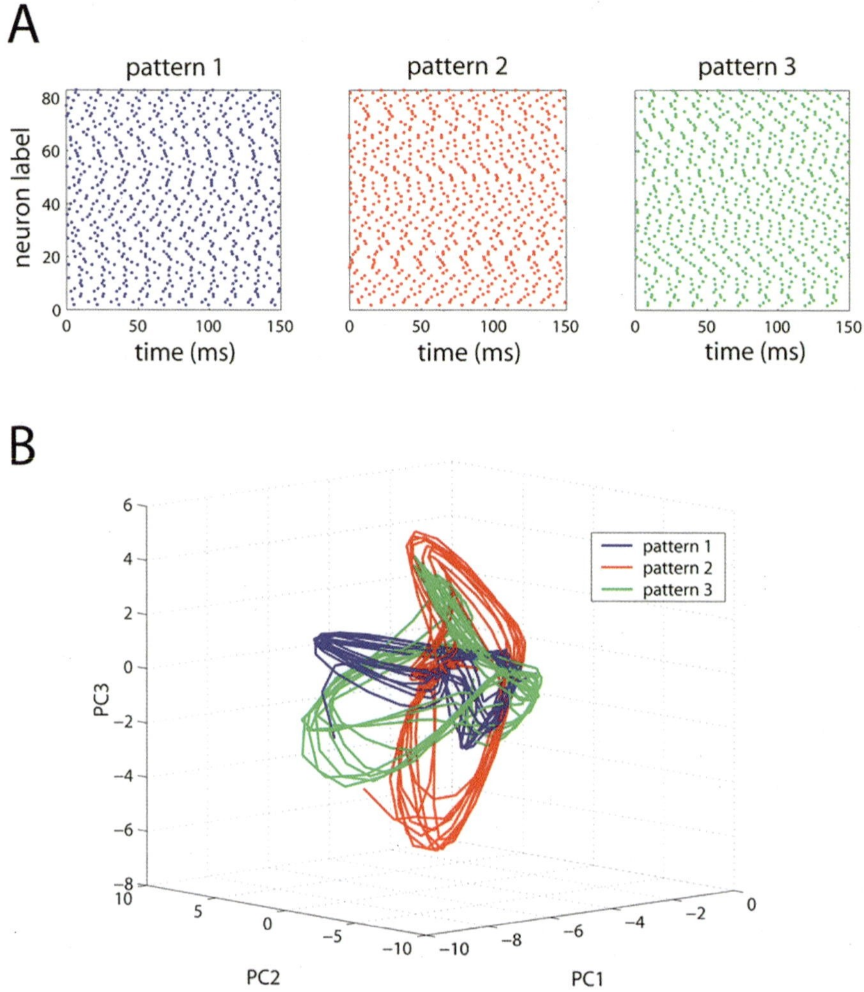

Fig. 5 Networks of neural oscillators with both excitatory and inhibitory connections can implement a latency code. (**A**) The same network converges to different spatiotemporal patterns in response to different initial conditions associated with different stimuli. (**B**) Representation of those patterns in a multidimensional space where each dimension is the activity of one oscillator and then projected down onto the first three principal components (PC) for visualization. The trajectories of different patterns are clearly distinct

of the oscillations with respect to background noise is determined by the phase–response curve. The stability of periodic motion can be quantified as the Liapunov exponent, λ of the oscillators. As in any other dynamical system, the larger the absolute value of a negative Liapunov exponent the more stable the dynamics. Indeed, the Liapunov exponent measures the rate of convergence/divergence of trajectories of the dynamical system starting from different initial conditions. Several authors have applied this approach recently to neural oscillators (Goldobin and Pikovsky,

2005; Teramae and Tanaka, 2004; Tateno and Robinson, 2007). As expected, the Liapunov exponent λ depends on the phase–response curve, $Z(\varphi)$. In particular, assuming that the oscillations are perturbed by background noise that is Gaussian and white with variance σ^2, one has

$$\lambda = -\frac{\sigma^2}{2T}\int_0^T \left[\frac{dZ(\varphi)}{d\varphi}\right]^2 d\varphi. \tag{20}$$

A continuous $Z(\phi)$ generally results in a negative Liapunov exponent (20), which is the common scenario in real neurons, although it is not always the case in some simplified neuron models, like the integrate-and-fire model. A negative Liapunov exponent has important consequences for neuroscience, explaining the existence of spike-time reliability (Bryant and Segundo, 1976; Mainen and Sejnowski, 1995; Movshon, 2000). In effect, when a rapidly fluctuating signal is presented to a neuron several times, the neuron will respond consistently in all trials: if one overlays the spike patterns from different trials, virtually all spike-times are preserved and the spikes from different trials overlap almost perfectly. Since the neuron fires with a roughly constant frequency due to the rapid fluctuations that drive the neuron, it can be described as an oscillator which is perturbed by some background noise that accounts for the variability across trials. A negative Liapunov exponent with large magnitude implies that in different trials the neuron's dynamics converge quickly to the same point of the limit cycle after a transient deviation due to background noise. As a result, the spike-times will be preserved across different trials despite the presence of irreproducible, background noise. Formula (20) can be interpreted as the average steepness of the phase–response curve, as it is the integral of the square of its derivative. Thus, the steeper the modulation of the phase–response curve, the more stable the oscillations. Finally, it is worth recalling that at least in simplified neural models the Liapunov exponent can be positive, impairing spike-time reliability (Teramae and Tanaka, 2004; Goldobin and Pikovsky, 2006).

3.5 Optimal Stimulus of a Neural Oscillator

We now consider an important link between dynamical system theory, in particular, the phase oscillator approximation, and information theory. Neurons can be described not only as dynamical systems, e.g., phase oscillators, but also as "filters" or black-boxes with a transfer function that transform stimuli in spike trains. Let $x(t)$ be the input to the neurons and $y(t)$ the output, i.e., the spike trains. Then using a Volterra expansion the relationship between $x(t)$ and $y(t)$ is approximated by

$$y(t) = \int_0^\infty K_1(s)x(t-s)ds + \int_0^\infty\int_0^\infty K_2(s,s')x(t-s)x(t-s')dsds' + \ldots$$

In a first-order approximation, i.e., if the spikes are sufficiently separated from each other so that a given spike has a negligible effect on the generation of successive spikes as compared to the effect of the input, one may retain only the first term

$$y(t) = \int_0^\infty K_1(s)x(t-s)ds.$$

The convolution kernel or transfer function, $K_1(s)$ happens to be the spike-triggered average reversed in time (Rieke et al., 1997), i.e., the average waveform of a white-noise input preceding a spike, plotted backwards in time. In studies of information theory applied to neuroscience, the spike-triggered average is usually referred to as the optimal stimulus of the neuron, as it is the most likely input trajectory triggering a spike. Recently, Ermentrout and collaborators have shown that the spike-triggered average is closely related to the phase–response curve in neurons that spike regularly, i.e., in neural oscillators (Ermentrout et al., 2007). In particular, they showed that

$$Z(t) = -\int_0^t K_1(T-t)dt \quad \Leftrightarrow \quad \frac{dZ(t)}{dt} = -K_1(T-t).$$

This has important consequences for computational neuroscience. If the phase–response curve is known, the optimal stimulus can be easily calculated. Reciprocally, the phase-response curve can be estimated from the spike-triggered average of the neuron (Ermentrout et al., 2007).

3.6 Optimal Timescale for Spike-Time Reliability

We now retake the study of spike-time reliability combining oscillator theory and basic information theory. Spike-time reliability can be regarded as a special case of stochastic synchronization. In the former case, a single neuron receives highly correlated, i.e., almost identical inputs in successive trials and responds with consistent spike-times across trials. In the latter case, two or more identical neurons receive correlated inputs and trigger synchronous spikes. The higher the correlation of the inputs, the more spikes are synchronized across neurons. The ability to generate reliable spikes across trials depends on the characteristic timescale of the inputs. Whereas constant inputs do not generate reliable spikes, fast fluctuating inputs do. In fact, there is an optimal timescale of the inputs that maximizes spike-time reliability (Galán et al., 2008). Let $x_1(t)$ and $x_2(t)$ be the inputs to two identical neurons, and let $y_1(t)$ and $y_2(t)$ be their spike trains. Then, following up on the previous paragraph, we write

$$\begin{cases} y_1(t) = \int_0^\infty K(s)x_1(t-s)ds \\ y_2(t) = \int_0^\infty K(s)x_2(t-s)ds \end{cases}$$

where $K(s)$ is the kernel obtained either from the spike-triggered average or from the phase–response curve. The inputs $x_i(t)$ consist of a fluctuating signal, $I(t)$ which is identical for both neurons, plus some background fluctuations modeled as uncorrelated, white noise

$$x_i(t) = I(t) + \eta_i(t),$$

with $\langle \eta_1(t)\eta_2(t-s)\rangle = 0$ and $\langle \eta_i(t)\eta_i(t-s)\rangle = \sigma_\eta^2 \delta(t-s)$. The timescale of the fluctuating signal $I(t)$ is characterized by its autocorrelation time, τ

$$\langle I(t)I(t-s)\rangle = \sigma_1^2 \exp(-|s|/\tau).$$

The ratio of the standard deviations of the signal, σ_I and the noise σ_η represents the signal-to-noise ratio of the stimulus. The degree of synchronization between neurons, or equivalently, the spike-time reliability across trials is quantified as the correlation coefficient of the spike trains

$$R \equiv \frac{\int_{-\infty}^{\infty} y_1(t)y_2(t)dt}{\sqrt{\int_{-\infty}^{\infty} y_1(t)y_1(t)dt \int_{-\infty}^{\infty} y_2(t)y_2(t)dt}} = \frac{\int_{-\infty}^{\infty} y_1(t)y_2(t)dt}{\int_{-\infty}^{\infty} y_1^2(t)dt}. \qquad (21)$$

where we have used, in the denominator, the fact that both neurons are identical. Upon substitution of the previous definitions in (21) one finally obtains (Galán et al., 2008)

$$R(\tau) = \frac{\sigma_1^2 \int_0^\infty \exp(-u/\tau)Q(u)du}{\sigma_1^2 \int_0^\infty \exp(-u/\tau)Q(u)du + \sigma_\eta^2 Q(0)},$$

with $Q(u) = \int_0^\infty K(s)K(s+u)ds$. Note that in the absence of background noise, $\sigma_\eta = 0$, spike-time reliability and stochastic synchronization are perfect, $R = 1$. On the other hand, in the absence of signal, $\sigma_I = 0$ spike-time reliability and stochastic synchronization are obviously, $R = 0$. Note also that R depends explicitly on the spike-triggered average, or equivalently on the phase–response curve of the neurons via $K(s)$, which in turn are determined by the internal degrees of freedom of the single-neuron dynamics. As shown in Galán et al. (2008), kernels $K(s)$ of real neurons and of realistic neural models maximize the curve $R(\tau)$ at approximately,

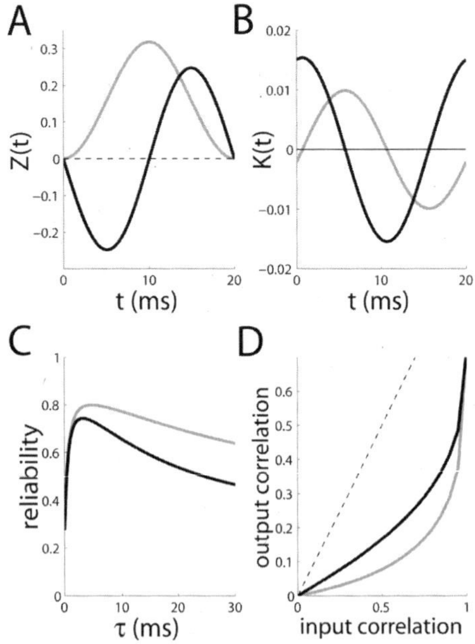

Fig. 6 Spike-time reliability, optimal stimulus, and stochastic synchronization in two canonical neural oscillators (*grey lines*, type I; *black lines*, type II). (**A**) Phase–response curves of the oscillators. (**B**) Transfer functions or kernels of the oscillators, which are their optimal stimulus reversed in time. (**C**) Reliability as a function of the timescale τ of the input signal in the presence of background noise (signal-to-noise ratio = 40%). Note that regardless of the oscillator type, there is a timescale that maximizes spike-time reliability in the range of few milliseconds. (**D**) Synchronization induced by correlated noise (very fast random fluctuations) of two identical type I oscillators (*grey*) and two identical type II oscillators (*black*). Note that the synchronization (output correlation) obtained is always smaller than the input correlation (*dashed lines* represent the identity $y = x$)

$\tau_{\text{opt}} = 10\%$ of the mean interspike interval. This implies that neurons firing in the beta/gamma frequency band are optimally reliable to signals with an autocorrelation time from 2 to 6 ms (Fig. 6C), which interestingly, coincides with the timescale of fast synaptic transmission.

3.7 Stochastic Synchronization

In the previous paragraph we have seen that spike-time reliability, stochastic synchronization, and their optimal timescale depend on the phase–response curve. It is therefore worth studying in more detail the peculiarities of the phase response contributing to these phenomena, within the framework of the phase oscillator approximation (Galán et al., 2007a, b). To this end, we start with two uncoupled oscillators driven by partially correlated stochastic inputs

$$\begin{cases} \dfrac{d\varphi_1}{dt} = \omega + Z(\varphi_1)\eta_1(t) \\ \dfrac{d\varphi_2}{dt} = \omega + Z(\varphi_2)\eta_2(t) \end{cases} \tag{22}$$

where $\eta_i(t)$ are zero-mean, white-noise stochastic inputs, $<\eta_i(t)\eta_i(t-\tau)> = \sigma_i{}^2\delta(\tau)$, that are spatially correlated $<\eta_1(t)\eta_2(t)> = c$, thus, the correlation coefficient of the inputs is $r = c/(\sigma_1\sigma_2)$. System (22) can be readily simulated to investigate synchronization as a function of the input correlation, r (Galán et al., 2007b). Alternatively, one can apply stochastic theory and solve the Fokker–Planck equation of system (22). In this case, the Fokker–Planck operator reads

$$\begin{aligned} L_{FP} \equiv &- \omega_1 \frac{\partial}{\partial_{\varphi_1}} - \omega_2 \frac{\partial}{\partial_{\varphi_2}} + \\ &+ \frac{\sigma_1^2}{2} \frac{\partial^2}{\partial_{\varphi_1^2}} Z^2(\varphi_1) + \frac{\sigma_2^2}{2} \frac{\partial^2}{\partial \varphi_2^2} Z^2(\varphi_2) + c \frac{\partial^2}{\partial_{\varphi_1}\partial_{\varphi_2}} Z(\varphi_1) Z(\varphi_2). \end{aligned}$$

And the stationary probability density function, $P(\phi_1,\phi_2)$ satisfies $L_{FP}P(\varphi_1, \varphi_2) = 0$, which can be solved efficiently with the finite-element method (Galán et al., 2007b). Using this approach, one can calculate $P(\phi_1,\phi_2)$ for different input correlation levels and for prototypical phase responses $Z(\varphi) = N(1 + \cos(\varphi + \pi))$ (type I neurons) and $Z(\varphi) = -M \sin\varphi$ (type II neurons), being N and M normalization constants, such that the integral of the absolute value of the phase–response curves is one. Then, to quantify the increase of stochastic synchronization as a function of the input correlation, we compute the cross-correlation coefficient, R of the phases from $P(\phi_1,\phi_2)$ as

$$R \equiv \frac{\iint\limits_{\Omega} \varphi_1\varphi_2 P(\varphi_1, \varphi_2) d\varphi_1 d\varphi_2 - \iint\limits_{\Omega} \varphi_1 P(\varphi_1, \varphi_2) d\varphi_1 d\varphi_2 \iint\limits_{\Omega} \varphi_2 P(\varphi_1, \varphi_2) d\varphi_1 d\varphi_2}{\iint\limits_{\Omega} \varphi_1^2 P(\varphi_1, \varphi_1) d\varphi_1 d\varphi_1 - \left[\iint\limits_{\Omega} \varphi_1 P(\varphi_1, \varphi_1) d\varphi_1 d\varphi_1\right]^2}.$$

As shown in Fig. 6D, R monotonically increases with increasing correlation of the stochastic inputs in both cases, but type II neurons are more efficient at synchronizing than type I are for each r, except for the limits $r \to 0$ and $r \to 1$ where both are equal.

3.8 Response Dynamics of a Population of Neural Oscillators

To investigate the modulation of firing rates in neural populations in response to pulse-like stimuli, Shea-Brown and collaborators have combined statistics with the phase oscillator approximation (Brown et al., 2004). In particular, they consider the probability density $\rho(\varphi, t)$ of a population of uncoupled phase oscillators

$$\frac{d\rho(\varphi,t)}{dt}=-\frac{\partial}{\partial\varphi}\left(\rho(\varphi,t)\frac{d\varphi}{dt}\right)=-\frac{\partial}{\partial\varphi}\left[\rho(\varphi,t)(\omega+Z(\varphi)I(t))\right].$$

Then, they study how the phase response of the neurons and the input-pulses affect the temporal modulation of the firing rates (flux of the probability density). In their detailed study, the authors report that the magnitude and timing of maximal population response depends strongly on both neuron type and stimulus duration relative to the baseline period. Among other results, they observe that for neuron models having a phase–response curve that takes negative values, the greatest deviations from baseline firing rates can occur significantly after the stimulus presentation. As a result, populations of such neurons could be used in detecting termination of sensory stimuli (off-responses).

3.9 Phase–Response Curves in EEG Studies of Epilepsy

So far, we have considered examples in which the oscillators represent single neurons. Phase models, however, are not limited to this case but they can also be applied to investigate interactions between populations of coherent oscillators. An interesting example from neuroscience is the interaction between thalamus and cortex. In effect, thalamus and cortex display spontaneous, slow (4–10 Hz) oscillations that are especially prominent during certain phases of sleep and in a type of pathological, epileptic activity, called absence seizures or *petit mal*. Pérez Velázquez and collaborators have recently estimated the phase–response curve of the thalamus and cortex during absence seizures in a rat model of this disease (Pérez Velázquez et al., 2007). Interestingly, cortex and thalamus do not synchronize but phase-lock with a stable phase lag of approximately 12 ms, thalamus ahead. This lag is not likely due to a conduction delay, which in this system is estimated to be 2 ms. An alternative explanation is that the system settles in a phase-locked, stable state, as in the theory of Sect. 3.1. To investigate this possibility, the authors studied the thalamo-cortical interactions using the Kuramoto model (14). Because the synaptic coupling between both areas is fast, the interaction functions can be replaced by the phase–response curves recorded in the experiments. Then, by fitting the phase–response curves and the experimentally observed phase lag to the Kuramoto model, the authors estimate the coupling strengths between the oscillators. The fit reveals that the thalamus drives the cortex twice as strongly as the cortex to the thalamus. This result is consistent with the classical role attributed to the thalamus as a gate of information flow to the cortex and the hypothesis of thalamic augmentation response (Morison and Dempsey, 1943). One disadvantage of the phase oscillator approximation, however, is that it only allows for the calculation of coupling strengths during epochs of oscillatory activity in both thalamus and cortex, which in the case of absence seizures is obviously a pathological state of the brain. Thus, one cannot conclude that the estimated coupling is anatomical and permanent, or functional and modulated by the state of the brain; a question, that must be addressed with complementary approaches.

4 Summary and Outlook

After a brief mathematical introduction to oscillator theory, we have provided several examples of the phase oscillator approximation in neuroscience. These applications orbit around the concepts of phase resetting and the phase–response curve. We have seen how this single-neuron property allows us to infer features of neural dynamics like spike-time reliability, synchronization, and phase-locking. These phenomena are thought to underlie encoding and information-processing capabilities of neural systems. Thus, the determination of phase–response curves from real neurons can provide valuable information about neural function.

A natural extension of the studies overviewed here will be an investigation into how the phase–response curve is affected by parameters of the neuron, such as the activation and inactivation of ionic conductances, and their spatial distribution along dendrites and soma. In effect, it is expected that channel blockers, as well as agonists and antagonists of receptors as well as neuromodulators affect the phase–response curve significantly and, therefore, the neural properties previously described. These studies could then contribute to understanding how and why neural synchronization is modulated by the state of awareness, e.g., during tasks requiring attention, or during sleep. In addition, they may shed light on the mechanisms underlying a pathological excess of neural synchronization, as in epilepsy, and a pathological defect of neural synchronization, as in attention deficits associated with schizophrenia and short-term memory loss.

In conclusion, the phase oscillator approximation transcends mathematics and physics to become a powerful analytical tool to study brain function.

Acknowledgments The author thanks Dr. Christopher G. Wilson for helpful comments. This work has been supported by the Mount Sinai Health Care Foundation and the Alfred P. Sloan Foundation.

References

Brody CD, Hopfield JJ (2003) Simple networks for spike-timing-based computation, with application to olfactory processing. Neuron 37:843–852.

Brown E, Moehlis J, Holmes P (2004) On the phase reduction and response dynamics of neural oscillator populations. Neural Comput 16:673–715.

Bryant HL, Segundo JP (1976) Spike initiation by transmembrane current: a white-noise analysis. J Physiol 260:279–314.

Canavier CC, Butera RJ, Dror RO, Baxter DA, Clark JW, Byrne JH (1997) Phase response characteristics of model neurons determine which patterns are expressed in a ring circuit model of gait generation. Biol Cybern 77:367–380.

Engel AK, Singer W (2001) Temporal binding and the neural correlates of sensory awareness. Trends Cogn Sci 5:16–25.

Ermentrout B (1996) Type I membranes, phase resetting curves, and synchrony. Neural Comput 8:979–1001.

Ermentrout B (2002) Simulating, Analyzing, and Animating Dynamical Systems: A Guide to XPPAUT for Researchers and Students. Philadelphia: SIAM.

Ermentrout GB, Galan RF, Urban NN (2007) Relating neural dynamics to neural coding. Phys Rev Lett 99:248103.

Foffani G, Uzcategui YG, Gal B, Menendez de la PL (2007) Reduced spike-timing reliability correlates with the emergence of fast ripples in the rat epileptic hippocampus. Neuron 55: 930–941.

Galán RF, Ermentrout GB, Urban NN (2005) Efficient estimation of phase-resetting curves in real neurons and its significance for neural-network modeling. Phys Rev Lett 94:158101.

Galán RF, Ermentrout GB, Urban NN (2006) Predicting synchronized neural assemblies from experimentally estimated phase-resetting curves. Neurocomputing 69:1112–1115.

Galán RF, Ermentrout GB, Urban NN (2007a) Reliability and stochastic synchronization in type I vs. type II neural oscillators. Neurocomputing 70:2102–2106.

Galán RF, Ermentrout GB, Urban NN (2007b) Stochastic dynamics of uncoupled neural oscillators: Fokker-Planck studies with the finite element method. Phys Rev E Stat Nonlin Soft Matter Phys 76:056110.

Galán RF, Ermentrout GB, Urban NN (2008) Optimal time scale for spike-time reliability: theory, simulations, and experiments. J Neurophysiol 99:277–283.

Galán RF, Sachse S, Galizia CG, Herz AVM (2004) Odor-driven attractor dynamics in the antennal lobe allow for simple and rapid olfactory pattern classification. Neural Comput 16:999–1012.

Goldobin DS, Pikovsky AS (2005) Synchronization of self-sustained oscillators by common white noise. Phys A – Stat Mech Appl 351:126–132.

Goldobin DS, Pikovsky A (2006) Antireliability of noise-driven neurons. Phys Rev E Stat Nonlin Soft Matter Phys 73:061906.

Hodgkin AL, Huxley AF (1952) A quantitative description of membrane current and its application to conduction and excitation in nerve. J Physiol 117:500–544.

Hopfield JJ, Brody CD (2001) What is a moment? Transient synchrony as a collective mechanism for spatiotemporal integration. Proc Natl Acad Sci U S A 98:1282–1287.

Hoppensteadt FC, Izhikevich EM (1997) Weakly Connected Neural Networks. New York: Springer-Verlag.

Kupper R, Gewaltig MO, Korner U, Korner E (2005) Spike-latency codes and the effect of saccades. Neurocomputing 65:189–194.

Kuramoto Y (1984) Chemical Oscillations, Waves, and Turbulence. Berlin: Springer Verlag.

Mainen ZF, Sejnowski TJ (1995) Reliability of spike timing in neocortical neurons. Science 268:1503–1506.

Mancilla JG, Lewis TJ, Pinto DJ, Rinzel J, Connors BW (2007) Synchronization of electrically coupled pairs of inhibitory interneurons in neocortex. J Neurosci 27:2058–2073.

Morison RS, Dempsey EW (1943) Mechanism of thalamocortical augmentation and repetition. Am J Physiol 138:0297–0308.

Movshon JA (2000) Reliability of neuronal responses. Neuron 27:412–414.

Netoff TI, Banks MI, Dorval AD, Acker CD, Haas JS, Kopell N, White JA (2005) Synchronization in hybrid neuronal networks of the hippocampal formation. J Neurophysiol 93:1197–1208.

Oprisan SA, Canavier CC (2002) The influence of limit cycle topology on the phase resetting curve. Neural Comput 14:1027–1057.

Oprisan SA, Prinz AA, Canavier CC (2004) Phase resetting and phase locking in hybrid circuits of one model and one biological neuron. Biophys J 87:2283–2298.

Pérez Velázquez JL, Galán RF, Domínguez LG, Leshchenko Y, Lo S, Belkas J, Guevara Erra R (2007) Phase response curves in the characterization of epileptiform activity. Phys Rev E Stat Nonlin Soft Matter Phys 76:061912.

Preyer AJ, Butera RJ (2005) Neuronal oscillators in aplysia californica that demonstrate weak coupling in vitro. Phys Rev Lett 95:138103.

Rieke F, Warland D, de Ryuter van Stevenick R, Bialek W (1997) Spikes. Exploring the neural code. Cambridge, Mass: MIT Press.

Rinzel J, Ermentrout B (1998) Analysis of neural excitability. In: Methods in Neuronal Modeling (Koch C, Segev I, eds), pp. 251–291. Cambridge, Mass: MIT Press.

Schafer C, Rosenblum MG, Abel HH, Kurths J (1999) Synchronization in the human cardiorespiratory system. Phys Rev e 60:857–870.

Shadlen MN, Movshon JA (1999) Synchrony unbound: a critical evaluation of the temporal binding hypothesis. Neuron 24:67–25.

Singer W, Gray CM (1995) Visual feature integration and the temporal correlation hypothesis. Annu Rev Neurosci 18:555–586.

Spors H, Wachowiak M, Cohen LB, Friedrich RW (2006) Temporal dynamics and latency patterns of receptor neuron input to the olfactory bulb. J Neurosci 26:1247–1259.

Tateno T, Robinson HP (2007) Phase resetting curves and oscillatory stability in interneurons of rat somatosensory cortex. Biophys J 92:683–695.

Teramae J, Tanaka D (2004) Robustness of the noise-induced phase synchronization in a general class of limit cycle oscillators. Phys Rev Lett 93:204103.

Tsubo Y, Takada M, Reyes AD, Fukai T (2007) Layer and frequency dependencies of phase response properties of pyramidal neurons in rat motor cortex. Euro J Neurosci 25:3429–3441.

Uhlhaas PJ, Singer W (2006) Neural synchrony in brain disorders: relevance for cognitive dysfunctions and pathophysiology. Neuron 52:155–168.

Vida I, Bartos M, Jonas P (2006) Shunting inhibition improves robustness of gamma oscillations in hippocampal interneuron networks by homogenizing firing rates. Neuron 49:107–117.

Winfree AT (2001) The Geometry of Biological Time. New York: Springer.

From Synchronisation to Networks: Assessment of Functional Connectivity in the Brain

Cornelis J. Stam

1 Introduction

There is increasing evidence from basic research that statistical correlations between electroencephalographic (EEG) and magnetoencephalographic (MEG) rhythms recorded from different brain areas, either 'at rest' or during a specific task, contain a wealth of information on functional interactions between brain regions (Buzsaki, 2006; Varela et al., 2001). In fact, synchronisation of oscillatory brain activity in various frequency bands may be one of the key mechanisms used by the brain to integrate information processed in multiple specialised local brain areas. Many new signal analysis tools have become available in recent years to study such synchronisation patterns and their disruption in various neurological and neuropsychiatric diseases. The study of functional interaction patterns in the brain is even beginning to show some remarkable similarities between the brain and other complex networks, such as the Internet (Reijneveld et al., 2007). In this chapter, we give a brief introduction of concepts underlying functional connectivity and network analysis in EEG and MEG, the major tools that are used, their methodological aspects and shortcomings and the application to neurological disorders. The introduction is focused on presenting the concepts and ideas behind the methods; for technical details, we refer to more extensive review papers.

2 Synchronisation Analysis

2.1 The Concept of 'Functional Connectivity'

Visual analysis of EEG and MEG time series allows a first crude assessment of correlations between different channels; however, this analysis is most effective in the

C.J. Stam (✉)
Department of Clinical Neurophysiology, VU University Medical Center, Amsterdam, The Netherlands
e-mail: CJ.Stam@VUmc.nl

J.L. Perez Velazquez, R. Wennberg (eds.), *Coordinated Activity in the Brain*, Springer Series in Computational Neuroscience 2, DOI 10.1007/978-0-387-93797-7_5,

case of large, usually abnormal epileptiform or periodic transients. For an accurate assessment of correlations between different time series, some form of quantitative analysis is required. A key concept here is that statistical interdependencies between time series of brain activity (EEG, MEG, but also fMRI BOLD signals) recorded from different, spatially separated brain regions reflect to some extent actual functional interactions between the underlying brain regions. This concept has become known as 'functional connectivity' (Aertsen et al., 1989; Lee et al., 2003). The notion of functional connectivity is most frequently used in the fMRI literature, but it is of course equally relevant for clinical neurophysiology. A related concept is that of 'effective connectivity' which is based upon an explicit model of the underlying network, and which aims to identify causal interactions between the brain regions studied (Dynamic causal modelling; Stephan et al., 2007). We will discuss the issue of correlation versus causality later on, but limit ourselves mostly to functional connectivity, which is a model-free concept. The important thing at this stage is that any measure of the statistical correlations between time series can thus be used to probe some aspect of functional connectivity. Later we will show how synchronisation analysis can be used as the starting point for network analysis.

Statistical measures of correlations between time series can be divided into linear and nonlinear measures. Below, we briefly introduce some examples of each of these two categories. For an in-depth treatment of linear and nonlinear correlation measures, we refer to an excellent review paper by Pereda et al. (2005).

2.2 Linear Measures

The simplest measure of correlation between two time series is the correlation coefficient of the amplitudes. This is the same correlation coefficient that is still used in statistics to determine the positive or negative correlation between two variables. The correlation coefficient is hardly used in time series analysis anymore since it has a number of disadvantages which can be solved easily by other methods: (i) the magnitude of the correlation may depend upon a time lag between the two time series; this requires that the correlation is computed for an (a priori unknown) range of time lags to find the highest value; (ii) the correlation coefficient does not take into account that correlations between time series may depend upon frequency band; (iii) both time series need to be stationary; (iv) the correlation coefficient only captures linear correlations between the time series. Problem (iv) and, to some extent, problem (iii) can be dealt with by using nonlinear measures which are discussed in the next section. The first two problems have been solved effectively by a method called coherence analysis, which has become the standard technique for linear correlation analysis.

Intuitively one can think of coherence as a (squared) correlation coefficient which is computed separately for each of the frequencies in the signals. The time lag between the signals is dealt with by a phase difference which is determined also separately for each frequency. Thus coherence analysis results in a graph which gives, for each frequency in the signal, the strength of the coupling (between 0 and 1). There are several different techniques to do coherence analysis. The most well known is based upon a discrete Fourier transform. A Fourier transform is taken from both time series. The Fourier transform of one channel is then multiplied with the complex conjugate of the Fourier transform of the other channel, resulting in a cross spectrum. The squared modulus of the cross spectrum (C) is then divided by the product of the power spectra (P) of both time series x and y to give the normalised coherence as a function of frequency f. In formula:

$$Coh_{xy}(f) = \frac{\left|C_{xy}(f)\right|^2}{P_{xx}(f)P_{yy}(f)} \tag{1}$$

To compute the coherence in this way, multiple epochs have to be analysed and the results have to be averaged. It is important to realise that even for uncorrelated time series coherence will typically not be zero. The exact value depends upon the number of epochs used in the computation: the more epochs, the closer coherence approaches zero if there is no coupling. Furthermore, this approach to coherence analysis has the same characteristics as the digital spectral analysis upon which it is based: the spectral resolution depends upon epoch length, and the highest frequency that can be analysed is determined by the sample frequency. For an excellent introduction into coherence analysis we refer to Nunez et al. (1997).

2.3 Nonlinear Measures

While linear measures of correlation are relatively straightforward and well understood, it is a priori very unlikely that they will be able to capture all the complex and nonstationary interdependencies between time series recorded from different brain regions. In the last few decades, enormous progress has been made in the mathematics of nonlinear dynamical systems. This has also lead to new insights in the way in which nonlinear, aperiodic and chaotic systems can synchronise their activity (Boccaletti et al., 2002). This progress in theoretical understanding has also resulted in the development and implementation of a whole range of new techniques for nonlinear time series analysis (Stam, 2005, 2006). In particular, several new ways to determine nonlinear statistical interdependencies between time series have been proposed (Pereda et al., 2005). These techniques can be divided into two main groups based upon (i) phase synchronisation and (ii) generalised synchronisation. Here we briefly discuss each of these approaches, referring to the review papers mentioned above for background information and mathematical detail.

The notion that regular oscillators can synchronise their dynamics when they are weakly coupled is already quite old and goes back to the seminal observations of Huygens on swinging pendulum clocks. However, it was discovered only recently that irregular and chaotic oscillators can also display a kind of phase synchronisation (Rosenblum et al., 1996). The basic idea here is that the phase difference between two oscillators remains *bounded.* If Φ_1 and Φ_2 are the phases of two time series and Φ is the phase difference or relative phase, n to m (with n and m some integers) phase synchronisation is defined as follows:

$$\left|\phi_{n,m}\right| = |n\phi_1 - m\phi_2| < \text{const} \tag{2}$$

Defined in this way, phase synchronisation can be applied to noisy, nonstationary and chaotic signals. Usually the analysis is restricted to the case of $n = m = 1$. To compute the phase synchronisation between two time series, it is necessary first to obtain a time series of instantaneous phase values. This can be done with the analytic signal concept and the Hilbert transformation (for technical details, see Pereda et al., 2005). Basically the original signal is represented as a complex signal with amplitude and phase values for each time point. From the phase time series, a distribution of phase differences can be obtained. The properties of this distribution (flat: no coupling; one or more peaks: coupling) then determine the strength of the phase synchronisation. Quantification can be done by taking the Shannon information entropy or the circular variance of the phase difference distributions. Phase synchronisation is a very elegant alternative to coherence since it can handle linear as well as nonlinear coupling, it does not require stationary data and it does not depend upon amplitude or power, in contrast to coherence. Furthermore, it can be implemented easily and is computationally fast. However, there is still controversy whether the method requires narrow band filtering of the data or can also be applied to broadband time series.

An alternative approach to synchronisation between nonlinear systems is based upon the concept of 'generalised synchronisation' introduced by Rulkov et al. (1995). In this case we do not consider the time series themselves, but the attractors reconstructed from them in some suitable 'state space'. Such an attractor is a series of vectors in an m-dimensional space representing the dynamics of the system. Attractors can be reconstructed from time series using the technique of time-delay embedding. The concept of generalised synchronisation implies that the state of dynamical system Y is a function of the state of dynamical system X: $Y = F(X)$. Here X is the driver and Y is the response system. The attractiveness of this concept is its generality. For instance, it does not require narrow band signals or a well-defined phase. However, reconstructing attractors from time series can be quite complex and requires a lot of parameters to be set. Also, violations of stationarity may be more problematic than in the case of phase synchronisation.

Several methods have been proposed to determine the strength, and sometimes also the direction, of generalised synchronisation. One of the simplest measures is the synchronisation likelihood, *SL* (Stam and van Dijk, 2002; Montez et al., 2006).

The *SL* is the likelihood that Y_j will be close to Y_i given that X_j is close to X_i (with $Y_{i,j}$ and $X_{i,j}$ vectors on attractors reconstructed from two time series); 'close to' being defined as an intervector distance shorter than a cut-off distance. The formal definition of the synchronisation likelihood between two systems at time *i* is

$$SL_i = \frac{1}{Np_{\text{ref}}} \sum_{j=i-w2|j-i|>w1}^{j=i+w2} \theta\left(r_{x,i} - \left|X_i - X_j\right|\right) \theta\left(r_{y,i} - \left|Y_i - Y_j\right|\right) \quad (3)$$

Here *N* is the total number of vectors over which the summation is done, P_{ref} a small number $0 < P_{\text{ref}} << 1$ which is used to choose the cut-off distance *r*, *w*2 a window which can be thought of as the epoch length over which the computation of *SL* is done, *w*1 the Theiler correction and the part right of the summation is a product of two Heaviside functions. The range of *SL* is $P_{\text{ref}} < SL < 1$. P_{ref} is the value of *SL* in the case the time series are not synchronised; 1 is the value of *SL* in the case of perfect synchronisation. By averaging over all *i* (or a suitable subset of all *i*) the mean *SL* between two time series can be computed. By skipping the averaging step *SL* can be computed as a function of time.

2.4 Methodological Aspects and Pitfalls

The large and still growing number of techniques currently available for assessing statistical correlations between time series has not made the implementation of functional connectivity analysis any easier. In fact for each clinical or experimental application one has to choose a proper method, taking into account its strengths and weaknesses and the specific demands of the task. There have been a few attempts to directly compare different techniques, usually with equivocal results (Ansari-Asl et al., 2006; David et al., 2004; Quian Quiroga et al., 2002). The mundane conclusion is that an ideal technique does not exist, and choosing the proper one in a specific case requires a bit of 'Fingerspitzengefuhl'. Here we discuss a few methodological aspects which need to be taken into account when choosing a method and interpreting the results.

Perhaps the single most important problem in assessing functional connectivity from multichannel EEG and MEG data is volume conduction. It is almost unavoidable that two EEG electrodes or MEG sensors, even when they are not really nearby, will pick up activity from a common source, giving rise to spuriously high correlations between the time series. These correlations are called 'spurious' since they do not reflect true functional interactions between different sources. Several solutions have been proposed in the literature. Nunez calculated the influence of volume conduction (and also the effect of an active reference in EEG, a related problem) on coherence in a model of the EEG (Nunez et al., 1997). He proposed to correct coherence estimates for the theoretical influence of volume conduction to obtain a 'reduced coherence'. If the activity of the common source is known, one can also use a partial coherence. Nolte has proposed to use only the imaginary component of the coherence since this is not affected by common sources (Nolte et al., 2004). Phase synchronisation has been shown to be influenced by volume conduction and the contribution of the common reference electrode in scalp EEG recordings as well (Guevara et al., 2005). The solution could be to use a modified estimator of phase synchronisation, the phase lag index (PLI) which ignores all coupling with zero phase lag since volume conduction and active reference effects will only show up as zero lag correlations (Stam et al., 2007a). Unfortunately, both the imaginary coherence and the phase lag index will also ignore true interactions with zero phase lag, giving perhaps an overly pessimistic assessment of functional interactions.

Another approach to the problem is to do the analysis of functional connectivity in source space rather than signal space. A well-known example of this approach is the dynamic imaging of coherent sources (DICS, Gross et al., 2001). A problem with such approaches may be that the assumptions of the specific source model used, for instance independence of the sources, may interfere with the subsequent assessment of functional connectivity. Hadjipapas et al. have shown that weakly coupled sources can be reliably reconstructed using SAM (synthetic aperture magnetometry) analysis of MEG data (Hadjipapas et al., 2005). Another popular source model, LORETA (low-resolution electromagnetic tomography analysis), has also been used to study functional interactions at the source level (Lehmann et al., 2006). The multitude of available models here reflects one of the major problems with this approach: there is no unique solution to the inverse problem, and the choice of a particular model is arbitrary and likely to reflect the outcome of the analysis.

A different problem concerns the choice between linear or nonlinear measures of correlation. Linear measures, in particular coherence, are well-known, generally available and require a minimal number of parameters (such as number and length of epochs) to be set. However, coherence analysis will miss the nonlinear part of the coupling and requires stationary data. Nonlinear measures are more general, since they will capture linear as well as nonlinear coupling. Some methods, such as phase synchronisation, are also suitable for nonstationary data. However, many nonlinear techniques, especially those based upon generalised synchronisation, are quite complex and require a lot of parameters to be chosen. This hampers their application in more routine applications. Nonlinear techniques may be more suitable for MEG than EEG, since MEG has been shown to contain more nonlinear information than EEG (Stam et al., 2003). Also, EEG and MEG recordings of epileptic activity are more likely to have nonlinear properties which cannot be fully characterised by linear techniques.

Recently, there has been an increased interest to go beyond functional connectivity and to determine directional, causal interactions between time series. Some of these methods are based upon the concept of Granger causality, originally derived from economics. Granger causality exists when the predictability of one time series is improved by having information of the past of another time series. The idea of Granger causality has been implemented in both linear and nonlinear algorithms (Chavez et al., 2003; Kaminski et al., 2001). Phase synchronisation has also been used to determine causal interactions between time series (Rosenblum and Pikovsky,

2001). These new methods for detecting directional and possibly causal interactions between different parts of the brain hold great potential for future clinical applications, for instance detecting abnormal driving of the brain by an epileptic focus.

3 Graph Analysis of Brain Networks

3.1 Modern Network Theory

At many levels of organisation, the brain can be considered as a complex network of interacting subunits. This is equally true at the level of interconnected neurons as well as at the level of interacting mesoscopic neural systems. While many linear and nonlinear methods, some of which were discussed above, can be used to investigate the level of interaction between pairs of such units, such an approach may not be sufficient to grasp the full complexity of the brain as a complex network. For instance, determining synchronisation between all possible pairs of MEG sensors in a particular frequency band will produce huge matrices of correlation data, which are difficult to interpret and to handle statistically. In the case of a full voxel-wise analysis of fMRI resting-state data the situation is even worse. This problem has been handled by averaging the data using task-related or anatomy-based considerations. An example of this approach where the pair-wise correlations between MEG channels were grouped in short and long distance regional patterns can be found in Stam et al. (2006). Alternatively, methods such as independent component analysis can be used to reduce the high-dimensional data sets to a limited number of sources. A problem here may be the fact that such approaches may interfere with the assessment of the interactions between brain regions. Here we describe an alternative approach based upon the concept of complex networks.

Networks have been known and studied in science for a long time in a branch of mathematics called graph theory. A graph is an abstract representation of a network in the form of a set of nodes (vertices) and connections (edges). The famous mathematician Leonard Euler (1707–1783) started graph theory when he solved the problem of the seven bridges of Königsberg by representing it in the form of a graph. A major advance in graph theory was the development of random graph models (Erdos and Renyi, 1960; Solomonov and Rapoport, 1951). Random graphs are characterised by the fact that edges between vertices exist with a probability p which is the same for all pairs of edges. Random graphs have interesting properties. For instance, random graphs show a phase transition towards a globally connected system as a function of increasing p. However, for a long time neither regular nor ER (Erdos Renyi)-type random graphs were particularly good models for many interesting types of networks encountered in nature. In particular, classical graph theory was not able to explain two interesting properties of social networks.

Social networks are often characterised by a high level of clustering. For instance, if A has acquaintances B and C, then B and C are more likely to know each other as well. At the same time social networks also display the property of short path lengths. This means that, even in a very large network, one can get from any network point (person) to any other network point in a very small number of steps (a step being for instance a pair of persons who know each other). This phenomenon has been investigated formally by Stanley Milgram (Milgram, 1967). He sent a large number of letters to randomly chosen persons in the USA. The purpose of the experiment was to get the letter to a target person in Boston. However, subjects were only allowed to send the letter to other persons they knew on a first name basis. Surprisingly, many of the letters did reach the target person in Boston, and they did so on average in only six steps. Milgram coined this property of social networks the 'small-world' phenomenon. Social networks are thus highly clustered and have the small-world property; this combination of properties could not be explained by classical graph theory.

The solution of this problem, and a breakthrough in thinking about complex networks in general, came from a brief paper published in *Nature* in 1998 by two physicists: Duncan Watts and Steve Strogatz (1998). Strogatz, a physicist with a background in nonlinear dynamics, was studying synchronisation in large systems of coupled oscillators. While synchronisation between pairs of oscillators was increasingly understood, theoretical understanding of large systems of oscillators was still limited. One of the best-studied models was that of Kuramoto which considers a fully connected array of phase oscillators (Strogatz, 2000). In this model, the onset of synchronisation (which takes the form of a phase transition) was well understood. However, the fully connected state was not a good model of interactions in real systems of coupled oscillators where connectivity is typically sparse. Watts, as a postdoc, was given the assignment to study synchronisation of large systems of oscillators using more realistic coupling patterns. What he came up with was a model of a network on a ring, where each vertex is coupled to its k nearest neighbours (with $k << N$ to make the network sparsely connected) (Fig. 1). In this regular network, clustering is high, but path lengths (average number of steps to go from one vertex to any other in the network) are long. Such systems do not synchronise very well. Now the brilliant idea was to introduce a rewiring probability p. With probability p an edge is chosen and rewired to a randomly chosen other vertex in the network. For increasing p, the network changes shape, and becomes a (sparsely connected) random network for $p = 1$. The random network with $p = 1$ has a very short path length, but also has a very low clustering. The really interesting behaviour of the model occurs for intermediate values of p, especially for small $p > 0$. Introduction of only a few random rewirings hardly affects the clustering which stays high, but it results in a very rapid drop in the path length, which approached the low value of random networks already for very small values of p. These intermediate networks that have both high clustering and short path lengths were called 'small-world' networks by Watts and Strogatz. The authors showed that several real complex networks (the central nervous system of *C. elegans*, a network of actors and a network of power plants in the USA) are small-world networks. Moreover, they showed that small-world networks are highly effective in transporting information; in fact, sparsely connected networks

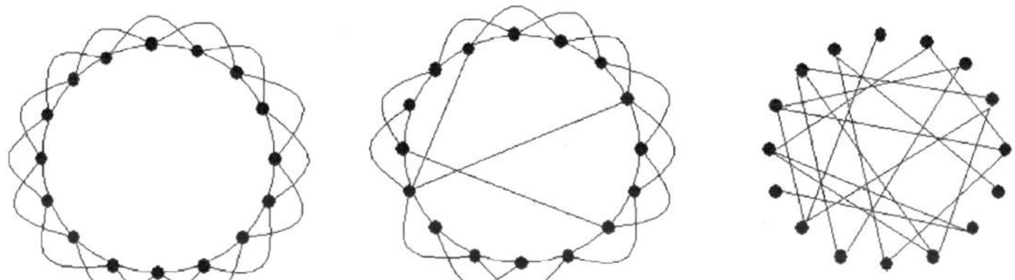

Fig. 1 Three basic network types in the model of Watts and Strogatz. The leftmost graph is a ring of 16 vertices ($N = 16$), where each vertex is connected to four neighbours ($k = 4$). This is an ordered graph which has a high clustering coefficient C and a long path length L. By choosing an edge at random and reconnecting it to a randomly chosen vertex, graphs with increasingly random structure can be generated for increasing rewiring probability p. In the case of $p = 1$, the graph becomes completely random, and has a low clustering coefficient and a short path length. For small values of p, the so-called small-world networks arise, which combine the high clustering coefficient of ordered networks with the short path length of random networks (from Stam and Reijneveld, 2007; copyright by the authors)

with a small-world configuration synchronise much better than regular lattice-like networks.

The discovery of small-world networks by Watts and Strogatz was rapidly followed by another breakthrough. Barabasi and Albert proposed a model of a growing network, where new edges were attached with a bias towards vertices with a high degree (the degree of a vertex is the number of edges it has) (Barabasi and Albert, 1999). They showed that such a model would converge to a state where the probability distribution of the degrees, $P(k)$ has the form of a power law: $P(k) \sim k^{\gamma}$, with the scaling component $\gamma = 3$. They suggested that several complex networks encountered in nature, for instance, the Internet and the World Wide Web might be examples of such 'scale-free' networks. The discovery of small-world and scale-free networks initiated an enormous interest in the study of complex networks, ranging from metabolic and gene networks to the Internet and social systems.

Modern network theory has now also been applied to study the nervous system, both at the level of anatomical connectivity and at the level of functional connectivity, that is, patterns of synchronisation between different components of the nervous system. Below, we first introduce some of the basic measures to characterise graphs. Next we show how graph analysis can be applied to neurophysiological data sets. Further details of modern network theory can be found in several reviews (Albert and Barabasi, 2002; Amaral and Ottino, 2004; Boccaletti et al., 2006; Newman, 2003). Application of graph theory to neuroscience has also been reviewed (Stam and Reijneveld, 2007).

3.2 Measures to Characterise Graphs

A graph is an abstract representation of a network in the form of nodes (or vertices) with connections (or edges). In the simplest case, two vertices are either connected (an edge between them exists) or not (an edge between them does not exist). Such graphs are called unweighted or binary graphs. If the strengths of connections are considered explicitly each edge is given a weight. The corresponding graphs are weighted graphs. Furthermore, it is possible to distinguish between graphs where the direction of the interaction between vertices is taken into account (directed graphs) and graphs where the relationship between vertices is symmetrical and bidirectional (undirected graphs). The pattern of connections in a graph is described by the adjacency matrix a_{ij}. If two vertices i and j are connected a_{ij} =1; otherwise a_{ij} is 0. In the case of weighted graphs, the weight of each existing edge is given by a weight matrix w_{ij}. The degree k of a vertex is the number of edges it has with other vertices, its set of neighbours. One can also consider the average degree K of the whole graph.

Graphs can now be characterised by several measures which allow to distinguish between different basic types. We will restrict ourselves to unweighted, undirected graphs unless indicated otherwise. The first measure is the clustering coefficient C. The clustering coefficient C_i of a vertex i is the probability that its neighbour vertices are also connected to each other. In other words, it is the ratio of the actual number of edges between its neighbours and the maximum possible number of edges between its neighbours. The clustering coefficient of all vertices can be averaged to give a graph clustering coefficient C. The clustering coefficient is a measure of local structure in the graph, and is related to resilience to random attacks on the network: loss of a few edges will have less effect if all the neighbours are connected to each other. The next important graph measure is the path length L. For unweighted graphs, this is defined as the smallest number of edges that have to be travelled to get from one vertex to another vertex. The path length is averaged over all possible pairs of vertices to give the average shortest path of the graph. This path length is a measure of global integration of the network. Finally, the degree distribution $P(k)$ can be considered. The degree distribution $P(k)$ is the probability that a randomly chosen vertex will have degree k.

With these three basic measures, it is possible to distinguish some of the most important types of graphs. The first type is a regular or lattice-like graph. This is characterised by a high clustering coefficient, a long path length and a degree distribution which is a delta function ($P(k) = 1$ for one value of k and $P(k) = 0$ for all other values of k). The regular graph corresponds to the case of $p = 0$ in the model of Watts and Strogatz (Fig. 1). The second type is the Erdos and Renyi (ER) type of random graph. Here, the likelihood that an edge exists between a pair of vertices is given by a likelihood p. Such graphs have a low clustering coefficient and a short path length. The degree distribution is a Poisson distribution, with a small variation around an average value. This type of graph corresponds to the case of a rewiring probability of one in the model of Watts and Strogatz. The third type of graph is the small-world network. This type of graph has a high clustering coefficient (comparable to a regular graph), a short path length (comparable to a random graph) and a Poisson-like degree distribution. It corresponds to a rewiring probability of $0 < p << 1$ in the model of Watts and Strogatz. The fourth basic type of graph is the scale-free graph. This is characterised by a degree distribution of the form $P(k) \sim k^{\gamma}$. Such scale-free graphs have very short path lengths, even shorter than ER-type random graphs (Cohen and Havlin, 2003). Sometimes scale-free graphs are considered as a subtype of small-world networks, but it should be noted that, in contrast to WS-type small-world networks, scale-free networks usually have a vanishing small clustering coefficient for the limit $N \rightarrow \infty$.

In addition to the classic graph measures (clustering coefficient, path length, degree distribution) mentioned above, many other measures have been introduced recently and the set of tools to describe graphs is still increasing (Boccaletti et al., 2006). Here we describe a number of the more important new measures. First, in many cases – especially the study of empirical networks – it is convenient to compare a measure to a reference network, usually a random network with the same number of vertices, nodes and degree. For instance, the clustering coefficient and the path length can be computed for a graph and also for an ensemble of random networks derived from this graph. This results in a normalised clustering coefficient $\gamma = C/\langle\text{C-random}\rangle$ and path length $\lambda = L/\langle\text{L-random}\rangle$, where $\langle\,\rangle$ denotes an ensemble average. The ratio between γ and λ is the small-world index introduced by Humphries et al. (2006) and Humphries and Gurney (2008). Although this approach of normalisation by comparison with an ensemble of random networks is most commonly used in combination with clustering coefficients and path lengths, it can of course also be used in combination with other measures.

A generalisation of the clustering coefficient is the notion of a graph motif (Milo et al., 2002). A motif is a small subgraph with a specific configuration. The clustering coefficient is in fact an analysis of the motif of a triangle; this can be generalised to more complex types of motifs. Another important graph measure is the vertex or edge 'centrality'. This measure is the ratio of the number of shortest paths between a set of vertices that cross through a vertex (or edge) divided by the total number of shortest paths between that set of vertices. Centrality measures indicate the relative importance of a vertex or edge and predict the consequences of its loss for network integrity. Another measure of network structure is the degree correlation. A degree correlation exists when the probability that a vertex i will have degree k depends upon the degree of the vertices to which it is connected. A simple reduced way to quantify this is to consider the correlation coefficient of degrees of connected vertices. Graphs with a positive degree correlation are called 'assortative'; graphs with a negative degree correlation are called 'disassortative'. Remarkably, social networks tend to be assortative, while biological and technological networks tend to be disassortative (Newman, 2003). Finally, the 'synchronisability' of a graph can be analysed with graph spectral analysis (Boccaletti et al., 2006).

Analysis of weighted graphs presents special problems. The usual measures developed for unweighted graphs cannot be directly applied to weighted graphs. One solution, which is often used in the application of graph analysis to neurophysiological data, is to convert the weighted graph to an unweighted graph by considering a threshold. This approach, which is described more fully below, is not without problems. The alternative solution is to consider measures which are specifically developed to deal with weighted graphs. Weighted versions of both the clustering coefficient and path length have been described (Barthelemy et al., 2005; Barrat et al., 2005; Newman, 2003). Related to this, one can consider the so-called local and global efficiencies, which are based upon the inverse of the path lengths (Latora and Marchiori, 2001, 2003). The concept of degree is also more complex in weighted networks since both the adjacency matrix a_{ij} and the weight matrix w_{ij} have to be considered. Ramasco and Goncalves introduced a measure of weight dispersion for weighted networks, which indicates the correlation between weights of individual vertices (Ramasco and Goncalves, 2007). Positive weight correlations are possibly related to the existence of 'super highways' and efficient communication in the network.

3.3 Application to Functional Connectivity: Principles and Problems

Application of graph analysis to neuroanatomical data is relatively straightforward and has already been done in the original paper of Watts and Strogatz where the full connectivity pattern of the brain of *C. elegans* was studied. The principal idea is to decide what the basic elements of the network are (neurons or anatomical regions) and to determine the connectivity matrix of all these elements (synapses between neurons or anatomical pathways between brain regions). The resulting connectivity matrices are usually analysed as unweighted, undirected graphs.

Application of graph analysis to neurophysiological data, such as multichannel EEG or MEG recordings or BOLD signals from fMRI studies, is less straightforward and requires a few intermediate steps. To illustrate one approach, we consider the scheme in Fig. 2. We start with a multichannel recording of neurophysiological activity, in this case an epoch of 21 channel EEG. After filtering the correlation or synchronisation between all possible pairs of channels is determined. This can be done with any of the linear or nonlinear measures described in Sect. 2. The result is a matrix of synchronisation or correlation values, as indicated in *B*. To convert this matrix to an unweighted graph, we consider each EEG channel to be a vertex and we assume that an edge exists between pairs of channels/vertices if the synchronisation strength between these two channels exceeds a threshold value. Preferably the threshold should be chosen such that the resulting graph is sparsely connected, but not so sparse that it becomes disconnected. Next, graph measures of choice can be computed from this graph, in the case of this example, the clustering coefficient C and the path length L. For comparison, an ensemble of random graphs is created from the original data, by shuffling the correlation matrix in *B*, preserving the total number of vertices and edges, as well as the symmetry of the matrix. In some cases such as the analysis of motifs it may be necessary to preserve the degree distribution as well (Sporns and Zwi, 2004). The average C and L of this ensemble are also calculated. The final measures are the ratio between $C/\langle C\text{-random}\rangle$ and $L/\langle L\text{-random}\rangle$ ($\langle\ \rangle$ indicates ensemble average).

While this approach allows in principle to apply the formalism of graph theory to multichannel neurophysiological data, we should realise there are several problems and limitations that should be taken into account. First, graph theory deals with the statistical properties of large networks, usually consisting of several hundreds to thousands of nodes. Several theorem's of graph theory may not be valid for small networks, consisting of only a few dozen of vertices such as the EEG network in the example. In particular, it may not be possible to derive meaningful degree distributions for networks with $N < 100$. Other network measures may be more robust in the case of small networks. Second, in the example, network analysis was applied to raw EEG recordings. It is conceivable that correlations between EEG channels will be influenced by the reference electrode and volume conduction (see also Sect. 3). This could bias network measures derived from such an analysis. An alternative could be either to use synchronisation measures that are less sensitive to volume conduction and reference electrode effects, such as the imaginary coherence or phase lag index or to perform the analysis on signal from reconstructed source analysis (De Vico Fallani et al., 2007a, b). Third, the choice of a threshold to convert the correlation matrix to an unweighted graph will affect the degree of the graph and many other measures that are computed from it. As an extreme example, with a threshold of zero the resulting graph will be fully connected with a clustering coefficient of 1 and a path length of 1. The choice of the threshold is thus important, but no straightforward criterion exists. One approach is to base the threshold on the significance of the correlations. Only correlations which exceed a certain significance criterion are converted to edges. This approach does not solve all problems since the choice of a significance criterion is itself a bit arbitrary, and the resulting graph is not necessarily an optimal representation of the network. Alternatively, one could consider a whole range of threshold values and study the behaviour of graph measures such as C and L as a function of threshold (for an early example, see Stam et al., 2007b). While less arbitrary, this approach introduces another problem when graph measures have to be compared between different conditions or groups. Here we have to multiply statistical tests, one for each value of the threshold, to compare conditions or groups, which may result in type I errors. The problem with thresholds can be avoided by analysing the original correlation matrix as a weighted graph (De Vico Fallani et al., 2007a, b). This restricts the analysis to graph measures which can be applied to weighted graphs, but does avoid the arbitrary choice of a threshold and the multiple statistical comparisons problem. Fourth, reconstructing sparsely connected unweighted networks from correlation matrices almost always results in a few vertices becoming disconnected from the graph. Leaving such disconnected vertices out of the calculation produces a bias in the graph measures, in particular path lengths. One solution is to consider inverse path lengths, where a path between disconnected vertices is given a value of infinity (and the inverse a value of zero); this approach based upon the concept of efficiencies is usually a good solution to the problem of disconnected points (Ponten et al., 2007). Fifth, problems may arise when sets of networks are compared which differ in the mean strength of

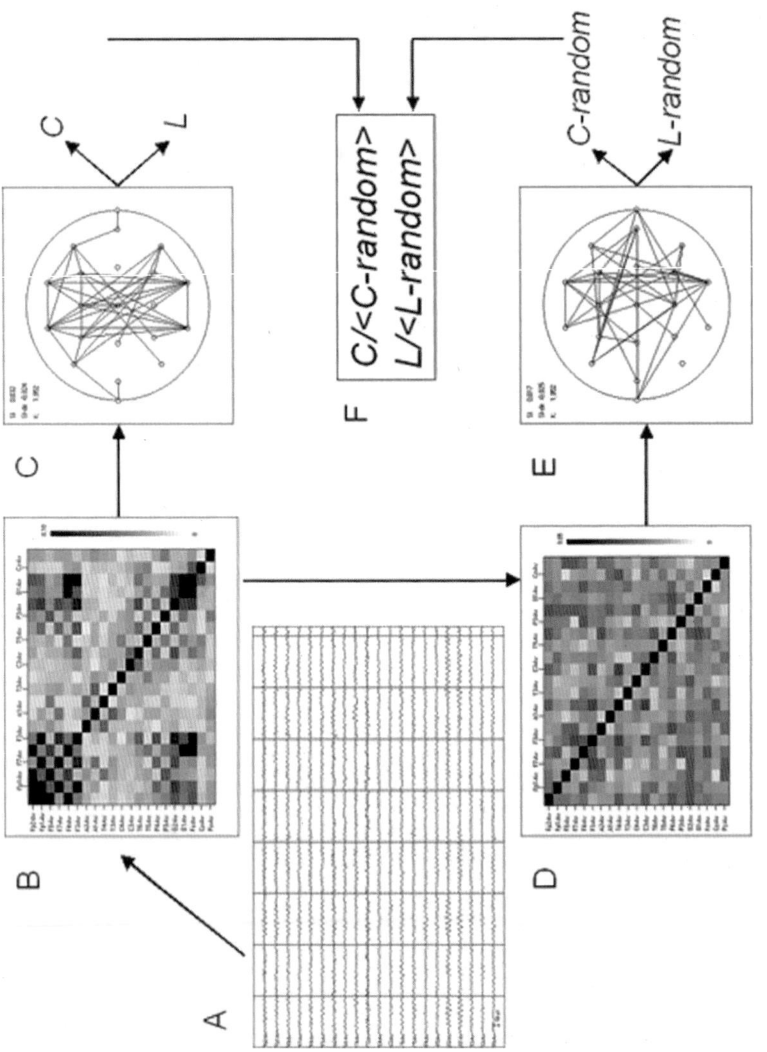
A
B
C
D
E
F
C
L
C/<C-random>
L/<L-random>
C-random
L-random

Fig. 2 Schematic illustration of graph analysis applied to multichannel recordings of brain activity (fMRI, EEG or MEG). The first step (panel **A**) consists of computing a measure of correlation between all possible pairs of channels of recorded brain activity. The correlations can be represented in a correlation diagram (panel **B**, strength of correlation indicated with *black white* scale). Next a threshold is applied, and all correlations above the threshold are considered to be edges connecting vertices (channels). Thus, the correlation matrix is converted to a unweighted graph (panel **C**). From this graph, various measures such as the clustering coefficient C and the path length L can be computed. For comparisons, random networks can be generated by shuffling the cells of the original correlation matrix of panel **B**. This shuffling preserves the symmetry of the matrix and the mean strength of the correlations (panel **D**). From the random matrices graphs are constructed, and graph measures are computed as before. The mean values of the graph measures for the ensemble of random networks are determined. Finally, the ratio of the graph measures of the original network and the mean values of the graph measures of the random networks can be determined (panel **F**) (from Stam and Reijneveld, 2007; copyright by the authors)

the correlation. In the case of a fixed threshold, networks with stronger correlations will be represented as graphs with more edges; this will affect many graph measures computed from these graphs. Thus the analysis will mix information on the topology of the network with information on the strengths of the correlation, making the interpretation of the analysis difficult. A solution is to consider a fixed degree K instead of a fixed threshold; here the threshold is adapted such that all reconstructed graphs will have the same degree. This eliminates the effect of differences in correlation strength. Any remaining differences can only be due to differences in network topology. An example of this approach can be found in Ponten et al. (2007). To some extent, normalisation by comparison with ensembles of random networks can also be used to solve this problem. We should realise that the 'fixed k' approach will result in unweighted networks where the edges do not have the same meaning in all networks. The weighted approach also suffers from the bias caused by differences in correlation strength; here a 'fixed k' solution is not possible and only normalisation by ensembles of random networks can be used.

4 Examples of Applications

4.1 Synchronisation

Many studies have used one or more of the linear or nonlinear correlation measures described above to study functional connectivity patterns in healthy subjects or various types of neurological or psychiatric disease. Overviews of much current work can be found in several review papers (Reijneveld et al., 2007; Schnitzler and Gross, 2005; Uhlhaas and Singer, 2006; Varela et al., 2001). For the sake of illustration, we briefly discuss applications in healthy subjects, patients with Alzheimer's disease and epilepsy.

Traditionally, EEG and MEG have been used in cognitive neuroscience to study the temporal aspects of brain responses in cognitive tasks using either event-related potentials or ERD/ERS (event-related dysynchronisation/event-related synchronisation) techniques. Now these techniques have been extended by analyses of functional connectivity in various frequency bands, either during tasks or in a 'resting-state' condition. Two interesting findings in healthy subjects deserve mentioning: (i) There is increasing evidence that synchronisation in the gamma band, especially in response to stimulus perception, may carry a special significance. In line with animal studies in the early 1990s, EEG studies have shown that gamma band synchronisation may be closely connected to perception, attention and conscious awareness. In a famous *Nature* paper, Rodriguez et al. have shown that conscious awareness of the face in Mooney pictures corresponds with large-scale changes in gamma band synchronisation (Rodriguez et al., 1999). Although these findings are still controversial, a close connection of gamma band synchronisation and consciousness is quite likely. (ii) Theta band rhythms play a major role in the hippocampus. There is now increasing evidence that theta band activity can be found in a wide range of other brain areas, most notably the frontal regions, that are directly or indirectly connected to the hippocampus. Of interest, theta band synchronisation seems to be closely associated with working memory processes. For instance, Sarnthein et al. were able to show that theta band synchronisation of the EEG specifically increases during the retention phase of a working memory paradigm (Sarnthein et al., 1998). These observations have been confirmed in other studies (Stam et al., 2002). These studies on gamma and theta band synchronisation nicely illustrate how EEG/MEG functional connectivity analysis may help to elucidate the functional significance of these EEG rhythms in healthy subjects.

Synchronisation in various frequency bands is a delicate process in healthy subjects which requires amongst others intact anatomical connectivity and functioning of neurotransmitter systems. These requirements may not be met in neurological disorders, resulting in a pathological disruption of synchronisation. One of the most intensely studied neurological disorders is Alzheimer's disease. Alzheimer's disease is the most frequent cause of dementia in the Western population. Alzheimer's disease is characterised by widespread neuronal loss, starting in the medial temporal areas and spreading to other, mostly associational cortical areas. Moreover, major neurotransmitters systems such as the cholinergic projections from the nucleus basalis of Meynert are also affected. Both neuronal loss and lack of acetylcholine might impair functional connectivity between distributed brain areas. There is now accumulating evidence from both EEG and MEG studies of functional connectivity in Alzheimer's disease that there is indeed a loss of coupling especially in the alpha and beta bands; changes in the lower frequency bands are more controversial (for review, see Jeong, 2004). However, in a recent MEG study it was shown that loss of connectivity in the alpha and beta bands could coexist with pathologically increased parieto occipital connectivity. Perhaps the changes in connectivity occurring in a neurodegenerative disease like Alzheimer's disease are more complex than a simple 'loss of connections', and rather reflect a different, pathological reorganisation of the topology of functional brain networks as discussed in the next section (Stam et al., 2007b).

Epileptic seizures have often been considered as classic examples of pathologically increased synchronisation. A nice illustration is the generalised spike-and-wave discharges during absence seizures. It has become clear that synchronisation related to epileptic seizures is often highly nonlinear and requires nonlinear techniques for optimal characterisation. Interest in nonlinear analysis of seizures increased enormously when the first reports on seizure prediction using nonlinear EEG analysis appeared in 1998 (Lehnertz and Elger, 1998; Martinerie et al., 1998). Subsequently a huge number of studies have tried to replicate and improve these results, often using some sort of nonlinear correlation measures (for review, see Stam, 2005). One conclusion is that the relation between seizures and synchronisation is quite complex; seizures often start with a decrease rather than an increase of synchronisation. Studies performed on intracranial recordings suggest in fact that increased synchronisation may characterise the end rather than the beginning of the seizure (Schindler et al., 2007). Increases in synchronisation occurring late in the seizure may also be accompanied by a reorganisation of the topology of brain networks towards a more normal 'small-world' pattern, while abnormal, more random interictal networks could be a risk factor for initiating seizures, as discussed below (Ponten et al., 2007). Synchronisation in relation to epilepsy is currently a very active field of research, and also the field that is most likely to produce clinically useful applications for seizure focus identification, guidance of epilepsy surgery and, possibly in the future, seizure prediction.

4.2 Network Analysis

4.2.1 MRI

Modern magnetic resonance imaging (MRI) techniques allow the characterisation of both structural as well as functional brain networks. He et al. used correlation matrices of cortical thickness derived from MRI to reconstruct brain networks in 94 healthy subjects. The resulting networks displayed high clustering, short path lengths and exponentially truncated degree distributions, consistent with small-world networks (He et al., 2007). A disadvantage of this approach is that it requires a large number of subjects to obtain reliable statistics. Alternatively, graph analysis can be applied to networks reconstructed from diffusion MRI. Hagmann et al. applied this approach to two human subjects and demonstrated a small-world pattern with an exponential degree distribution (Hagmann et al., 2007). Another application of graph analysis to diffusion-weighted MRI has been described by Iturria-Medina et al. (2007). Schmitt et al. showed that the small-world features of MRI-based structural brain networks may have a strong genetic base (Schmitt et al., 2008).

Graph analysis can also be applied to BOLD (blood oxygen level dependent) signals derived from functional MRI. Dodel et al. were the first to suggest graph analysis of fMRI data (Dodel et al., 2002). In a series of studies characteristics of resting-state fMRI-derived networks were characterised (Salvador et al., 2005a, b; Achard et al., 2006). These studies confirmed the presence of small-world properties independent of frequency. Although hubs (high-degree nodes) could be identified in the networks, the authors state that the networks are not truly scale-free, but have exponentially truncated degree distributions (Achard et al., 2006). In contrast, another study reported a scale-free degree distribution for fMRI data and no effect of different tasks on graph measures (Eguiluz et al., 2005). A scale-free degree distribution was also reported in a later study (Cecchi et al., 2007). The analysis of the latter two studies was based upon a voxel-wise approach, whereas the other studies analysed BOLD time series from 90 regions of interest. Achard and Bullmore, using neural efficiency measures, showed that elderly subjects have a reduced efficiency of networks (Achard and Bullmore, 2007). Administration of a dopamine antagonist reduced efficiency in healthy subjects. Dosenbach et al. demonstrated the presence of two distinct task-control network based upon graph analysis of resting-state fMRI data (Dosenbach et al., 2007). Age-related changes reflected by reduced connectivity degree in motor networks were reported by Wu et al. (2007).

A few studies have used MRI-based graph analysis to study changes in brain networks in neuropsychiatric disease. He et al. showed that networks based upon cortical thickness correlations were altered in Alzheimer patients, displaying increased clustering, longer path lengths and increased vulnerability to targeted attacks (He et al., 2008). Achard et al. reported loss of small-world features in a single subject with a brain stem lesion (Achard et al., 2006). Resting-state fMRI analysis in a group of children with attention-deficit/hyperactivity disorder (ADHD) showed a change of network topology consisting of increased local and decreased global efficiency compared to controls (Wang et al., 2008). This would imply a shift towards a more regular network topology in ADHD patients. Liu et al. studied resting-state functional connectivity in 31 schizophrenia patients and 31 healthy controls (Liu et al., 2008). The authors found disrupted small-world topology in the schizophrenia patients, correlating with disease duration.

4.2.2 EEG/MEG

Probably the first attempt to apply the concepts of graph theory to functional rather than anatomical connectivity was the study by Klaas Stephan et al. (2000). This study was based upon published information about the functional interrelations of brain areas in experimental animals, derived from strychnine lesions. The authors showed for the first time evidence for a small-world structure of functional connectivity in primate cerebral cortex. Small-world networks have now been demonstrated in vivo at the level of neuronal networks in monkeys (Yu et al., 2008).

Recently, a rapidly increasing number of studies have studied functional brain networks from the perspective of graph theory (Bassett and Bullmore, 2006; Reijneveld et al., 2007; Sporns and Honey, 2006). The first indications for a small-world pattern of functional connectivity in healthy humans were reported in an MEG study in five subjects in 2004 (Stam, 2004). Here, synchronisation likelihood of resting-state MEG filtered in different frequency bands was used to obtain correlations matrices, which were converted to unweighted graphs after thresholding. Functional networks for intermediate frequencies from 8 to 30 Hz were regular, while networks for lower and higher frequencies showed small-world features. Using wavelet decomposition of MEG recordings in 22 healthy subjects, Bassett et al. found evidence for small-world networks across different frequency bands (Bassett et al., 2006). Furthermore, they found evidence that brain networks were operating close to an order/disorder transition. Performance of a motor task was associated with the emergence of long-range connections in beta and gamma bands. EEG studies, although much more limited in the number of channels from which to reconstruct networks, have confirmed the presence of small-world functional networks in healthy subjects (Micheloyannis et al., 2006a, 2007; Ray et al., 2007; Smit et al., 2007; Stam et al., 2007b).

Of interest, the topology of functional brain networks is influenced amongst others by genetic factors, age and task demands. Smit et al. showed in a large cohort of twins that, after correction for influences of synchronisation strength, clustering coefficients and path lengths have a strong heritability across different frequency bands (Smit et al., 2007). The study of Micheloyannis et al., which compared children aged 8–12 years with young adults, showed a relative loss of small-world organisation of adults (Micheloyannis et al., 2009). Somewhat surprisingly, a lower intelligence quotient (IQ) in otherwise healthy subjects was correlated with a more outspoken small-world organisation of functional brain networks compared to high IQ subjects (Micheloyannis et al., 2006a).

Brain state also affects the topology of functional networks. During sleep a transition towards a more outspoken small-world organisation with increased clustering coefficients was reported (Ferri et al., 2007, 2008). A range of studies have shown changes in network topology during the performance of motor or cognitive tasks (Astolfi et al., 2007; Basset et al., 2006; De Vico Fallani et al., 2007a, b; Micheloyannis et al., 2006a, b, 2007; Ray et al., 2007). However, the patterns of changes in these studies were quite variable, depending upon the specific task and the details of the analysis.

Finally, changes in functional brain networks have been demonstrated in several neuropsychiatric disorders. De Vico Fallani et al. showed an increase in theta, alpha and beta band local efficiency in spinal cord-injured patients; these changes were interpreted as an indication of compensation mechanisms (De Vico Fallani et al., 2007b). In patients with brain tumours, widespread changes in functional connectivity were found, accompanied by a loss of small-world features in theta, beta and gamma bands (Bartolomei et al., 2006). In Alzheimer's disease, the normal small-world pattern was shown to be disrupted in the beta band (Stam et al., 2007b). A similar loss of small-world features has been reported in patients with schizophrenia, especially during performance of a working memory task (Micheloyannis et al., 2006b). Reconstructing networks from nonlinear interactions in EEG recordings Rubinov et al. also showed a loss of small-world features in schizophrenia patients (Rubinov et al., 2007).

Several modelling studies suggested a possible relationship between network architecture and the propensity to propagate epileptic seizures (Netoff et al., 2004; Percha et al., 2005; Srinivas et al., 2007). Abnormalities of network topology during an epileptic seizure were first reported by Wu et al. (2006). They showed a tendency towards more small-world-like networks during the seizure. In a more extensive study based upon intracranial recordings in patients with temporal lobe seizures a similar pattern of changes, with increases of both clustering coefficients and path lengths in the late phase of the seizure was described (Ponten et al., 2007). On the basis of these findings, it was conjectured that the interictal networks could actually be more random, with a very low threshold for synchronisation, and the 'regularisation' during seizures might present a mechanism by which seizure spread comes to a halt. These findings seem to be in agreement with the more random interictal networks in brain tumour patients who have a very high incidence of epilepsy (Bartolomei et al., 2006).

5 Summary and Conclusion

Multichannel EEG and MEG recordings contain a wealth of information on functional interactions between brain areas. Proper analysis of these interactions requires the use of quantitative techniques. Statistical interdependencies between time series can be characterised with linear techniques such as coherence or nonlinear techniques based upon the concepts of phase synchronisation or generalised synchronisation. Important methodological issues in applying these techniques are the influence of volume conduction and, in the case of EEG, the active reference; both may produce spuriously high correlations. This problem can be dealt with by computing 'reduced coherence', by using only the imaginary part of coherence or by the phase lag index. Alternatively, functional connectivity may be assessed in source rather than signal space. In general, linear techniques are more straightforward and easy to use, but nonlinear techniques can capture more information, especially in MEG recordings and in the case of epileptic phenomena.

Going beyond synchronisation, the nature of complex structural and functional networks can be studied using the framework of graph theory. This approach has gained enormous momentum after the discovery of small-world and scale-free networks, which can be characterised by measures such as clustering coefficients, path lengths, degree distribution and an increasing number of new measures. Network analysis can be applied to matrices of functional connectivity derived from any type of synchronisation analysis. However, this approach is still quite new and many methodological issues, such as the proper choice of a threshold to construct unweighted graphs, remain open.

Analysis of functional connectivity has demonstrated the significance of gamma and theta bands for consciousness and working memory. In neurodegenerative disease, connectivity may be decreased as well as abnormally organised spatially. During epileptic seizures, synchronisation may undergo complex changes depending upon the phase of the seizure and accompanied by a reorganisation of functional brain networks. Network analysis has shown that both anatomical as well as functional networks in the human brain have typical properties of small-world networks, such as high clustering and short path lengths. Controversy exists with respect to the question whether brain networks are truly scale-free, of whether their degree distributions are exponentially truncated power laws. The topology of complex networks has a strong genetic basis, changes during aging, and is affected to some extent by state and task demands. The structure of complex brain networks is disrupted in various neuropsychiatric conditions such as Alzheimer's disease, schizophrenia, ADHD, brain tumour and epilepsy. Major challenges for graph analysis of functional connectivity are to obtain a proper understanding of the processes, genetic and epigenetic, that give rise to these complex brain networks, the processes that determine the plasticity of these networks in adults, and the mechanisms by which disease disturbs the optimal network patterns giving rise to cognitive dysfunction and epileptic seizures.

References

Achard S, Salvador R, Whitcher B, Suckling J, Bullmore E (2006) A resilient, low-frequency, small-world human brain functional network with highly connected association cortical hubs. J Neurosci 26: 63–72.

Achard S, Bullmore E (2007) Efficiency and cost of economical brain functional networks. PloS Comput Biol 3: e17.

Aertsen AMHJ, Gerstein GL, Habib MK, Palm G (1989) Dynamics of neuronal firing correlation: modulation of 'effective connectivity'. J Neurophysiol 61: 900–917.

Albert R, Barabasi AL (2002) Statistical mechanics of complex networks. Rev Modern Phys 74: 47–97.

Amaral LAN, Ottino JM (2004) Complex networks. Augmenting the framework for the study of complex systems. Eur Phys J B 38: 147–162.

Ansari-Asl K, Senhadji L, Bellanger J-J, Wendling F (2006) Quantitative evaluation of linear an nonlinear methods characterizing interdependencies between brain signals. Phys Rev E 74: 031916.

Astolfi L, De Vico Fallani F, Cincotti F, Mattia D, Marciani MG, Bufalari S, Salinari S, Colosmimo A, Ding L, Edgar JC, Heller W, Miller GA, He B, Babiloni F (2007) Imaging functional brain connectivity patterns from high-resolution EEG and fMRI via graph theory. Psychophysiology 44: doi: 10.1111/j.1469-8986.2007.00556.x.

Barabasi AL, Albert R (1999) Emergence of scaling in random networks. Science 286: 509–512.

Barrat A, Bathelemy M, Pastor-Satorras R, Vespignani A (2005) The architecture of complex weighted networks. Proc Natl Acad Sci USA 101: 3747–3752.

Barthelemy M, Barrat A, Pastor-Satorras R, Vespignani A (2005) Characterization and modelling of weighted networks. Phys A 346: 34–43.

Bartolomei F, Bosma I, Klein M, Baayen JC, Reijneveld JC, Postma TJ, Heimans JJ, van Dijk BW, de Munck JC, de Jongh A, Cover KS, Stam CJ (2006) Disturbed functional connectivity in brain tumour patients: evaluation by graph analysis of synchronization matrices. Clin Neurophysiol 117: 2039–2049.

Bassett DS, Bullmore E (2006) Small-world brain networks. Neuroscientist 12: 512–523.

Bassett DS, Meyer-Linderberg A, Achard S, Duke T, Bullmore E (2006) Adaptive reconfiguration of fractal small-world human brain functional networks. Proc Natl Acad Sci 103: 19518–19523.

Boccaletti S, Kurths J, Osipov G, Valladares DL, Zhou CS (2002) The synchronization of chaotic systems. Phys Rep 366: 1–101.

Boccaletti S, Latora V, Moreno Y, Chavez M, Hwang D-U (2006) Complex networks: structure and dynamics. Phys Rep 424: 175–308.

Buzsaki G (2006) Rhythms of the brain. Oxford University Press, Oxford.

Cecchi GA, Rao AR, Centeno MV, Baliki M, Apkarian AV, Chialvo DR (2007) Identifying directed links in large scale functional networks: application to brain fMRI. BMC Cell Biol 8(Suppl I): S5 doi: 10.1186/1471-2121-8-S1-S5.

Chavez M, Martinerie J, Le van Quyen M (2003) Statistical assessment of nonlinear causality: application to epileptic EEG signals. J Neurosci Methods 124: 113–128.

Cohen R, Havlin S (2003) Scale-free networks are ultrasmall. Phys Rev Lett 90: art 058701.

David O, Cosmelli D, Friston KJ (2004) Evaluation of different measures of functional connectivity using a neural mass model. NeuroImage 21: 659–673.

De Vico Fallani F, Astolfi L, Cincotti F, Mattia D, Tocci A, Marciani MG, Colosimo A, Salinari S, Gao S, Cichocki A, Babiloni F (2007a). Extracting information from cortical connectivity patterns estimated from high resolution EEG recordings: a theoretical graph approach. Brain Topogr 19: 125–136.

De Vico Fallani F, Astolfi L, Cincotti F, Mattia D, Marciani MG, Salinari S, Kurths J, Gao S, Cichocki A, Colosimo A, Babiloni F (2007b). Cortical functional connectivity networks in normal and spinal cord injured patients: evaluation by graph analysis. Hum Brain Mapp 28(12): 1334–1346.

Dodel S, Hermann JM, Geisel T (2002) Functional connectivity by cross-correlation clustering. Neurocomputing 44–46: 1065–1070.

Dosenbach NUF, Fair DA, Miezin FM, Cohen AL, Wenger KK, Dosenbach RAT, Fox MD, Snyder AZ, Vincet JL, Raichle ME, Schlaggar BL, Petersen SE (2007) Distinct brain networks for adaptive and stable task control in humans. Proc Natl Acad Sci USA 104: 11073–11078.

Eguiluz VM, Chialvo DR, Cecchi GA, Baliki M, Apkarian AV (2005) Scale-free brain functional networks. Phys Rev Lett 94: 018102.

Erdos P, Renyi A (1960) On the evolution of random graphs. Publi Math Inst Hung Acad Sci 12: 17–61.

Ferri R, Rundo F, Bruni O, Terzano MG, Stam CJ (2007) Small-world network organization of functional connectivity of EEG slow-wave activity during sleep. Clin Neurophysiol 118: 449–456.

Ferri R, Rundo F, Terzano MG, Stam CJ (2008) The functional connectivity of different EEG bands moves towards small-world network organization during sleep. Clin Neurophysiol doi: 10.1016/j.clinph.2008.04.294.

Gross J, Kujala J, Hamalainen M, Timmermann L, Schnitzler A, Salmelin R (2001) Dynamic imaging of coherent sources: studying neural interactions in the human brain. Proc Natl Acad Sci USA 98: 694–699.

Guevara R, Perez Velazquez JL, Nenadovic V, Wennberg R, Senjanovic G, Dominguez LG (2005) Phase synchronization measurements using electroencephalographic recordings. What can we really say about neuronal synchrony? Neuroinformatics 3: 301–313.

Hadjipapas A, Hillebrand A, Holliday IE, Singh K, Barnes G (2005) Assessing interactions of linear and nonlinear neuronal sources using MEG beamformers: a proof of concept. Clin Neurophysiol 116: 1300–1313.

Hagmann P, Kurant M, Gigandet X, Thiran P, Wedeen J van, Meuli R, Thiran J-P (2007) Mapping human whole-brain structural networks with diffusion MRI. PloS ONE 2: e597.

He Y, Chen ZJ, Evans AC (2007) Small-world anatomical networks in the human brain revealed by cortical thickness from MRI. Cereb Cortex 17(10): 2407–2419.

He Y, Chen Z, Evans A (2008) Structural insights into aberrant topological patterns of large-scale cortical networks in Alzheimer's disease. J Neurosci 28: 4756–4766.

Humphries MD, Gurney K, Prescott TJ (2006) The brainstem reticular formation is a small-world, not scale-free network. Proc R Soc B 273: 503–511.

Humphries MD, Gurney K (2008) Network 'small-world-ness': a quantitative method for determining canonica network equivalence. PloS ONE 3: e0002051.

Iturria-Medina Y, Canales-Rodriguez EJ, Melie-Garcia L, Valdes-Hernandez PA, Martinez-Montes E, Aleman-Gomez Y, Sanchez-Bornot JM (2007) Characterizing brain anatomical connections using diffusion weighted MRI and graph theory. NeuroImage 36: 645–660.

Jeong J (2004) EEG dynamics in patients with Alzheimer's disease. Clin Neurophysiol 115: 1490–1505.

Kaminski M, Ding M, Truccolo WA, Bressler SL (2001) Evaluating causal relations in neural systems: Granger causality, directed transfer function and statistical assessment of significance. Biol Cybern 85: 145–157.

Latora V, Marchiori M (2001) Efficient behavior of small-world networks. Phys Rev Lett 87: 198701.

Latora V, Marchiori M (2003) Economic small-world behavior in weighted networks. Eur Phys 32: 249–263.

Lee L, Harrison LM, Mechelli A (2003) A report of the functional connectivity workshop, Dusseldorf 2002. NeuroImage 19: 457–465.

Lehmann D, Faber PL, Gianotti LRR, Kochi K, Pascual-Marqui RD (2006) Coherence and phase locking in the scalp Eeg and between LORETA model sources, and microstates as putative mechanisms of brain temporo-spatial functional organization. J Physiol Paris 99(1): 29–36.

Lehnertz K, Elger CE (1998) Can epileptic seizures be predicted? Evidence from nonlinear time series analysis of brain electrical activity. Phys Rev Lett 80: 5019–5022.

Liu Y, Zhou Y, He Y, Hao Y, Song M, Yu Ch, Liu H, Kiu Z, Jiang T (2008) Disrupted small-world networks in schizophrenia. Brain 131: 945–961.

Martinerie J, Adam C, Quyen M Le Van, Baulac M, Clemenceau S, Renault B, Varela FJ (1998) Epileptic seizures can be anticipated by non-linear analysis. Nat Med 4: 1173–1176.

Micheloyannis S, Pachou E, Stam CJ, Vourkas M, Erimaki S, Tsirka V (2006a) Using graph theoretical analysis of multi channel EEG to evaluate the neural efficiency hypothesis. Neurosci Lett 402: 273–277.

Micheloyannis S, Pachou E, Stam CJ, Breakspear M, Bitsios P, Vourkas M, Erimaki S, Zervakis M (2006b). Small-world networks and disturbed functional connectivity in schizophrenia. Schizophr Res 87: 60–66.

Micheloyannis S, Vourkas M, Tsirka V, Karakonstantaki E, Kanatsouli K, Stam CJ (2009) The influence of ageing on complex brain networks: a graph theoretical analysis. Hum Brain Mapp 30: 200–208.

Milgram S (1967) The small world problem. Psychol Today 2: 60–67.

Milo R, Shen-Orr S, Itzkovitz S, Kashtan N, Chklovskii D, Alon U (2002) Network motifs: simple building blocks of complex networks. Science 298: 824–827.

Montez T, Linkenkaer-Hansen K, van Dijk BW, Stam CJ (2006) Synchronization likelihood with explicit time-frequency priors. Neuroimage 33: 1117–1125.

Netoff TI, Clewley R, Arno S, White JA (2004) Epilepsy in small-world networks. J Neurosci 24: 8075–8083.

Newman MEJ (2003) The structure and function of complex networks. SIAM Rev 45: 167–256.

Nolte G, Wheaton OBL, Mari Z, Vorbach S, Hallett M (2004) Identifying true brain interaction from EEG data using the imaginary part of coherency. Clin Neurophysiol 115: 2292–2307.

Nunez PL, Srinivasan R, Westdorp AF, Wijesinghe RS, Tucker DM, Silberstein RB, Cadusch PJ (1997) EEG coherency I: statistics, reference electrode, volume conduction, Laplacians, cortical imaging, and interpretation at multiple scales. Electroen Clin Neurol 103: 499–515.

Percha B, Dzakpasu R, Zochowski M (2005) Transition from local to global phase synchrony in small world neural network and its possible implications for epilepsy. Phys Rev E 72: 031909.

Pereda E, Quian Quiroga R, Bhattacharya J (2005) Nonlinear multivariate analysis of neurophysiological signals. Progress in Neurobiology 77: 1–37.

Ponten SC, Bartolomei F, Stam CJ (2007) Small-world networks and epilepsy: graph theoretical analysis of intracerebrally recorded mesial lobe seizures. Clin Neurophysiol 118: 918–927.

Quian Quiroga R, Kraskov A, Kreuz T, Grassberger G (2002) Performance of different synchronization measures in real data: a case study on electroencephalographic signals. Phys Rev E Stat Nonlin Soft Matter Phys 65(4 Pt 1): 041903.

Ramasco JJ, Goncalves B (2007) Transport on weighted networks: when the correlation are independent of the degree. Phys Rev E 76: 066106.

Ray C, Ruffini G, Marco-Pallarex J, Fuentemilla L, Grau C (2007) Complex networks in brain electrical activity. Europhys Lett 79: 38004.

Reijneveld JC, Ponten SC, Berendse HW, Stam CJ (2007) The application of graph theoretical analysis to complex networks in the brain. Clin Neurophysiol 118: 2317–2331.

Rodriguez E, George N, Lachauz JP, Martinerie J, Renault B, Varela FJ (1999) Perception's shadow: long distance synchronization of human brain activity. Nature 397: 430–433.
Rosenblum MG, Pikovsky AS, Kurths J (1996) Phase synchronization of chaotic oscillators. Phys Rev Lett 76: 1804–1807.
Rosenblum MG, Pikovsky AS (2001) Detecting direction of coupling in interacting oscillators. Phys Rev E 64: 045202.
Rubinov M, Knock SA, Stam CJ, Micheloyannis S, Harris AWF, Williams LM, Breakspear M (2007) Small-world properties of nonlinear brain activity in schizophrenia. Hum Brain Mapp in press.
Rulkov NF, Sushchik MM, Ysimring LS, Abarbanel HDI (1995) Generalized synchronization of chaos in directionally coupled chaotic systems. Phys Rev E 51: 980–994.
Salvador R, Suckling J, Coleman MR, Pickard JD, Menon D, Bullmore E (2005a) Neurophysiological architecture of functional magnetic resonance images of human brain. Cereb Cortex 15: 1332–1342.
Salvador R, Suckling J, Schwarzbauer Ch, Bullmore E (2005b) Undirected graphs of frequency-dependent functional connectivity in whole brain networks. Phil Trans R Soc B 360: 937–946.
Sarnthein J, Petsche H, Rappelsberger P, Shaw GL, von Stein A (1998) Synchronization between prefrontal and posterior association cortex during human working memory. Proc Natl Acad Sci USA 95: 7092–7096.
Schindler K, Leung H, Elger CE, Lehnertz K (2007) Assessing seizure dynamics by analysing the correlation structure of multichannel intracranial EEG. Brain 130: 65–77.
Schmitt JE, Lenroot RK, Wallace GL, Ordaz S, Taylor KN, Kabani N, Greenstein D, Lerch JP, Kendler KS, Neale MC, Giedd JN (2008) Identification of genetically mediated cortical networks: a multivariate study of pediatric twins and siblings. Cereb Cortex doi:10.1093/cercor/bhm211.
Schnitzler A, Gross J (2005) Normal and pathological oscillatory communication in the brain. Nat Rev Neurosci 6: 285–296.
Smit DJ, Stam CJ, Posthuma D, Boomsma DI, de Geus EJ (2007) Heritability of 'small-world' networks in the brain: a graph theoretical analysis of resting-state EEG functional connectivity. Hum Brain Mapp [Epub ahead of print].
Solomonov R, Rapoport A (1951) Connectivity of random nets. Bull Math Biophys 13: 107–117.
Sporns O, Zwi JD (2004) The small world of the cerebral cortex. Neuroinformatics 2: 145–162.
Sporns O, Honey CJ (2006) Small world inside big brains. Proc Natl Acad Sci 51: 19219–19220.
Srinivas KV, Jain R, Saurav S, Sikdar SK (2007) Small-word network topology of hippocampal neuronal network is lost, in an in vitro glutamate injury model of epilepsy. Eur J Neurosci 25: 3276–3286.
Stam CJ (2004) Functional connectivity patterns of human magnetoencephalographic recordings: a 'small-world' network? Neurosci Lett 355: 25–28.
Stam CJ (2005) Nonlinear dynamical analysis of EEG and MEG: review of an emerging field. Clin Neurophysiol 116: 2266–2301.
Stam CJ (2006) Nonlinear brain dynamics. Nova Science, New York.
Stam CJ, van Dijk BW (2002) Synchronization likelihood: an unbiased measure of generalized synchronization in multivariate data sets. Phys D 163: 236–241.
Stam CJ, van Cappellen van Walsum AM, Micheloyannis S (2002) Variability of EEG synchronization during a working memory task in healthy subjects. Int J Psychophysiol 46: 53–66.
Stam CJ, Breakspear M, van Cappellen van Walsum AM, van Dijk BW (2003) Nonlinear synchronization in EEG and whole-head MEG recordings of healthy subjects. Hum Brain Mapp 19: 63–78.
Stam CJ, Jones BF, Manshanden I, van Cappellen van Walsum AM, Montez T, Verbunt JPA, de Munck JC, van Dijk BW, Berendse HW, Scheltens P (2006) Magnetoencephalographic evaluation of resting-state functional connectivity in Alzheimer's disease. Neuroimage 32: 1335–1344.

Stam CJ, Reijneveld JC (2007) Graph theoretical analysis of complex networks in the brain. Nonlin Biomed Phys 1: 3.

Stam CJ, Nolte G, Daffertshofer A (2007a) Phase lag index: assessment of functional connectivity from multichannel EEG and MEG with diminished bias from common sources. Hum Brain Mapp 28: 1178–1193.

Stam CJ, Jones BF, Nolte G, Breakspear M, Scheltens PH (2007b). Small-world networks and functional connectivity in Alzheimer's disease. Cereb Cortex 17: 92–99.

Stephan KE, Hilgetag C-C, Burns GAPC, O'Neill MA, Young MP, Kotter R (2000) Computational analysis of functional connectivity between areas of primate cerebral cortex. Phil Trans R Soc Lond B 355: 111–126.

Stephan KE, Harrison LM, Kiebel SJ, David O, Penny WD, Friston KJ (2007) Dynamic causal models of neural system dynamics: current state and future extensions. J Biosci 32: 129–144.

Strogatz SH (2000) From Kuramoto to Crawford: exploring the onset of synchronization in populations of coupled oscillators. Phys D 143: 1–20.

Uhlhaas PJ, Singer W (2006) Neural synchrony in brain disorders: relevance for cognitive dysfunctions and pathophysiology. Neuron 52: 155–168.

Varela F, Lachaux J-P, Rodriguez E, Martinerie J (2001) The brainweb: phase synchronization and large-scale integration. Nat Rev Neurosci 2: 229–239.

Wang L, Zhu C, He Y, Zang Y, Cao Q, Zhang H, Zhong Q, Wang Y (2008) Altered small-world brain functional networks in children with attention-deficit/hyperactivity disorder. Hum Brain Mapp [Epub ahead of print].

Watts DJ, Strogatz SH (1998) Collective dynamics of 'small-world' networks. Nature 393: 440–442.

Wu H, Li X, Guan X (2006) Networking property during epileptic seizure with multi-channel EEG recordings. J. Wang et al. (Eds.): ISNN 2006, Lecture Notes Computer Science 3976, pp. 573–578.

Wu T, Zang Y, Wang L, Long X, Hallett M, Chen Y, Li K, Chan P (2007) Aging influence on functional connectivity of the motor network in the resting state. Neurosci Lett 442: 164–168.

Yu S, Huang D, Singer W, Nikolic D (2008) A small world of neuronal synchrony. Cereb Cortex [Epub ahead of print].

The Size of Neuronal Assemblies, Their Frequency of Synchronization, and Their Cognitive Function

Johannes Sarnthein and Astrid von Stein

1 Introduction

In many aspects of cognitive neuroscience it is useful to assume that the main task of cognition is the guidance of action. Cognition evolves in a constantly changing environment in an adaptive and context-dependent way. Cognitive operations are successful if they generate not "correct" representations of external stimuli, but actions that are adapted to the specific situation (von Stein, 1999; Engel et al., 2001). Hence, the human brain is very flexible in its decision to assign a behavioral response to a perceived external stimulus and the same stimulus may elicit one of a variety of motor responses depending on the internal brain state and the behavioral context assigned to the stimulus.

In perception, one distinguishes between two types of processing. Bottom-up processing is driven by properties inherent to the stimulus, e.g., color or contrast for visual stimuli. At the same time, the internal brain state and the behavioral context influence perception (top-down). Bottom-up processing is constantly subject to top-down modifications that enhance task-relevant, and prune irrelevant, information (Ahissar and Hochstein, 2004). These computations are very fast and reliable. In this way, efficient top-down resources in the brain allow to create predictions about future situations and to constantly match expectations against signals from the environment.

Synchronization of neuronal activity has been proposed to define functional neuronal assemblies (Eckhorn et al., 1988; Gray et al., 1989; Singer and Gray, 1995). Here synchronization emerges as an intrinsic property of cognitive processing. In this dynamical definition, neuronal synchronization is not only due to properties inherent to the stimulus (bottom-up) but also modified top-down by patterns of ongoing activity (Arieli et al., 1996). Since bottom-up processing takes place in

J. Sarnthein (✉)
Neurochirurgie, Universitäts Spital Zürich, 8091 Zürich, Switzerland
e-mail: johannes.sarnthein@usz.ch

J.L. Perez Velazquez, R. Wennberg (eds.), *Coordinated Activity in the Brain*, Springer Series in Computational Neuroscience 2, DOI 10.1007/978-0-387-93797-7_6,

primary sensory areas, top-down interactions with higher areas involve a large-scale synchronization of neuronal activity in assemblies extending over several brain areas (von Stein et al., 2000; von Stein and Sarnthein, 2000; Engel et al., 2001; Varela et al., 2001; Fries, 2005; Palva and Palva, 2007). For communication in the brain, the synaptic path length of connectivity is critical and effective connectivity determines the possible routes of information between areas. Since many aspects of brain activity are governed largely by rhythms (Buzsaki, 2006), the frequency of the oscillation limits how far information gets transmitted in one cycle. Hence, fast oscillations favor local activity as it is recruited for bottom-up processing. The interaction between distant neuronal assemblies as required in top-down control becomes manifest in activity with longer time periods and lower frequencies.

In this chapter, we will first discuss hierarchy in the brain and general considerations on the local field potential (LFP) and the electroencephalogram (EEG) as measures for synchronization within and between neuronal assemblies. We will then present an algorithm to compute power spectral density and magnitude-squared coherence, the quantities that we use to assess synchronization. Finally, in extending the experimental evidence presented in our earlier publication (von Stein and Sarnthein, 2000), we will review some recent examples from the literature where bottom-up and top-down processing are mediated by synchronization within and between brain areas at different frequencies.

2 Hierarchy in the Brain

Brain regions are thought to be ordered in a hierarchical fashion. Functionally, a hierarchy is derived in the sense that abolishing activity in lower areas will silence higher areas. Silencing higher areas will, however, only modulate the activity of lower areas.

Between brain areas, if area A connects anatomically to area B in the brain, then there usually are reciprocal connections from B to A. Anatomical connections are labeled as ascending and descending, feedforward and feedback, or bottom-up or top-down. In the sensory system, patterns of long-range cortical connectivity are used to define bottom-up and top-down pathways. Anatomical projections from the specific nuclei of the thalamus (e.g., LGN in vision) form synapses in the granular layer. Between cortical areas, anatomical connections terminate with characteristic laminar distributions. Either they form synapses mainly in the granular layer (bottom-up type) or in the supra- and infragranular layers (top-down type); thus, cortical areas can be assigned to different levels of the hierarchy of the visual system (Salin and Bullier, 1995). Traditionally, the description starts at the site of sensory input, so that perception is described as a bottom-up hierarchy from lower to higher stages of processing. Best documented is vision, where the successive stages of elaboration of the visual stimuli from the retina to the various visual areas have been extensively studied.

Alternatively, the description can also start at the behavioral relevance of a stimulus. The behavioral relevance affects internal activity of the brain, like expectation, preparation, or attention. The internal brain activity occurs simultaneously to the sensory input and corresponds to activity typically from the frontal lobes or the limbic system or in associative cortices, but far removed from the primary sensory areas. The influence of this activity on lower stages of processing is referred to as top-down. There is psychophysical and physiological evidence for active top-down participation even in basic stages of sensory perception. In these terms, the top-down connections generate predictions and the bottom-up connections convey prediction error or mismatch signals. More generally speaking, both incoming and internal activity are integrated in a large-scale network (Varela et al., 2001).

3 Synchronized Brain Activity Measured by LFP and EEG

3.1 Local Scale Synchronization

On a spatial scale of less than 2 mm, the dense packing and the mixture of neurons in a cortical column are prone to oscillate in a synchronous mode. The functional characteristics of a cortical module are instantiated in a simple basic excitatory–inhibitory network, referred to as a canonical microcircuit (Douglas and Martin, 2004). In a microcircuit, pyramidal axon collateral branches ascend back to superficial layers and other branches distribute excitation in the horizontal plane, forming a strongly recurrent excitatory network. The strong recurrent excitation matches the paucity of thalamic input – very few of the synapses on pyramidal cells are thalamocortical synapses: less than 10–20% in the input layers of cortex and less than 5% across the entire depth of cortex (Douglas and Martin, 2007). The amplification of the input signal caused by this positive feedback loop is controlled by a network of inhibitory interneurons. These can receive excitatory and inhibitory synapses and have only local connections. Activation of a microcircuit sets in motion a sequence of excitation and inhibition in every neuron of the module, rather than a sequential activation of separate neurons at different hypothetical processing stages. The nonlinearities in the microcircuits may generate substantial dissociations between subthreshold population activity and the spiking of pyramidal cells (Belitski et al., 2008). Re-excitation of microcircuits is tightly controlled by local inhibition and the time course of excitation–inhibition is far longer than the synaptic delays involved. Therefore the magnitude and timing of local mass action arise as properties of microcircuits.

This tight interconnectivity is not restricted to such a small area, as collaterals are known to extend over several millimeters, linking larger patches of neural tissue by reciprocal connections (Bosking et al., 1997). Consequently, in columns of the primary visual cortex separated by 2–7 mm, which have non-overlapping receptive fields, neurons that share similar feature properties tend to synchronize (Gray, 1999). In this way local integration occurs over a local network distributed over an area of $\sim$1 cm^2 through monosynaptic connections with conduction delays of a few milliseconds (Girard et al., 2001).

3.2 Measuring Electrical Dipole Fields from the Brain

When a neuron receives synaptic input at its dendrite, the postsynaptic membrane potential will deviate from the resting membrane voltage. Compared to the fast action potentials, membrane potential changes associated with postsynaptic potentials are much smaller in amplitude. However, they last for tens of milliseconds, which is important for the generation of LFPs and EEG recordings. For the transmembrane potential to change in a given neuron, there must be a transmembrane current, i.e., a flow of ions across the membrane. A current from the intracellular space to the extracellular space creates a current source and the opposite a current sink. Extracellular field potentials add up linearly throughout the extracellular tissue due to electric superposition, representing the weighted sum of all current sinks and current sources along the geometry of multiple cells. Tissue impedance is frequency independent, allowing the description of cortex as an ohmic resistor. As such, the tissue can be described by its specific electrical resistance (Logothetis et al., 2007). Furthermore, the resistive properties of the gray matter are largely isotropic, i.e., its resistivity is the same along each direction. Because of their longer duration, postsynaptic potentials have a much higher chance to occur in a temporally overlapping way than do the very short action potentials. In addition, postsynaptic potentials are displayed by many more neurons than are spikes because only a very small minority of neurons reaches the spike threshold at a certain time. For these reasons, the contribution of action potentials to the LFP and especially to the scalp EEG is negligible. Both LFP and EEG capture local network activity (e.g., recurrent excitatory–inhibitory loops of the cortical microcircuits) and also global activity (e.g., generated by cortico-cortical or thalamocortical loops).

The mean extracellular potential, which is measured with electrodes, critically depends on the local geometrical arrangements of neurons. For cells with diametrically opposite orientations, currents of equal magnitude but opposite polarity will generate potentials that tend to cancel each other. As a result, information is potentially lost. Three arrangements produce a characteristic mean extracellular potential: the open field, the closed field, and the open–closed field. The first is encountered when the neurons are organized in a laminar array, with their dendrites facing in one direction and the somata in the other. Typical examples of open field are the neocortex, the cerebellum, and the hippocampus. Simultaneous activation of the dendrites of neurons in this arrangement generates strong dipole layers whose field potentials are easily captured by electrodes. The second arrangement produces dipoles with spherical symmetry. Measurements from outside the sphere record zero potential. The third arrangement is a mixture, where measurements outside the sphere record changes in the open field component only. For example, this kind of LFP records the activity from a small number of neurons ($\sim$5) in the thalamus (Sarnthein and Jeanmonod, 2007, 2008).

Several approaches have been explored to estimate the upper limit of the spatial extent of LFP summation in cortex. Indirect calculation by computing the phase coherence of LFP as a function of inter-electrode distance in experiments with simultaneous multiple-electrode recordings gave an upper limit of 0.5–3 mm from the electrode tip (Juergens et al., 1999). In another approach, cortex was severed until a change in LFP was noticeable and an upper limit of 0.1 mm was found (Engel et al., 1990). Coherence analysis shows a frequency-dependent coupling region of 2.9 mm (2–8 Hz), 2.4 mm (8–15 Hz), and 2.0 mm (20–60 Hz) (Goense and Logothetis, 2008).

3.3 Neuronal Synchronization as a Basis for LFP and EEG Measurements

It has become clear that EEG and LFP signals are manifestations of the synchronous activity of neuronal assemblies. Next, one can estimate the proportion of neuronal elements that have to be synchronously active for a signal to emerge (Hari et al., 1997; Nunez, 1999). Suppose some EEG phenomenon is due to source generation in cortical pyramidal cells that form a dipole layer. Let us imagine that the dipole layer is composed of a mixture of M coherent sources with synchronized activity and N incoherent sources with activity at random phases. All sources are located within roughly the same distance of a scalp electrode. The potentials due to individual coherent sources combine by linear superposition to produce the summed contribution to the scalp potential. In contrast, the contribution of the incoherent sources is due only to statistical fluctuation. Let the amplitude of each source be A and the variance σ^2. The maximal signal produced by the synchronous sources is MA. According to the central limit theorem, the variance of signals produced by the incoherent sources is $N\sigma^2$. The signal-to-noise ratio is then $MA\Big/\left(\sigma\sqrt{N}\right) \propto M\Big/\sqrt{N}$. Thus, the relative contribution of coherent to incoherent sources is estimated to be $M\Big/\sqrt{N}$. For example, approximately 10^7 pyramidal cells line up in parallel within a 1 cm^2 portion of a cortical gyrus. If only 1% are coherent, the relative contribution to scalp potential of these minority neurons would then be approximately $10^5\Big/\sqrt{10^7}$, or about 30 times greater than that of the 99% neurons that produce incoherent sources. This estimation neglects the effect of different distances between sources and scalp electrode and also the degree of synchrony in the group of coherent sources. Still, the general idea is valid that EEG scalp potentials are due primarily from combined contributions from neuronal sources with high synchrony.

The relative importance of coherent sources can be much less with LFP measurements using small electrodes in the brain. If we assume that fewer than 100 neurons are close enough to contribute substantially to the measured potential and that 10% are coherent, the relative contributions of coherent and incoherent sources are about equal. These examples show the critical importance of spatial scale in the study of EEG and LFP.

In the case of scalp EEG, the localization of the source of the cortical generators is estimated by many distinct approaches (e.g., Pascual-Marqui et al., 1994; Scherg and Berg, 1996; Gross et al., 2001). Evoked potentials are generated by repetitive presentation of a stimulus and subsequent time-domain averaging of the EEG signal time-locked to the stimulus. The averaging eliminates those contributions to the EEG, which are generated by intrinsic, ongoing brain activity. The sources of evoked potentials are successfully localized by fitting a small number of electric dipole sources (Scherg and Berg, 1996). In the absence of time-domain averaging, the EEG is dominated by intrinsic brain activity independent of the external stimulus. This activity stems from fluctuating patterns of synchronization and desynchronization within neuronal assemblies in the cortical tissue. It reflects the instantaneous state of cortical networks (Arieli et al., 1996). Since intrinsic fields have low amplitude, they are effective only in close proximity to the generating neuronal assemblies and they influence the EEG mainly at individual scalp electrodes. Conversely, we can assume that scalp EEG is generated in the cortical tissue underlying the electrode. It has been estimated that the signal at one electrode stems from within a radius of $\sim$5 cm on the scalp corresponding to an area of $\sim$10 cm^2 of cortex (Nunez, 1999). This area matches roughly with the size of a cortical area. Therefore, in a first approximation, changes in the synchronization of neuronal activity of a neuronal assembly representing a cortical area are reflected in changes in EEG amplitude in the EEG signal from one electrode on the scalp.

3.4 Synchronization Between LFP and EEG Signals

The synchronization between the signals from two electrodes can be investigated by spectral coherence. The coherence spectrum describes the phase constancy between two signals in a frequency range, which in turn is a measure for synchronization. Here synchronization is not used in the strict sense, but to describe oscillations in constant phase. Enhanced synchronization between two signals is taken as enhanced cooperation between the neuronal generators of the signals.

The analysis of coherence between two scalp EEG signals aims at characterizing the interaction between two neuronal assemblies as it is realized by axonal transmission of action potentials between member neurons of the assemblies. However, high coherence between two scalp EEG signals may also contain a substantial contribution of volume conduction of ionic currents through the extracellular space. The following considerations help to identify the contribution due to axonal transmission: (1) Volume conduction should be the same for all cognitive states and can be eliminated by subtraction of a baseline state. (2) Volume conduction is frequency independent (Logothetis et al., 2007) and occurs instantaneously, i.e., with zero lag independent of frequency. (3) Volume conduction causes enhanced EEG amplitude at both recording sites. Enhanced coherence in one brain state with concomitant reduction in EEG amplitude makes a strong contribution of volume conduction unlikely. (4) Volume conduction falls off with the distance between scalp electrodes and should be small for distances exceeding the spatial resolution (5 cm) of a scalp EEG electrode. Therefore, if carefully applied, coherence indicates the degree of synchronization of neuronal activity between cortical areas.

3.5 What Does Synchronization of Neuronal Activity Mean for Cognitive Processing?

For the neural implementation of top-down influences in the brain, models have been proposed that are based on the synchronization of neural activity (von Stein et al., 2000; von Stein and Sarnthein, 2000; Engel et al., 2001; Varela et al., 2001; Fries, 2005; Palva and Palva, 2007). Here cognitive function is explained in terms of synchronous behavior of large neuronal assemblies that are dynamically bound within and across subsystems (Eckhorn et al., 1988; Gray et al., 1989; Singer and Gray, 1995). The assemblies are defined primarily by the synchronous activity of their member neurons. The models assume that neural synchrony with precision in the millisecond range is crucial for object representation, response selection, and sensorimotor integration. Synchrony is supposed to enhance the saliency of neural responses because correlated discharges have a much stronger impact on receiving neuronal groups than temporally disorganized inputs. The central idea is that synchronization can select neuronal responses for further joint processing. In this way, synchronization can be used to encode information about the relatedness of neural signals. Importantly, synchronization in this model can be intrinsically generated (not imposed on the system by external stimuli), and modulated by intrinsic signals that reflect experience, contextual influences, and action goals. Furthermore, synchronization can be modified within a fraction of a second while anatomical connections stay unchanged on that timescale. In this context, top-down control is exerted by the influence of ongoing patterns of activity on the processing of sensory signals (Arieli et al., 1996) as a result of continuous large-scale interactions between higher order and lower order brain areas.

If a message shall be communicated efficiently to a receiving neuron, the sending neuron's output has to be timed such that it arrives at the receiving neuron when that neuron is excitable (Fries, 2005). The following requirements have to be met to enable this prospective timing: (1) The fluctuations of excitability in the receiving neuron must be predictable. This requirement is met by the physiologically occurring neuronal oscillations (Kopell et al., 2000) that are sufficiently regular to allow prediction of the next excitability peak. (2) The spike traveling time from the sending to the receiving neuron must be reliable. (3) The sending neuron has to emit output such that it arrives when the receiving neuron is excitable. If the excitability of the receiving neuron is indeed predictable because it oscillates, then this requirement is met if the sending neuron also oscillates and if the two neurons are phase-locked to each other. In extension of this reasoning, synchronization in neuronal assemblies is realized through phase-locked oscillatory activity of the member neurons. Precise timing is important in this scheme of communication. The frequency of the synchronous oscillations, the relative phase between oscillations in two neurons, and the conduction delay need to match. For oscillatory activity, the frequency of the oscillation limits how far information gets transmitted in one cycle of the oscillation.

For local neuronal assemblies, synchronous activity has mostly been found in the gamma frequency range (30–100 Hz). Here the term "local" refers to spatially adjacent neurons as well as monosynaptically connected neurons, e.g., in cortical areas of both hemispheres connected via the corpus callosum. While this corresponds to cycle lengths of 10–30 ms, the local conduction delays are around 1–3 ms. Hence, fast oscillations favor local activity and local activity is recruited for bottom-up processing.

Large-scale synchronization concerns neural assemblies which are farther apart in the brain (>1 cm, transmission delays >8–10 ms over polysynaptic pathways). For example, neurons in occipital and frontal lobes are separated by dozens of milliseconds in transmission time. In this case, synchronization is mediated by distant connections, either cortico-cortical fibers or thalamocortical reciprocal pathways (Llinas et al., 1998). These pathways correspond to bottom-up and top-down connections that link different levels of the hierarchical network in different brain regions. Lower frequency bands are more robust to spike timing delays and thus may be better suited for longer range coupling between multiple distant areas. Therefore, the interaction between distant neuronal assemblies as required in top-down control becomes manifest in activity with longer time periods and lower frequencies.

4 Methods of Spectral Analysis

In the following we will apply spectral analysis to data recorded from a patient with Parkinson's disease (Sarnthein and Jeanmonod, 2007). LFP was recorded from the thalamus and, simultaneously, EEG was recorded from the scalp. The patient was awake with eyes closed. A data length of 60 s and the code are available in the supplementary material for this chapter http://www.springer.com// 978-0-387-93796-0.

4.1 Power Spectral Density

For spectral estimates, the multitaper Fourier transform method has become popular in the neuroscience community (Jarvis and Mitra, 2001). This technique deals explicitly with the problems of bias and variance in spectral estimation and thereby allows us to optimize the trade-off between the two. In particular, signals can be faithfully described, where, like the EEG, the range of meaningful amplitudes spans several orders of amplitude.

We suppose a time series $X(t)$ that is a realization of a stationary process defined on the time interval $t = 1, 2, \ldots, N$ with sampling interval Δt (Percival and Walden, 1993). The multitaper spectral estimator $\hat{S}^{mt}(f)$ is an average of K direct spectral estimators

$$\hat{S}^{mt}(f) = \frac{1}{K}\sum_{k=0}^{K-1} \hat{S}_k^{mt}(f) \quad \text{for} \quad \hat{S}_k^{mt}(f) = \Delta t \left| \sum_{t=1}^{N} X(t)h_k(t)e^{-2\pi i f t \Delta t} \right|^2$$

where $h_k(t)$ is the data taper for the kth direct spectral estimator $\hat{S}_k^{mt}(f)$. The tapers form a sequence of orthogonal functions, defined on the time interval $t = 1, 2, ..., N$ with normalization $\sum_{t=1}^{N} h_k(t)^2 = 1$. A set of K orthogonal data tapers with good bias properties is given by portions of the *discrete spheroidal sequences* (*dpss* or *Slepian sequences*). They are parameterized by a bandwidth parameter W, such that there are $K < 2NW\Delta t$ basis functions with spectra that have >90% of their power within the frequency interval $[f - W, f + W]$ around f, the frequency of interest. The spectral concentration on the frequency interval ensures good bias properties. Thus, the use of Slepian functions minimizes the bias, whereas the averaging over multiple orthogonal tapers on the same data minimizes the variance.

In practice, we first select the *resolution bandwidth* $2W$. Typically, W is taken to be a small multiple $j > 1$ of the fundamental frequency $W = j/(N\Delta t)$ for $j = 2, 3$ or 4. If we make Wlarge, the number of tapers increases – hence we can make K large and decrease the variance of $\hat{S}^{mt}(f)$. On the other hand, the resolution of $\hat{S}^{mt}(f)$ decreases as W increases and fine features in the spectrum can inadvertently be smeared out. Second, we compute the spectral estimators $\hat{S}_k^{mt}(f)$ for $k = 0$ to $K - 1$ with $K < 2NW\Delta t$. Finally, we average together K of these spectral estimators, where K is determined by examining the individual spectral estimators for evidence of bias. The average across tapers reduces the variance in the multitaper spectral estimator $\hat{S}^{mt}(f)$ by a factor of $1/K$.

If we assume that the measured EEG and LFP time series are stationary within each brain state, we can break up the signal into epochs. The length N of the epoch determines the fundamental frequency $1/(N\Delta t)$ which is the smallest nonzero Fourier frequency. For the analysis of EEG corresponding to cognitive states, we are interested in frequencies above ~2 Hz, an epoch length of 2 or 5 s is appropriate if the state lasts several seconds. In the case where different brain states are induced by individual trials of a cognitive task, we analyze the group of all epochs corresponding to one brain state separately. We then perform multitaper spectral estimation for each epoch and finally take the mean to obtain the average power spectral density P_{xx} and similarly the cross-power spectral density P_{xy}. The EEG and LFP power spectra (Fig. 1A, B) show a strong alpha peak as expected for a wake subject with eyes closed. The alpha peak is shifted down to lower frequency, which is typical for patients with Parkinson's disease (Stoffers et al., 2007; Moazami-Goudarzi et al., 2008)

4.2 Coherence

The magnitude-squared coherence γ^2 for two signals, x and y, is equal to the average cross-power spectrum P_{xy} normalized by the averaged power spectra of the signals:

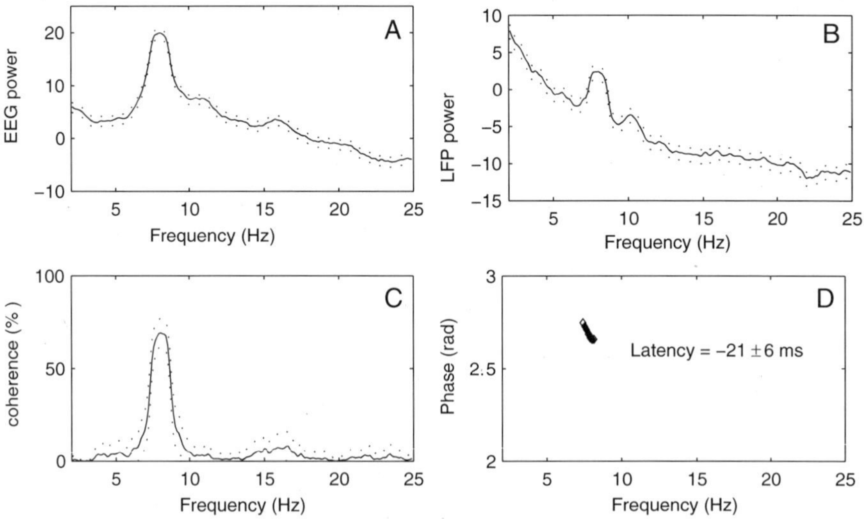

Fig. 1 Spectral analysis of EEG and LFP in a patient with Parkinson's disease. (**A, B**) Power spectral density (units $10\log_{10}\mu V^2\,\mathrm{Hz}^{-1}$) of scalp EEG at site C3 (**A**) and thalamic LFP (**B**). (**C**) The coherence between EEG and LFP is significant from 7 to 9 Hz in the theta band, *dotted line*: 95% confidence interval. (**D**) Phase φ between EEG and LFP. The slope dφ/df was transformed into latency. Window length 4 s, $W = 4,\ K = 3$. Adapted from Sarnthein and Jeanmonod (2007)

$$\gamma^2 = \frac{|P_{xy}|^2}{|P_{xx}P_{yy}|}.$$

Coherence measures the strength of the linear relationship between two signals at every frequency *f*. Its value lies between 0 and 100% and it estimates the degree to which phases and amplitudes at the frequency of interest are dispersed. $\gamma^2 = 0$ means phases and amplitudes are randomly dispersed among all epochs. Signals are perfectly coherent ($\gamma^2 = 100\%$) at a given frequency when they have both constant phase difference ϕ and constant amplitude ratio over the time considered. In this case, phases of signals x and y are identical in all epochs, i.e., the two signals are totally phase-locked at this frequency.

In principle, coherence mixes the effects of amplitude and phase in the interrelations between two signals. Therefore methods have been devised that investigate phase locking alone (e.g., phase coherence, phase-locking value [PLV]). Nevertheless, by taking into account the amplitude, coherence focuses on stronger signals, which may be less prone to noise and in some cases a more relevant description of physiological processes. Figure 1C shows the coherence spectrum between the scalp EEG and the thalamic LFP.

4.3 Phase and Latency

As a complement to coherence, the phase difference φ between LFP and EEG signals is contained in the polar form of the cross-power spectrum

$P_{xy}(f) = a(f)e^{i\varphi(f)}$. The phase spectrum $\phi(f)$ is plotted in Fig. 1D for the range of significant coherence.

In the description of wave propagation, the choice of a single frequency and phase value corresponds to the concept of phase speed; the choice of a range of frequencies and of the slope of the phase spectrum corresponds to the concept of group speed (Gotman, 1983). When transforming phase measurements into latency, ambiguity arises because 360° can always be added or subtracted from a phase value: a lead of +36° is also a lag of –324°. At 10 Hz this phase difference could be interpreted as a lead of 10 ms as well as a lag of 90 ms, which add up to the wave period of 100 ms. However, when using the slope of the phase over several frequencies, ambiguity becomes less likely. For example, the slope between two phase values at 10 Hz (36°) and 12 Hz (43.2°) corresponds to a lead of 10 ms; subtracting 360° from only one phase value leads to a lag of –490 ms where the period 500 ms comes from the difference between the two frequencies (2 Hz). In our calculation we chose a range of frequencies usually separated only by the resolution of the spectrum (0.25 Hz), making the latencies essentially unambiguous. In physical terms, the latency L_{xy} is defined as "group delay" or "envelope delay" and describes the time needed for wave packets to propagate between the two recording sites at group speed.

The latency L_{xy} between LFP and EEG signals was calculated from the slope of the phase spectrum $d\phi/df$. We chose a range of frequency points in the range of significant coherence, fitted a straight regression line and calculated the latency according to $L_{xy} = d\phi/df/2\pi$. The standard deviation is taken from the standard deviation of the regression. This estimate of latency bears on the implicit assumption that information flow between thalamic LFP and scalp EEG is mediated by propagating wave packets.

4.4 Confidence Intervals

The confidence intervals for power and coherence can be estimated on the basis of the χ^2 distribution (Bendat and Piersol, 1986). The power spectral density is then distributed as $\hat{P}_{xx}(f) = a\,\chi^2_{\upsilon}$ with constant a and υ degrees of freedom, which are determined by the numbers of epochs nep and tapers K. The 95% confidence interval around the estimated value $\hat{P}_{xx}$ for each frequency point is then

$$\frac{P_{xx}}{1+2\varepsilon} \leq \hat{P}_{xx} \leq \frac{P_{xx}}{1-2\varepsilon} \quad \text{with} \quad \varepsilon \approx \sqrt{1/(nepK)}.$$

For magnitude-squared coherence, the approach is similar, except that the width of the confidence interval depends on the magnitude of coherence:

$$\frac{\gamma^2}{1+2\varepsilon} \leq \hat{\gamma}^2 \leq \frac{\gamma^2}{1-2\varepsilon} \quad \text{with} \quad \varepsilon \approx \sqrt{\frac{2}{nepK}\frac{1-\gamma^2}{|\gamma|}}.$$

A more advanced approach to confidence intervals estimated with jackknife statistics is offered by the chronux toolbox (www.chronux.org).

4.5 Frequency Bands

Traditionally, LFP and EEG are classified in a number of specific frequency bands. Rhythmic EEG is subdivided into frequency bands known as delta (0–4 Hz), theta (4–8 Hz), alpha (8–12 Hz), beta (12–24 Hz), and gamma (24–80 Hz), which are typically characterized by different amplitudes. This classification was based on the correlation of each band with a distinct behavioral state. The oscillatory activity in these bands is associated with the thalamocortical loops and cortical networks. The definition of the bands is empirical and often the range of a particular band is determined arbitrarily.

5 Experimental Evidence for the Inverse Relationship Between Assembly Size and Synchronization Frequency

If neuronal assemblies are defined solely on their anatomical extension, high frequencies of synchronization have been suggested for small neuronal assemblies and low frequencies for assemblies with larger spatial extent (Nunez, 2000). This negative correlation between frequency of activity and spatial scale has been observed, e.g., for the alpha peak frequency, which decreases with head size of the subject (Nunez et al., 1978). In this chapter, however, we investigate the hypothesis that neuronal assemblies are not defined on only anatomical grounds, but are rather recruited functionally for a specific cognitive task (von Stein and Sarnthein, 2000). In this view, the size of an assembly depends less on the absolute distance between sites, but rather on the number of synapses involved in the interaction. In the following, we will present a few salient examples of experimental evidence, that interareal synchronization during tasks recruiting more than one brain area is a manifestation of cooperative activity in larger neuronal assemblies and that this synchronization appears at lower frequencies than local visual processing.

5.1 Local Synchronization in High (Gamma) Frequencies

The importance of neural synchrony for cortical processing was first observed in the visual cortex of anesthesized cats and the hypothesis was put forward that functional assemblies are defined by the synchronous activity of their member neurons (Eckhorn et al., 1988; Gray et al., 1989; Engel et al., 2001; Varela et al., 2001). In these experiments, visual stimulation elicited gamma activity in the LFP and MUA measured in primary visual cortex. This is an example of pure bottom-up activity, since top-down activity is eliminated by anesthesia.

Inspired by these results, we performed a simple experiment in humans while recording EEG over visual and temporal cortex (von Stein and Sarnthein, 2000). The subjects viewed parallel gratings where the number of bars per degree of visual field was varied. There was no behavioral response required and the experiment aimed to elicit bottom-up processing. The main finding was that EEG power increased in the lower gamma frequency band (24–35 Hz) with the number of bars per degree. Importantly, this increase was confined to primary visual cortex.

Two publications report gamma activity in central EEG as a function of sound intensity (Schadow et al., 2007a) and in parieto-occipital EEG as a function of visual stimulus contrast (Schadow et al., 2007b). The latter publication also discusses possible difficulties for eliciting gamma responses in humans (Juergens et al., 1999) and lists several references where local gamma activity in humans has been reported over the years. These findings are compatible with the notion that local neuronal assemblies involved in bottom-up processing are characterized by synchronization in higher frequency bands.

5.2 Interareal Interactions Mediated by Synchronization in Intermediate Frequencies

To investigate interareal interactions we devised a task which requires interactions between different, but close cortical areas. In the task, subjects had to combine percepts from different modalities to one object, which is known as supramodal processing (von Stein et al., 1999). Everyday objects were presented in three modalities: as written words, as spoken words, or as pictures. To reduce memory processes, subjects were instructed not to memorize the items. Instead of demanding an immediate behavioral response, subject's active participation and depth of perception was monitored after the experiment by testing incidental memory. The supramodal processing of inputs resulted in enhanced coherent activity between electrodes over the adjacent temporal and parietal cortices. The main synchronization took place in the beta frequency range (13–18 Hz). There was no increase in EEG power in either electrode, indicating that the number of active neurons contributing to the EEG signal remained constant in each cortical area. Thus, supramodal processing did not elicit a global activation of one specific area but rather enhanced synchronization between neurons of different neurons of different cortical areas, i.e., members of a distributed ensemble. While it was known previously that temporal and parietal brain regions are involved in supramodal processing, these results indicate that processing is mediated by synchronized activity between neuronal assemblies.

Integrative processing between the visual and the auditory modality has also been investigated by testing reaction times to audiovisual stimulation (Senkowski et al., 2006). Audiovisually presented targets are processed faster than auditory-alone or visual-alone target stimuli and beta activity correlated with reaction times, speaking for an association of beta oscillations and integrative multisensory processing. Synchronization in the beta frequency range appears also between adjacent frontal and temporal cortical areas (Ghuman et al., 2008). Repeated object classification lead to more efficient classifications for repeated compared with novel items and, at the same time, decreased neural response in prefrontal cortex and temporal cortex, but enhanced neural synchrony between these regions. In this priming paradigm, bottom-up and top-down interactions are mediated by synchrony in the beta band. The results suggest that enhanced synchronization reflects a process where neural and computational efficiency are increased by strengthening only those interareal interactions, which serve the subject's behavioral goals. With optimized communication between brain areas, top-down and bottom-up processes converge more rapidly on a parsing of the input stimulus, leading to a more rapid behavioral response. In this way, enhanced synchronization facilitates performance and reduces processing demands.

These experimental findings of interareal synchronization in the beta band are supported by theoretical considerations (Kopell et al., 2000) stating that beta frequency, as opposed to gamma frequency, is able to synchronize over long conduction delays, which corresponds to signals traveling more than just local distances in the brain.

5.3 *Top-Down Activity Mediated by Long-Distance Interactions in Low Frequencies*

We will now list some experiments which indicate that during top-down processing neuronal assemblies interact over long distances and that the interaction is mediated by oscillatory synchronization in the theta band.

Working memory (WM) is our ability to keep events actively "in mind" for short periods of time. Information retained in working memory is continuously updated as it is used to guide behavior. The retention interval in a working memory paradigm corresponds to a cognitive state without current sensory input. The mental rehearsal of memory items during retention is a state where a "sensory" experience is produced not by an external stimulus but by an internal process. This is an example for top-down activity. In this cognitive state, top-down activity is mediated by large-scale integration in the brain to produce an adequate behavioral response. We used a working memory paradigm to probe interactions between distant cortical areas in the context of top-down processing (Sarnthein et al., 1998). In this task, subjects had to retain in memory strings or abstract line drawings and had to reproduce them after each trial. It was known previously that frontal and parietal brain regions are involved in working memory retention. During the retention interval, synchrony was observed between the prefrontal and posterior association cortices in the theta range (4–7 Hz) for both types of stimuli. In agreement with the common dissociation of WM into verbal and visuospatial components (Baddeley, 1992), letter strings induced a pattern of left hemispheric dominance. During retention of abstract line drawings, both left and right hemispheres were involved, in agreement with other investigations of visuospatial memory tasks in humans. In a similar task, where letter strings had to be retained in memory, increased workload induced a strengthening of left-hemisphere frontal–posterior theta coherence (Payne and Kounios, 2009).

A similar fronto-temporo-parietal network was found active in an experiment on encoding and retrieval of visual memory (Sauseng et al., 2004; Sauseng and Klimesch, 2008). In this experiment, encoding required only the processing of visual information, whereas during retrieval verbal and visual information had to be recalled at the same time. During encoding, prefrontal brain sites were exclusively coupled with right posterior electrode sites, whereas during retrieval connectivity between anterior and bilateral temporo-parietal sites was found. This shows that theta phase coupling was strictly dependent on the modality (visual vs. verbal) of the encoded or retrieved material. Since frontal cortex was leading posterior cortical areas, it was suggested that frontal cortex monitors information stored in posterior brain areas. In this view, fronto-parietal theta coupling is relevant for the executive control (top-down) of working memory processes.

Attentional processes may provide another type of top-down activity. Here perception is not only driven bottom-up by external stimuli but also modulated by internal constraints such as expectancy or the current behavioral goal. In an experiment with awake behaving cats, the cortical mechanisms of top-down effects were investigated by analyzing the synchronization between neurons on different levels of the cortical hierarchy during perception of stimuli with varying behavioral significance (von Stein et al., 2000). Cats were trained to perform a go/no-go task and LFP recorded from three layers of cortex in three areas at different levels of the hierarchy of the visual system. In trials where cats had to change their behavior from attentive viewing to paw-press (go trials), interareal synchronization appeared in the 4–12 Hz theta/alpha frequency range and not in no-go trials, where behavior had to remain unchanged. The layer specificity of the interaction indicated a top-down directed interaction. For novel unexpected stimuli, processed in a bottom-up way, interactions occurred not in the 4–12 Hz range but in the gamma range. The emphasis of higher frequency synchrony during unexpected stimuli could reflect local enhancement of stimulus representations that are passed forward from sensory to frontal cortex. These results indicate that synchronization in the gamma range corresponds to bottom-up processes and synchronization between cortical areas in the theta/alpha range may mediate top-down processes.

Attention is directed in the interplay between the memory of task demands (top-down) and the salience of external stimuli (bottom-up). In an experimental paradigm, where monkeys had to switch between top-down and bottom-up attention, neurons were recorded in prefrontal and parietal cortices (Buschman and Miller, 2007). Although both frontal and parietal cortex were involved in attention, bottom-up signals appeared first in parietal cortex and top-down signals appeared first in frontal cortex. Synchronization between frontal and parietal areas was stronger in higher frequencies during bottom-up attention and stronger in lower frequencies during top-down attention. In the interplay of bottom-up and top-down processes, also the phases of gamma waves and theta waves display enhanced synchrony when humans attend a visual target (Sauseng et al., 2008). This is interpreted as matching between incoming visual information (bottom-up) and information stored in memory (top-down). In this context, synchrony at higher frequency bands might support the local interactions needed to enhance stimulus representations. The low-frequency synchrony of searching attention could reflect a widespread transmission of top-down signals on a larger anatomical scale.

5.4 Alpha Waves as Markers of Top-Down Activity

Even though alpha waves are readily seen in EEG by the naked eye (Berger, 1930), their functional role is a subject of current debate (Klimesch et al., 2007; Palva and Palva, 2007). Subjects with eyes closed often display a dominant frequency in their EEG power spectrum that is usually called alpha peak (Fig. 1A). The peak frequency may decrease down to the theta range in elderly subjects (Niedermeyer and Lopes da Silva, 1999) and patients with neurological diseases (e.g., Sarnthein et al., 2006; Stoffers et al., 2007; Moazami-Goudarzi et al., 2008) or medication (e.g., Salinsky et al., 2003). Nevertheless, We will retain the term "alpha peak" in this chapter in spite of its, sometimes, low frequency. Alpha waves of the scalp EEG can be highly synchronous with thalamic LFP in patients (Sarnthein and Jeanmonod, 2007, 2008) (Fig. 1C) and in animals (Lopes da Silva et al., 1973; Schürmann et al., 2000). In comparison of magnitude, cortico-cortical coherence in some cases matches but usually exceeds thalamocortical coherence (Sarnthein and Jeanmonod, 2007, 2008). The thalamic LFP, in turn, shows high correlation with rhythmic spiking of thalamocortical neurons induced by high-threshold calcium spikes (Hughes et al., 2004). As induced by pacemaker cells and without obvious role in cognitive processing, alpha waves have been termed a basic "idling rhythm" of the brain (Adrian and Matthews, 1934; Pfurtscheller et al., 1996). In view of the importance of background activity (Arieli et al., 1996), the state without external input appears as a state of pure top-down activity.

This view is supported by variations of alpha power in the EEG that are induced by a wide span of complex cognitive tasks investigating – among others – attention, memory, and expectation (Klimesch, 1999; Jensen et al., 2002; Babiloni et al., 2008; Michels et al., 2008; van Dijk et al., 2008). While patients were engaged in a working memory task, also the thalamic LFP was modulated (Sarnthein et al., 2005). Concerning its role in cognitive processing, alpha activity has been suggested to reflect either active inhibition (Klimesch et al., 2007) or it might contribute directly to cortical processing (von Stein et al., 2000; Palva and Palva, 2007), but in both cases alpha waves can be taken as a marker of top-down control of brain activity.

6 Conclusions

The experimental evidence listed above supports the hypothesis that cortico-cortical interactions may occur on different scales and that the complexity of the interaction influences the temporal dynamics within the neuronal assembly. Local interactions during visual processing involve gamma frequency activity as measured by the ongoing EEG generated by the cortex underlying the electrode. Interactions between different but adjacent cortical areas involve beta frequency activity, and long-range interactions during internal mental (top-down) processes involve interactions in a low theta/alpha frequency range. In the latter two cases, EEG coherence measures large synchronous cell assemblies, whose member neurons are distributed over cortical areas. These results are compatible with the notion that functional cortical assemblies are defined by the cognitive task in which they are engaged. Here the size of a functional cortical assembly is reflected by the EEG frequency and the size of the activated neuronal assembly is inversely related to the frequency of synchronization (von Stein and Sarnthein, 2000). Additionally, these studies point to the involvement of alpha/theta frequency synchronization in processes with a predominance of internal mental activity, i.e., top-down effects. By analyzing the frequency of activity, EEG may be used to distinguish between bottom-up and top-down processing. Taking these findings together, the low-frequency interactions during top-down processing may reflect complex computations, where information of many brain areas is integrated in interareal neuronal assemblies. Conversely, fast local processing of sensory information (bottom-up) occurs in the gamma band. This suggests that the brain may emphasize synchronization at different frequency bands for the dynamic modulation of interareal connections, which in turn engages the network best suited for the current task.

Acknowledgment Astrid von Stein has contributed substantially to the concepts and our experiments presented in this chapter (Sarnthein, 2005).

References

Adrian, E.D., Matthews, B.H., 1934. The interpretation of potential waves in the cortex. J Physiol 81, 440–471.

Ahissar, M., Hochstein, S., 2004. The reverse hierarchy theory of visual perceptual learning. Trends Cogn Sci 8, 457–464.

Arieli, A., Sterkin, A., Grinvald, A., Aertsen, A., 1996. Dynamics of ongoing activity: explanation of the large variability in evoked cortical responses. Science 273, 1868–1871.

Babiloni, C., Capotosto, P., Brancucci, A., Percio, C.D., Petrini, L., Buttiglione, M., Cibelli , G., Romani, G.L., Rossini, P.M., Arendt-Nielsen, L., 2008. Cortical alpha rhythms are related to the anticipation of sensorimotor interaction between painful stimuli and movements: a high-resolution EEG study. J Pain 9, 902–911.

Baddeley, A., 1992. Working memory. Science 255, 556–559.

Belitski, A., Gretton, A., Magri, C., Murayama, Y., Montemurro, M.A., Logothetis, N.K., Panzeri, S., 2008. Low-frequency local field potentials and spikes in primary visual cortex convey independent visual information. J Neurosci 28, 5696–5709.

Bendat, J.S., Piersol, A.G., 1986. Random Data. Wiley, New York.

Berger, H., 1930. Über das Elektroenkephalogramm des Menschen II. J Psychol Neurol 40, 160–179.

Bosking, W.H., Zhang, Y., Schofield, B., Fitzpatrick, D., 1997. Orientation selectivity and the arrangement of horizontal connections in tree shrew striate cortex. J Neurosci 17, 2112–2127.

Buschman, T.J., Miller, E.K., 2007. Top-down versus bottom-up control of attention in the prefrontal and posterior parietal cortices. Science 315, 1860–1862.

Buzsaki, G., 2006. Rhythms of the Brain. Oxford University Press, New York.
Douglas, R.J., Martin, K.A., 2004. Neuronal circuits of the neocortex. Annu Rev Neurosci 27, 419–451.
Douglas, R.J., Martin, K.A., 2007. Mapping the matrix: the ways of neocortex. Neuron 56, 226–238.
Eckhorn, R., Bauer, R., Jordan, W., Brosch, M., Kruse, W., Munk, M., Reitboeck, H.J., 1988. Coherent oscillations: a mechanism of feature linking in the visual cortex? Multiple electrode and correlation analyses in the cat. Biol Cybern 60, 121–130.
Engel, A.K., Fries, P., Singer, W., 2001. Dynamic predictions: oscillations and synchrony in top-down processing. Nat Rev Neurosci 2, 704–716.
Engel, A.K., König, P., Gray, C.M., Singer, W., 1990. Stimulus-dependent neuronal oscillations in cat visual cortex: inter-columnar interaction as determined by cross-correlation analysis. Eur J Neurosci 2, 588–606.
Fries, P., 2005. A mechanism for cognitive dynamics: neuronal communication through neuronal coherence. Trends Cogn Sci 9, 474–480.
Ghuman, A.S., Bar, M., Dobbins, I.G., Schnyer, D.M., 2008. The effects of priming on frontal-temporal communication. Proc Natl Acad Sci USA 105, 8405–8409.
Girard, P., Hupe, J.M., Bullier, J., 2001. Feedforward and feedback connections between areas V1 and V2 of the monkey have similar rapid conduction velocities. J Neurophysiol 85, 1328–1331.
Goense, J.B., Logothetis, N.K., 2008. Neurophysiology of the BOLD fMRI signal in awake monkeys. Curr Biol 18, 631–640.
Gotman, J., 1983. Measurement of small time differences between EEG channels: method and application to epileptic seizure propagation. Electroencephalogr Clin Neurophysiol 56, 501–514.
Gray, C.M., 1999. The temporal correlation hypothesis of visual feature integration: still alive and well. Neuron 24, 31–47, 111–125.
Gray, C.M., König, P., Engel, A.K., Singer, W., 1989. Oscillatory responses in cat visual cortex exhibit inter-columnar synchronization which reflects global stimulus properties. Nature 338, 334–337.
Gross, J., Kujala, J., Hamalainen, M., Timmermann, L., Schnitzler, A., Salmelin, R., 2001. Dynamic imaging of coherent sources: studying neural interactions in the human brain. Proc Natl Acad Sci USA 98, 694–699.
Hari, R., Salmelin, R., Makela, J.P., Salenius, S., Helle, M., 1997. Magnetoencephalographic cortical rhythms. Int J Psychophysiol 26, 51–62.
Hughes, S.W., Lorincz, M., Cope, D.W., Blethyn, K.L., Kekesi, K.A., Parri, H.R., Juhasz, G., Crunelli, V., 2004. Synchronized oscillations at alpha and theta frequencies in the lateral geniculate nucleus. Neuron 42, 253–268.
Jarvis, M.R., Mitra, P.P., 2001. Sampling properties of the spectrum and coherency of sequences of action potentials. Neural Comput 13, 717–749.
Jensen, O., Gelfand, J., Kounios, J., Lisman, J.E., 2002. Oscillations in the alpha band (9–12 Hz) increase with memory load during retention in a short-term memory task. Cereb Cortex 12, 877–882.
Juergens, E., Guettler, A., Eckhorn, R., 1999. Visual stimulation elicits locked and induced gamma oscillations in monkey intracortical- and EEG-potentials, but not in human EEG. Exp Brain Res 129, 247–259.
Klimesch, W., 1999. EEG alpha and theta oscillations reflect cognitive and memory performance: a review and analysis. Brain Res Brain Res Rev 29, 169–195.
Klimesch, W., Sauseng, P., Hanslmayr, S., 2007. EEG alpha oscillations: the inhibition-timing hypothesis. Brain Res Rev 53, 63–88.
Kopell, N., Ermentrout, G.B., Whittington, M.A., Traub, R.D., 2000. Gamma rhythms and beta rhythms have different synchronization properties. Proc Natl Acad Sci USA 97, 1867–1872.
Llinas, R., Ribary, U., Contreras, D., Pedroarena, C., 1998. The neuronal basis for consciousness. Philos Trans R Soc Lond B Biol Sci 353, 1841–1849.

Logothetis, N.K., Kayser, C., Oeltermann, A., 2007. In vivo measurement of cortical impedance spectrum in monkeys: implications for signal propagation. Neuron 55, 809–823.

Lopes da Silva, F.H., von Lierop, T.H.M.T., Schrijer, C.F., Storm van Leeuwen, W., 1973. Organization of thalamic and cortical alpha rhythms: spectra and coherences. Electroencephalogr Clin Neurophysiol 35, 627–639.

Michels, L., Moazami-Goudarzi, M., Jeanmonod, D., Sarnthein, J., 2008. EEG alpha distinguishes between cuneal and precuneal activation in working memory. Neuroimage 40, 1296–1310.

Moazami-Goudarzi, M., Sarnthein, J., Michels, L., Moukhtieva, R., Jeanmonod, D., 2008. Enhanced frontal low and high frequency power and synchronization in the resting EEG of parkinsonian patients. Neuroimage 41, 985–997.

Niedermeyer, E., Lopes da Silva, F.H., 1999. Electroencephalography: Basic Principles, Clinical Applications, and Related Fields, 2nd ed. Lippincott Williams & Wilkins, Philadelphia, PA.

Nunez, P.L., 1999. Neocortical Dynamics and Human EEG Rhythms. Oxford University Press, New York.

Nunez, P.L., 2000. Toward a quantitative description of large-scale neocortical dynamic function and EEG. Behav Brain Sci 23, 371–398; discussion 399–437.

Nunez, P.L., Reid, L., Bickford, R.G., 1978. The relationship of head size to alpha frequency with implications to a brain wave model. Electroencephalogr Clin Neurophysiol 44, 344–352.

Palva, S., Palva, J.M., 2007. New vistas for alpha-frequency band oscillations. Trends Neurosci 30, 150–158.

Pascual-Marqui, R.D., Michel, C.M., Lehmann, D., 1994. Low resolution electromagnetic tomography: a new method for localizing electrical activity in the brain. Int J Psychophysiol 18, 49–65.

Payne, L., Kounios, J., 2009. Coherent oscillatory networks supporting short-term memory retention. Brain Res Brain Res Rev 1247, 126–132.

Percival, D.B., Walden, A.T., 1993. Spectral Analysis for Physical Applications. Cambridge University Press, Cambridge.

Pfurtscheller, G., Stancak, A., Jr., Neuper, C., 1996. Event-related synchronization (ERS) in the alpha band – an electrophysiological correlate of cortical idling: a review. Int J Psychophysiol 24, 39–46.

Salin, P.A., Bullier, J., 1995. Corticocortical connections in the visual system: structure and function. Physiol Rev 75, 107–154.

Salinsky, M.C., Oken, B.S., Storzbach, D., Dodrill, C.B., 2003. Assessment of CNS effects of antiepileptic drugs by using quantitative EEG measures. Epilepsia 44, 1042–1050.

Sarnthein, J., 2005. Astrid von Stein (1961–2002) – obituary. Int J Psychophysiol 57, 79–79.

Sarnthein, J., Jeanmonod, D., 2007. High thalamocortical theta coherence in patients with Parkinson's disease. J Neurosci 27, 124–131.

Sarnthein, J., Jeanmonod, D., 2008. High thalamocortical theta coherence in patients with neurogenic pain. Neuroimage 39, 1910–1917.

Sarnthein, J., Morel, A., von Stein, A., Jeanmonod, D., 2005. Thalamocortical theta coherence in neurological patients at rest and during a working memory task. Int J Psychophysiol 57, 87–96.

Sarnthein, J., Petsche, H., Rappelsberger, P., Shaw, G.L., von Stein, A., 1998. Synchronization between prefrontal and posterior association cortex during human working memory. Proc Natl Acad Sci USA 95, 7092–7096.

Sarnthein, J., Stern, J., Aufenberg, C., Rousson, V., Jeanmonod, D., 2006. Increased EEG power and slowed dominant frequency in patients with neurogenic pain. Brain 129, 55–64.

Sauseng, P., Klimesch, W., 2008. What does phase information of oscillatory brain activity tell us about cognitive processes? Neurosci Biobehav Rev 32, 1001–1013.

Sauseng, P., Klimesch, W., Doppelmayr, M., Hanslmayr, S., Schabus, M., Gruber, W.R., 2004. Theta coupling in the human electroencephalogram during a working memory task. Neurosci Lett 354, 123–126.

Sauseng, P., Klimesch, W., Gruber, W.R., Birbaumer, N., 2008. Cross-frequency phase synchronization: a brain mechanism of memory matching and attention. Neuroimage 40, 308–317.

Schadow, J., Lenz, D., Thaerig, S., Busch, N.A., Frund, I., Herrmann, C.S., 2007a. Stimulus intensity affects early sensory processing: sound intensity modulates auditory evoked gamma-band activity in human EEG. Int J Psychophysiol 65, 152–161.

Schadow, J., Lenz, D., Thaerig, S., Busch, N.A., Frund, I., Rieger, J.W., Herrmann, C.S., 2007b. Stimulus intensity affects early sensory processing: visual contrast modulates evoked gamma-band activity in human EEG. Int J Psychophysiol 66, 28–36.

Scherg, M., Berg, P., 1996. New concepts of brain source imaging and localization. Electroencephalogr Clin Neurophysiol Suppl 46, 127–137.

Schürmann, M., Demiralp, T., Basar, E., Basar-Eroglu, C., 2000. Electroencephalogram alpha (8–15 Hz) responses to visual stimuli in cat cortex, thalamus, and hippocampus: a distributed alpha network? Neurosci Lett 292, 175–178.

Senkowski, D., Molholm, S., Gomez-Ramirez, M., Foxe, J.J., 2006. Oscillatory beta activity predicts response speed during a multisensory audiovisual reaction time task: a high-density electrical mapping study. Cereb Cortex 16, 1556–1565.

Singer, W., Gray, C.M., 1995. Visual feature integration and the temporal correlation hypothesis. Annu Rev Neurosci 18, 555–586.

Stoffers, D., Bosboom, J.L., Deijen, J.B., Wolters, E.C., Berendse, H.W., Stam, C.J., 2007. Slowing of oscillatory brain activity is a stable characteristic of Parkinson's disease without dementia. Brain 130, 1847–1860.

van Dijk, H., Schoffelen, J.M., Oostenveld, R., Jensen, O., 2008. Prestimulus oscillatory activity in the alpha band predicts visual discrimination ability. J Neurosci 28, 1816–1823.

Varela, F., Lachaux, J.P., Rodriguez, E., Martinerie, J., 2001. The brainweb: phase synchronization and large-scale integration. Nat Rev Neurosci 2, 229–239.

von Stein, A., 1999. Does the brain represent the world? Evidence against the mapping assumption. In: Riegler, A., Peschl, M., von Stein, A. (Eds.), Understanding Representation in the Cognitive Sciences. Kluwer, New York, pp. 161–167.

von Stein, A., Chiang, C., König, P., 2000. Top-down processing mediated by interareal synchronization. Proc Natl Acad Sci USA 97, 14748–14753.

von Stein, A., Rappelsberger, P., Sarnthein, J., Petsche, H., 1999. Synchronization between temporal and parietal cortex during multimodal object processing in man. Cereb Cortex 9, 137–150.

von Stein, A., Sarnthein, J., 2000. Different frequencies for different scales of cortical integration: from local gamma to long range alpha/theta synchronization. Int J Psychophysiol 38, 301–313.

Supplemental electronic material can be found at
http://www.springer.com/978-0-387-93796-0

Synchrony in Neural Networks Underlying Seizure Generation in Human Partial Epilepsies

Fabrice Bartolomei and Fabrice Wendling

1 Introduction

Focal (or partial) epilepsies are characterized by recurrent seizures generated in an abnormal region of the brain, the epileptogenic zone (EZ). Approximately 30% of cases are resistant to antiepileptic drugs. In this situation, surgical resection of the EZ is the only therapeutic option able to suppress seizures or, at least, to significantly reduce their frequency. The localization and the definition of the EZ are therefore crucial issues in epileptology and are addressed through detailed analysis of anatomo-functional data acquired in epileptic patients during pre-surgical evaluation. Among investigation methods used during this evaluation, intracerebral exploration remains the only way to directly record the electrophysiological activity (depth-EEG) from brain structures and to formulate hypotheses about their potential involvement in epileptogenic processes. In this context, a large number of studies have been dedicated to the analysis of depth-EEG signals. Based on the estimation of interdependences (i.e., statistical coupling) between signals recorded from distinct sites, some reports have demonstrated that the areas involved in the generation of seizures (defining the EZ) are characterized by synchronous oscillations at seizure onset (Bartolomei et al., 2005b; Bartolomei et al., 2001; Bartolomei et al., 2004b; Bartolomei et al., 1999; Duckrow and Spencer, 1992; Gotman and Levtova, 1996; Le Van Quyen et al., 1998; Lieb et al., 1987). The "synchronization" of activities recorded from brain structures is therefore an important phenomenon that may be used for identifying epileptogenic networks (i.e., promoting the initiation of seizures). Other studies based on nonlinear associations in multivariate signals (Guye et al., 2006) have also reported that long distance functional connectivity is dramatically altered during seizures, or indicated that the topology of networks changes as ictal activity develops (Ponten et al., 2007).

F. Bartolomei (✉)
INSERM, Faculté de Médecine, and Assistance Publique – Hôpitaux de Marseille, Hôpital de la Timone, Service de Neurophysiologie Clinique, Aix Marseille Université, U751, Marseille, F-13000, France
e-mail: fbartolo@medecine.univ-mrs.fr, fabrice.bartolomei@medecine.univ-mrs.fr

J.L. Perez Velazquez, R. Wennberg (eds.), *Coordinated Activity in the Brain*, Springer Series in Computational Neuroscience 2, DOI 10.1007/978-0-387-93797-7_7,

In this chapter, our objective is to present the work performed in our group dealing with the study of synchronization processes between distant structures during seizure genesis and development. We will also discuss data showing that abnormal changes in neural synchrony provoked by epileptic rhythms may be responsible of (or at least be in relation with) the production of ictal symptoms or changes in patient's behavior.

2 The Epileptogenic Zone: A Network of Hyperexcitable Structures

The EZ may be considered as a general notion encompassing probably different conditions. Regarding the organization of the EZ, two schematically different situations may be observed. In some cases, the EZ corresponds to a relatively restricted area of the brain. Seizure genesis takes place in a unique "functional area", a situation corresponding to the classical notion of an epileptogenic focus. The concept of epileptic focus and its related notions such as the symptomatic zone (see Rosenow and Luders, 2001) is by far the most popular in epileptology.

However, in a large number of cases (and usually in temporal lobe seizures), the seizure onset is characterized by discharges that rapidly involve several distinct, and distant, brain regions. In this situation, the model of the epileptogenic focus is too simplistic and can not accurately describe the spatial organization of the EZ. The propagation of seizures is also a very complex process. It is indeed usual to observe that partial seizures involve several cortical and subcortical structures during their time course, sometimes with a remarkable reproducibility from one seizure to the other, in a given patient (Wendling et al., 1997). Figure 1 shows one example of "distributed" areas generating seizures in a case of classical mesial temporal lobe seizure. In this example, the seizures start simultaneously from at least distinct regions, such as the hippocampus, the amygdala, and the temporal pole.

In a series of works based on the analysis of SEEG signals in a large number of patients, we have proposed that the EZ corresponds to a network of hyperexcitable structures sharing at least two main spatio-temporal characteristics:

- the ability to generate fast oscillations, in the beta or/and the gamma range, (classically called low voltage rapid discharges, LVRD).
- the ability to synchronize their activity at seizure onset and during the course of the seizure.

LVRDs constitute a characteristic electrophysiological pattern in focal seizures of human epilepsy characterized by a marked decrease of signal voltage (sometimes preceded by high-amplitude spikes) with a marked increase of signal frequency. They have long been observed in depth EEG signals. Rapid discharges generally appear in brain regions in keeping with the epileptogenic zone (Bancaud et al.,

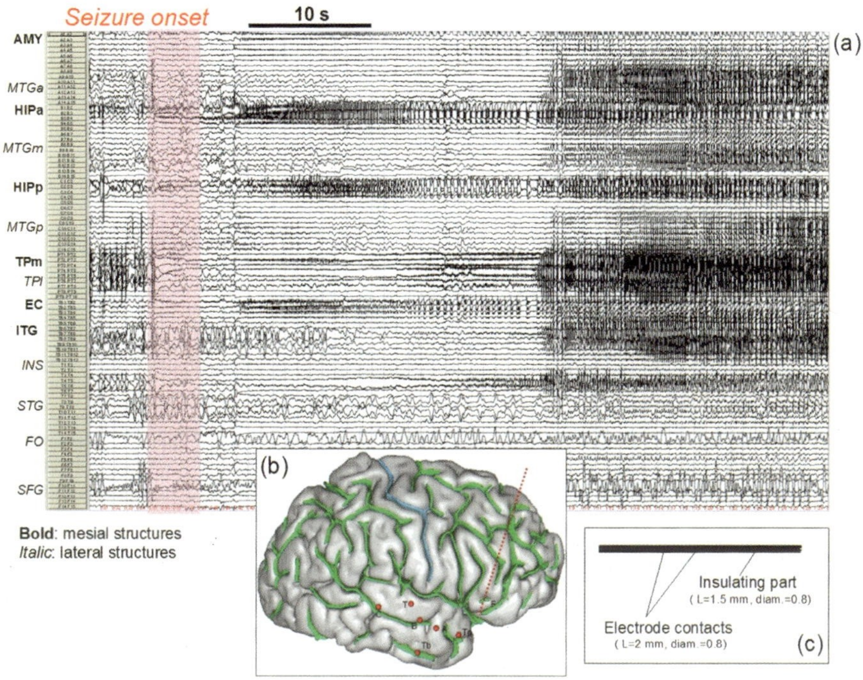

Fig. 1 (**a**) Example of mesial temporal lobe seizure and (**b**) corresponding SEEG scheme showing the electrodes exploring the temporal lobe structures. (**c**) Each electrode contains 15 contacts numbered from 1 (medial) to 15 (lateral). A: electrode exploring the amygdala (AMY, medial contacts) and the anterior part of the middle temporal gyrus (MTGa) (lateral contacts) B: electrode exploring the anterior hippocampus (HIPa, medial contacts) and the mid part of MTGm (lateral contacts); TB: Electrode exploring the entorhinal cortex (EC, medial contacts) and the anterior part of the inferior temporal gyrus (ITG, lateral contacts); C: electrode exploring the posterior hippocampus (HIPp, medial contacts) and the posterior part of MTG (MTGp, lateral leads); T: electrode exploring the insular cortex (INS, medial contacts) and the anterior part of the superior temporal gyrus (STG, lateral contacts) F: electrode exploring the frontal lobe (fronto-orbital region FO and superior frontal gyrus SFG). SEEG traces are represented in bipolar montages. Each line represents the recording between two adjacent contacts. Seizure starts from mesial structures, i.e., medial contacts of electrodes A (amygdla), B (Hippocampus), TP (temporal pole), and TB (entorhinal cortex). One can notice the involvement of several structures including neocortical regions during seizure propagation, approximately 30 s after the onset

1970) and are therefore considered as important marker providing substantial information about the EZ spatio-temporal organization (Alarcon et al., 1995; Wendling et al., 2003). In temporal lobe seizures, the LVRD are generally less rapid disclosing a frequency peak in the beta or low gamma range, typically from 20 to 40 Hz (Bartolomei et al., 2004b; Spencer et al., 1992).

3 Synchronization of Ictal Activity: From Normal Brain Synchrony to Epileptic "Binding"

As indicated above, the seizure onset characterized by LVRD involves often distant and functionally distinct brain sites almost simultaneously. Thus it could be hypothesized that a "synchronizing phenomenon" gives rise to the simultaneous start of fast oscillations. This hypothesis prompted us to study the spatio-temporal dynamics of these phenomena through measuring the interdependencies between generated signals.

It has been known for a long time that epileptic phenomena are associated with dramatic changes in brain synchrony mechanisms (Brazier, 1972) and subsequent studies have shown that seizures in humans are associated with the abnormal synchronization of distant structures (Bartolomei et al., 1999; Gotman and Levtova, 1996; Le Van Quyen et al., 1998). This synchronization can be quantified by measuring the interdependencies between signals recorded in different brain regions involved in the EZ. Numerous methods were proposed over the past decades, often categorized according to their ability to assess the linear (coherence, linear regression analysis, etc) or nonlinear (mutual information, nonlinear regression analysis, similarity between state-space trajectories reconstructed from observed signals) properties of the relationship. Readers may refer to Ansari-Asl et al. (2006) for the presentation of various kinds of methods and evaluation based on different models of relationships between signals.

Using these methods, it is therefore possible to study functional couplings between several brain regions involved or not at seizure onset. To us, the use of nonlinear approaches is probably best suited as it does not require assumptions on the nature or the relationship (Bartolomei et al., 2001). In this context, the so-called nonlinear regression analysis was introduced by Pijn and Lopes Da Silva (1993) in the middle of the 1990s. It provides a parameter, referred to as the nonlinear correlation coefficient h^2, which takes values in [0, 1]. Low values of h^2 denote that two signals X and Y under analysis are independent. On the other hand, high values of h^2 mean that the second signal Y may be explained by a transformation (possibly nonlinear) of the first signal X (i.e., both signals are dependent).

In addition, this method offers the possibility to study the direction of the coupling between neuronal populations which is an important parameter to determine the "leader region" responsible for the "driving" input in the system. In addition to the estimation of h^2, a second quantity has been proposed (Wendling et al., 2001) that brings information on the causal property of the association. This quantity, referred to as the direction index D, takes into account both the estimated time delay τ between signals X and Y (latency) and the asymmetry property of the nonlinear correlation coefficient h^2 (values of the h^2 coefficient are different if the computation is performed from X to Y or from Y to X). Values of parameter D range from –1.0 (X is driven by Y) to 1.0 (Y is driven by X). In the past 5 years, these methods have been mainly applied to the study of temporal lobe seizures (TLS) (Bartolomei et al., 2005b; Bartolomei et al., 2001; Bartolomei et al., 2004b; Bartolomei et al., 1999; Wendling et al., 2001). Results of the signal quantification demonstrate that the regions involved in the EZ establish preferential functional links during the seizure and favor the existence of a network organization of the EZ.

In addition, the study of the statistical relations existing between SEEG signals at seizure onset, allowed us to identify four sub-types of TLEs according to the interactions between mesial (amygdala–hippocampus–entorhinal cortex) and neocortical structures: mesial, mesial–lateral, and lateral–mesial and lateral (Bartolomei et al., 2001; Bartolomei et al., 1999). Mesial TLS are the most frequent forms of TLS and Fig. 2 shows one typical example, studied with the method of non-linear correlation. In this group, functional coupling between several regions belonging to the mesial structures is observed. Absence of coupling between mesial and neocortical structures at seizure onset is also a characteristic feature of these seizures.

A more general phenomenon is that the coupling between brain structures is maximal before the emergence of the LVRD and tends to decrease after, to then increase once more later during the seizure course. This phenomenon is illustrated in Fig. 2. This "preictal" synchronization has been particularly quantified in a group of mesial temporal lobe epilepsy (MTLE), by studying the interactions between entorhinal cortex, hippocampus, and amygdala (Bartolomei et al., 2004b). Two main patterns of transition between interictal to ictal state may be observed in these seizures. Some seizures start with a fast discharge and seem to be particularly under the control of the entorhinal cortex, while seizures starting by a

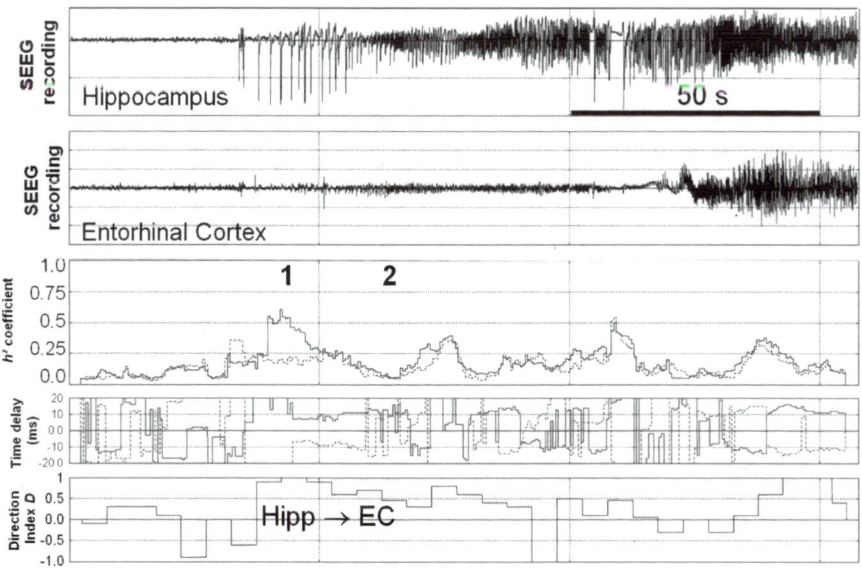

Fig. 2 A study of functional coupling between entorhinal cortex and hippocampus in a patient with mesial temporal lobe seizures. An increase of synchrony (1) during the phase of preictal spiking is observed as measured by the coefficient h^2 between the two structures. The direction index D indicates that the activity in the hippocampus is leading that recorded in the entorhinal cortex. Note that the rapid discharge is associated with a decrease of correlation

"preictal" periodic spiking are probably more likely triggered by the hippocampus. The desynchronization between signals observed after a first phase of synchrony was found to be particularly important when we studied the signal from neocortical sites in neocortical seizures (Wendling et al., 2003). Finally, the pattern of synchronization/desynchronization between regions forming the EZ appears to be a characteristic property of the EZ, that may be observed in many different forms of partial seizures (Schindler et al., 2007).

4 The Role of Synchronization/Desynchronization in Ictal Semiology

Video–SEEG recordings are classically used to make anatomo-electro-clinical correlations and to determine which regions of the brain are responsible for clinical symptoms. One of the most striking phenomena is that semiology generally appears when a large set of different structures are involved in the ictal discharge. In addition, the first clinical signs are observed after the seizure onset (typically several seconds) and are largely related to the propagation of the discharge.

We made the hypothesis that signs and symptoms during the seizure course could be related to changes and new interactions taking place between brain regions during the development of ictal rhythms. Unfortunately, so far very few studies have attempted to correlate symptoms to changes in brain synchronization. In recent years, we have suggested that some symptoms could be related to the uncontrolled activation of functionally normal neural networks by epileptic rhythms or, on the contrary, be related to the disruption of mechanisms underlying normal brain function.

The former situation can be seen in the building of some "elaborated" symptoms mimicking normal behaviors. One of the most remarkable examples is the "dreamy state" that encompasses two kinds of phenomena: an illusion of familiarity (déjà vu-déjà vécu) or a reminiscence of visual memories. For a long time, the anatomical localization and the mechanisms of these phenomena have been the object of impassioned debates (Bancaud et al., 1994). It is admitted today that these phenomena are related to epileptic discharges involving the memory systems of the mesial temporal lobe. An assumption formulated by Gloor (1990) and Halgren and Chauvel (1993) proposed that the discharge transiently rebuilds a neural network implicated in declarative memory. We recently showed that these "dysmnesic" phenomena were more frequent after stimulation of the rhinal cortices than after stimulation of the hippocampus or the amygdala (Bartolomei et al., 2004a). We particularly studied one of the patients from this series in whom the stimulation of perirhinal cortex provoked the recollection of vivid visual scenes. The interdependencies existing between signals generated at the time of the phenomenon of memory recollection were quantified by linear correlation between intracerebral EEG signals filtered into classical frequency sub-bands. We demonstrated that a transitory theta range synchronization between the mesial temporal lobe structures as well as a primary visual area was observed at the time of the reminiscence (Barbeau et al., 2005). These results are thus suggestive of the transient activation of a neural network normally involved in the recall of memories.

Another example is the humming/singing automatism. Some patients during temporal seizures exhibit a humming/singing behavior, highly reproducible from one seizure to another. We studied three patients explored by SEEG and presenting with such an automatism (Bartolomei et al., 2002). In comparison with the seizure onset period, we showed that a functional coupling (characterized by the coherence averaged over the frequency band) appears between the temporal superior gyrus and the prefrontal cortex during the humming phenomenon.

These two examples are remarkable in that they suggest that ictal activity, observable in areas remote from the EZ, is able to "reactivate" neural networks involved in some specific (memory for dreamy state, musical function) brain functions.

An opposite pattern was found to occur in patients presenting with complex emotional behaviors (seizures are marked by screaming, agitation, terror expression) (Bartolomei et al., 2005c). In these cases, the symptoms are more correlated with the rapid ictal discharge affecting different "para-limbic" regions of both frontal and temporal lobes. Study of correlations between signals have shown that these intense emotional alterations were associated with a decrease of synchrony (Wendling et al., 2003) between signals recorded from the neural networks known to be involved in emotional processing, and in particular a loss of synchrony between the orbito-frontal cortex and the amygdala. This disruption of functional connections could then result in the disruption of emotional regulation leading to the release of altered behavior.

5 Involvement of Thalamo-cortical Synchrony in Temporal Lobe Epilepsy (TLE) Seizures

The anatomy of TLE seizures has been particularly studied from human electroencephalographic (EEG) recordings as well as from animal studies using induced temporal lobe seizures. However, structures outside the temporal lobe might also play a role in temporal lobe seizures. A potentially important extra-temporal structure that could be involved early in the temporal organization of MTLE seizures is the midline thalamus (TH). The role of synchronized thalamo-cortical oscillations has been well demonstrated in animal models of absence seizures (Meeren et al., 2002). During partial seizures, the role of such interactions has been less studied. Electrophysiological recordings have shown that midline TH may be early involved in spontaneous seizures of chronic epileptic rats (Bertram et al., 2001). It has thus been suggested that the TH could act as a seizure amplifier and synchronizer of the ictal activity. Recently we have studied thalamic activity during TLE seizures (using intracerebral recordings, SEEG technique) to analyze the synchronization between signals from temporal lobe structures and the TH (Guye et al., 2006). Synchrony was studied using a statistical measure of SEEG signal interdependencies (nonlinear correlation). The results demonstrated an overall increase of synchrony between TH and temporal lobe structures during seizures. We found that values from seizure onset period were significantly higher than values from the background period ($p = 0.001$). Values at the end of seizure were significantly higher than values from the seizure onset ($p<0.0001$). Several indices were also defined in order to correlate some clinical features to the degree of coupling between cortical structures and TH. In patients with mesial TLEs, a correlation was found between the degree of thalamo-cortical synchrony and the presence of an early loss of consciousness but not with other clinical parameters. In addition, surgical prognosis seemed to be better in patients with low values of thalamo-cortical couplings at the seizure onset. This report (Guye et al., 2006) demonstrated that the TH and remote cortical structures synchronize their activity during TLE seizures and suggested that the extension of the epileptogenic network to the TH is a potential important factor determining surgical prognosis. Figure 3 shows an example of high synchrony between cortical temporal lobe structures and thalamus.

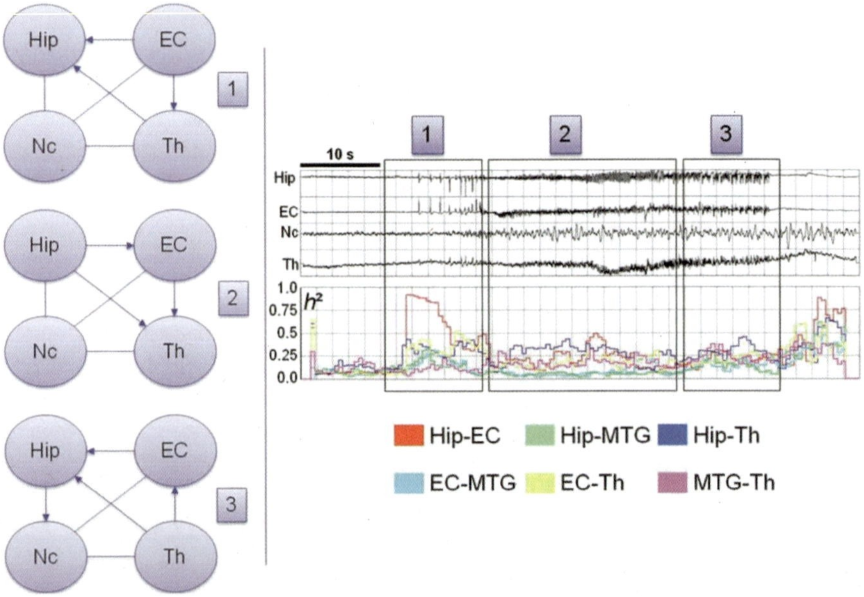

Fig. 3 Nonlinear correlation is studied between temporal lobe structures (Hip: hippocampus, EC: entorhinal cortex, NC: temporal neocortex) and the midline thalamus (TH). Large increase of correlation (determined by the coefficient h^2) between temporal lobe and thalamus is observed as soon as the seizure starts (1)

6 The Epileptogenic Zone Discloses Altered Synchrony During the Interictal State

Despite the classical notion stipulating that epilepsy is associated with abnormally high synchrony within and between neuronal populations, there are in fact very few works that have proved that the epileptic brain is the seat of altered synchronization between distant neuronal assemblies during the interictal state. The study reported in Mormann et al. (2000) showed a local increase in interictal synchrony using the mean phase coherence in a group of 17 MTLE patients. A recent study estimated the interdependencies between signals recorded by subdural grids in nine patients during pre-surgical evaluation with neocortical epilepsy (Schevon et al., 2007). Inter-electrode synchrony was quantified using the mean phase coherence algorithm and revealed areas of elevated local synchrony that may be a marker of epileptogenic cortex.

In a recent study (Bettus et al., 2008), we characterized the functional coupling between mesial structures during the interictal state by comparing the MTLE group to a "control" group including patients with extra-temporal seizures. Results showed that functional coupling was enhanced in the epileptogenic zone in this group of patients (MTLE). This effect was significant for the theta, alpha, beta, and gamma bands and to a large part independent of the occurrence of interictal spiking. The functional significance of an increase of EEG synchrony between structures forming the EZ is unclear. We can assume that even during the interictal state, abnormal long-range reinforced EEG connectivity between structures forming the EZ network is a fundamental characteristic of this region. Of particular interest, in a recent study using the amygdala kindling model in the rat (an animal model of secondary epileptogenesis), an increase of coherence values between amygdala and frontal cortex was observed. Therefore, enhanced connectivity appears to be a marker of the kindling phenomenon at the EEG level (Blumenfeld et al., 2007). This result suggests that this particular state can "facilitate" seizure occurrence by "priming" the system to synchronize during seizures.

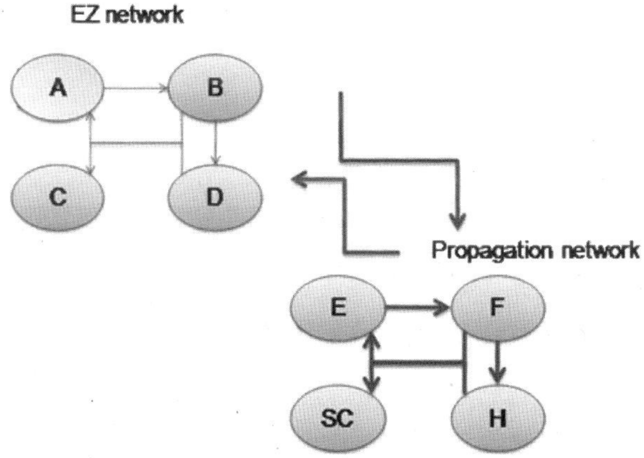

Fig. 4 A model of the organization of partial seizures. The brain structures are represented by letters (A, B, etc). See text for details

7 A General Scheme for the Anatomo-functional Organization of Human Partial Seizures

A general scheme of the organization of human partial seizures can be outlined from the aforementioned studies and is illustrated in Fig. 4 (Bartolomei et al., 2005a). Seizures are generated in an initial network of epileptogenic brain structures (A, B, C, D) (epileptogenic zone network, EZ network) whose activity is transiently synchronized and afterward desynchronized, with the appearance of fast oscillations. This network may include a lesion (A, green circle). During a second ictal phase, other cortical (E, F, H) and subcortical structures (SC) are the seat of slower rhythmic modifications that are synchronized (propagation network). The emergence of a particular clinical semiology in the course of the seizure depends on these phenomena which can in certain cases "mimic" a normal cerebral process or on the contrary provoke a major rupture in normal cerebral functioning.

References

Alarcon G, Binnie CD, Elwes RD, Polkey CE. Power spectrum and intracranial EEG patterns at seizure onset in partial epilepsy. Electroencephalogr Clin Neurophysiol 1995; 94: 326–37.

Ansari-Asl K, Senhadji L, Bellanger JJ, Wendling F. Quantitative evaluation of linear and nonlinear methods characterizing interdependencies between brain signals. Phys Rev E Stat Nonlin Soft Matter Phys 2006; 74: 031916.

Bancaud J, Angelergues R, Bernouilli C, Bonis A, Bordas-Ferrer M, Bresson M, et al. Functional stereotaxic exploration (SEEG) of epilepsy. Electroencephalogr Clin Neurophysiol 1970; 28: 85–6.

Bancaud J, Brunet-Bourgin F, Chauvel P, Halgren E. Anatomical origin of déjà vu and vivid 'memories' in human temporal lobe epilepsy. Brain 1994; 117: 71–90.

Barbeau E, Wendling F, Regis J, Duncan R, Poncet M, Chauvel P, et al. Recollection of vivid memories after perirhinal region stimulations: synchronization in the theta range of spatially distributed brain areas. Neuropsychologia 2005; 43: 1329–37.

Bartolomei F, Barbeau E, Gavaret M, Guye M, McGonigal A, Régis J, et al. Role of the rhinal cortices in déja vu-déjà vécu and reminiscences of memories. Neurology (accepté pour publication) 2004a.

Bartolomei F, Chauvel P, Wendling F. Spatio-temporal dynamics of neuronal networks in partial epilepsy. Rev Neurol (Paris) 2005a; 161: 767–80.

Bartolomei F, Khalil M, Wendling F, Sontheimer A, Regis J, Ranjeva JP, et al. Entorhinal cortex involvement in human mesial temporal lobe epilepsy: an electrophysiologic and volumetric study. Epilepsia 2005b; 46: 677–87.

Bartolomei F, Trebuchon A, Gavaret M, Regis J, Wendling F, Chauvel P. Acute alteration of emotional behaviour in epileptic seizures is related to transient desynchrony in emotion-regulation networks. Clin Neurophysiol 2005c; 116: 2473–9.

Bartolomei F, Wendling F, Bellanger J, Regis J, Chauvel P. Neural networks involved in temporal lobe seizures: a nonlinear regression analysis of SEEG signals interdependencies. Clin Neurophysiol 2001; 112: 1746–60.

Bartolomei F, Wendling F, Regis J, Gavaret M, Guye M, Chauvel P. Pre-ictal synchronicity in limbic networks of mesial temporal lobe epilepsy. Epilepsy Res 2004b; 61: 89–104.

Bartolomei F, Wendling F, Vignal JP, Chauvel P, Liegeois-Chauvel C. Neural networks underlying epileptic humming. Epilepsia 2002; 43: 1001–12.

Bartolomei F, Wendling F, Vignal J, Kochen S, Bellanger J, Badier J, et al. Seizures of temporal lobe epilepsy: identification of subtypes by coherence analysis using stereo-electroencephalography. Clin Neurophysiol 1999; 110: 1741–54.

Bertram EH, Mangan PS, Zhang D, Scott CA, Williamson JM. The midline thalamus: alterations and a potential role in limbic epilepsy. Epilepsia 2001; 42: 967–78.

Bettus G, Wendling F, Guye M, Valton L, Regis J, Chauvel P, et al. Enhanced EEG functional connectivity in mesial temporal lobe epilepsy. Epilepsy Res 2008.

Blumenfeld H, Rivera M, Vasquez JG, Shah A, Ismail D, Enev M, et al. Neocortical and thalamic spread of amygdala kindled seizures. Epilepsia 2007; 48: 254–62.

Brazier MA. Spread of seizure discharges in epilepsy: anatomical and electrophysiological considerations. Exp Neurol 1972; 36: 263–72.

Duckrow R, Spencer S. Regional coherence and the transfer of ictal activity during seizure onset in the medial temporal lobe. Electroencephalogr Clin Neurophysiol 1992; 82: 415–22.

Gloor P. Experiential phenomena of temporal lobe epilepsy. Facts and hypotheses. Brain 1990; 113 (Pt 6): 1673–94.

Gotman J, Levtova, V. Amygdala-hippocampus relationships in temporal lobe seizures: a phase coherence study. Epilepsy Res 1996; 25: 51–7.

Guye M, Regis J, Tamura M, Wendling F, McGonigal A, Chauvel P, et al. The role of corticothalamic coupling in human temporal lobe epilepsy. Brain 2006; 129: 1917–28.

Halgren E, Chauvel P. Experiential phenomena evoked by human brain electrical stimulation. In: Devinsky O, Beric A and Dugali M, editors. Electrical and Magnetic Stimulation of the Brain and Spinal Cord. New York: Raven Press Ltd, 1993: 123–40.

Le Van Quyen M, Adam C, Baulac M, Martinerie J, Varela F. Nonlinear interdependencies of EEG signals in human intracranially recorded temporal lobe seizures. Brain Res 1998; 792: 24–40.

Lieb J, Hoque K, Skomer C, Song X. Interhemispheric propagation of human medial temporal lobe seizures: a coherence/phase analysis. Electroencephalogr Clin Neurophysiol 1987; 67: 101–19.

Meeren HK, Pijn JP, Van Luijtelaar EL, Coenen AM, Lopes da Silva FH. Cortical focus drives widespread corticothalamic networks during spontaneous absence seizures in rats. J Neurosci 2002; 22: 1480–95.

Mormann F, Lehnertz K, David P, Elger CE. Mean phase coherence as a measure for phase synchronization and its application to the EEG of epilepsy patients. Phys D 2000; 144: 358–69.

Pijn J, Lopes Da Silva F. Propagation of electrical activity: nonlinear associations and time delays between EEG signals. In: Zschocke and Speckmann, editors. Basic Mechanisms of the EEG. Boston: Birkauser, 1993.

Ponten S, Bartolomei F, Stam C. Small-world networks and epilepsy: Graph theoretical analysis of intracerebrally recorded mesial temporal lobe seizures. Clin Neurophysiol 2007; doi:10.1016/j.clinph.2006.12.002.

Rosenow F, Luders H. Presurgical evaluation of epilepsy. Brain 2001; 124: 1683–700.

Schevon CA, Cappell J, Emerson R, Isler J, Grieve P, Goodman R, et al. Cortical abnormalities in epilepsy revealed by local EEG synchrony. Neuroimage 2007.

Schindler K, Leung H, Elger CE, Lehnertz K. Assessing seizure dynamics by analysing the correlation structure of multichannel intracranial EEG. Brain 2007; 130: 65–77.

Spencer S, Guimaraes P, Katz A, Kim J, Spencer D. Morphological patterns of seizures recorded intracranially. Epilepsia 1992; 33: 537–45.

Wendling F, Badier J, Chauvel P, Coatrieux J. A method to quantify invariant information in depth-recorded epileptic seizures. Electroencephalogr Clin Neurophysiol 1997; 102: 472–85.

Wendling F, Bartolomei F, Bellanger JJ, Bourien J, Chauvel P. Epileptic fast intracerebral EEG activity: evidence for spatial decorrelation at seizure onset. Brain 2003; 126: 1449–59.

Wendling F, Bartolomei F, Bellanger J, Chauvel P. Interpretation of interdependencies in epileptic signals using a macroscopic physiological model of EEG. Clin Neurophysiol 2001; 112: 1201–18.

Detection of Phase Synchronization in Multivariate Single Brain Signals by a Clustering Approach

Axel Hutt and Matthias H.J. Munk

Abstract Analog signals of the cerebral cortex in behaving subjects frequently express strong oscillatory components. To investigate functional interactions among different areas of the cortex, it is biologically plausible to determine dependencies of oscillatory signals such as their phase relation both within and across areas. The chapter introduces a cluster approach algorithm to detect phase synchronization in single brain signals. The introduced synchronization index allows for the extraction of time windows, which exhibit strong phase synchronization in all examined time series. This kind of phase synchronization is highly non-stationary and is called mutual phase synchronization. Further the assessment of single trials with respect to the trial average revealed that a number of features in time–frequency space are common to different trials.

1 Introduction

Synchronization effects have been found in many natural systems (Pikovsky et al., 2001). In neuroscience, physiological studies of neocortical function have revealed that the most prominent type of neuronal interactions both within and across different areas is of synchronous nature (Salin and Bullier, 1995). Neuronal synchronization has been proposed to play an important role in sensory processing (Singer, 1999), attention (Steinmetz et al., 2000), memory (Tallon-Baudry et al., 2004) and signal transmission in cortex (Salinas and Sejnowski, 2001).

Several experimental studies have detected synchronization with zero phase difference, so-called in-phase synchronization, e.g. between hemispheres (Engel et al., 1991; Nowak et al., 1995), between cortex, lateral geniculate nucleus and retina (Castelo-Branco et al., 1998) and between different visual areas (Bressler, 1996; Eckhorn et al., 1988; Nelson et al., 1992; Frien et al., 1994; Knig et al., 1995).

A. Hutt (✉)
INRIA CR Nancy – Grand Est, CS20101, 54603 Villers-ls-Nancy Cedex, France
e-mail: axel.hutt@loria.fr

J.L. Perez Velazquez, R. Wennberg (eds.), *Coordinated Activity in the Brain*, Springer Series in Computational Neuroscience 2, DOI 10.1007/978-0-387-93797-7_8,

Further, neuronal in-phase synchronization is supposed to represent a possible solution for the visual binding problem, i.e. the neuronal connection of different perceptive features (Singer, 1999).

In order to quantify the synchronization effects typically experimental paradigms are used in which sensory stimuli evoke neuronal responses. These stimuli might represent sequences of auditory tones or visual figures. Currently there are two hypotheses on how the brain responds to such stimuli (Makeig et al., 2002; Shahl et al., 2004):

- The conventional so-called ERP hypothesis assumes the stimulus-locked neuronal response, which is identical each time the stimulus is presented. This assumption involves a resting state before the stimulus onset and thus allows for the average of the detected neuronal responses obtained from several stimulus realizations. Most current analysis methods apply the statistics over several realizations of experimental trials (Allefeld and Kurths, 2003; beim Graben et al., 2000). These approaches assume responses, which are time-locked to the stimulus and thus allow for averaging of trials. The great advantage of such methods is the easy implementation, while they need a large number of trials and neglect temporal jitters in single trials.
- The other hypothesis has been developed more recently and assumes a phase resetting of ongoing neuronal activity at stimulus onset (Arieli et al., 1995; Grindvald et al., 2002). This approach declines the average of single neuronal responses and proposes the analysis of single neuronal responses. In recent years several methods have been developed, which focus on the analysis of such single neuronal responses (Effern et al., 2000; Karjalainen and Kaipio, 1999; Laskaris and Ioannides, 2002; Ioannides et al., 2002; Lachaux et al., 1999; Pikovsky et al., 2000).

Since single experimental realizations in controlled neuropsychological experiments yield measurements which show prominent time shifts to each other and different scales in amplitude and time, the concept of ongoing activity promises new insights into the understanding of brain functions. Consequently this chapter aims to elucidate some difficulties and solution paths in the analysis of multivariate signals obtained from a single experimental realization.

The analysis of typical single brain signals obtained during neuropsychological experiments has to deal with many difficulties. Let us conceive an experimental setup of an in vivo multielectrode measurement of local field potentials in two brain areas simultaneously involving a large set of electrodes while engaging the subject in specific task. We point out three major problems illustrated by such an experimental set-up:

- The signals observed exhibit a high noise level, i.e. in most cases no prominent amplitude peaks indicate underlying neuronal processes. In technical terms brain signals exhibit a low signal-to-noise ratio. Hence it is difficult to separate task-related activity from random activity in the brain, e.g. caused by synaptic failures, ion-channel fluctuations and others (Koch, 1999). To separate the noise and signal, several separation methods assume a linear superposition of random and task-related activity (Liu et al., 2001; Quian Quiroga and Garcia, 2003), while others consider nonlinear (Hutt et al., 1999; Hutt, 2002) or dynamical (Prusseit and Lehnertz, 2007) signal–noise interactions.

- The signal is high dimensional, with the dimension equal to the number of detectors. This aspect makes the analysis very complex since most synchronization indices reflect the phase synchronization between pairs of time series, and thus the number of synchronization indices is very high comparing all-to-all relations. In fact one has $N(N-1)/2$ pairs out of N time series. One solution to this problem is the dimensionality reduction of the data on a relevant subspace (Gratton et al., 1989; MaKeig et al., 1997; Hutt and Riedel, 2003) or the investigation of a low-dimensional statistical variable based on a data model, such as a phase synchronization index.
- The dynamics of the signals under study is transient in time by virtue of the processing of the external incoming stimulus. This aspect denies all methods based on the stationarity of a signal and necessitates instantaneous approaches (Boashash, 1992; Lachaux et al., 1999; Tass et al., 1998; Le Van Quyen et al., 2001; Lehmann and Skrandies, 1980; Hutt and Riedel, 2003; Hutt, 2004).

The current work attacks the dimensionality problem considering temporal transients, while the noise aspect is not considered. However, noise can be reduced by a typical denoising as a pre-processing step, see, e.g., Effern et al. (2000).

In recent years further methods have been developed, which focus on the analysis of single neuronal responses (Effern et al., 2000; Karjalainen and Kaipio, 1999; Laskaris and Ioannides, 2002; Ioannides et al., 2002; Lachaux et al., 1999; Pikovsky et al., 2000). In this context, the method, which has been applied most frequently to detect phase synchronization, computes instantaneous circular variances in a sliding window, (Haig et al., 2000). However, the major disadvantage of this approach is the arbitrary choice of the fixed width of the sliding window. This point is crucial as neuronal responses show non-stationary behaviour with varying timescales in time and thus the corresponding analysis necessitates adaptive time windows.

This study considers the phase synchronization in single trials, while the time window of the statistical evaluation is extracted from the data. The subsequent sections introduce the notion of the instantaneous phase, the definition of phase synchronization and the applied clustering algorithm. Section 3 shows the results obtained from the applications to experimental local field potentials. Here, the study compares results obtained from single trials with averages of results. Section 4 discusses the obtained results.

2 Methods

This section introduces the notion of instantaneous phases and the phase synchronization in time series and introduces the cluster approach.

2.1 Phase Definitions

Let us assume a zero-mean time series $s(t)$ of T time points $t = t_n = n\Delta$ with $n = 1, \ldots, T$ and the sampling interval Δ. In general, this signal can be decomposed into the instantaneous amplitude $A(t)$ and the phase $\phi(t)$, e.g. by $s(t) = A(t)\sin\phi(t)$. The phase is related to the instantaneous frequency $\nu(t)$ by $\phi(t) = \int \nu(t)dt$ and thus may be obtained from frequency analysis (Boashash, 1992), e.g. by wavelet analysis or Gabor transformation.

In the subsequent applications, the phase shall be defined by wavelet analysis following the approach of DeShazer et al. (2001). A complex signal $\bar{s}(t) = A(t)\exp(i\Phi(t))$ is computed by

$$\bar{s}(t) = \frac{2}{\sqrt{2\pi}} \int_0^\infty \tilde{s}(\nu)e^{-(\nu-\nu_0)^2/2\sigma_f^2} e^{-i2\pi\nu t} d\nu, \quad \nu_0 > 0 \tag{1}$$

with the centre frequency $\nu_0 > 0$ and the Fourier transform $\tilde{s}$ of the time series. The new time series $\bar{s}(t)$ is a complex number since ν_0 is chosen positive. In addition, $2\sigma_f$ may be interpreted as the width of the frequency band. Consequently $\bar{s}(t)$ may be interpreted being limited to the frequency band $[\nu_0 - \sigma_f, \nu_0 + \sigma_f]$. Then, the instantaneous phase and amplitude read

$$\Phi(t) = \arctan(I(\bar{s}(t))/R(\bar{s}(t)), A(t) = \sqrt{R(\bar{s}(t))^2 + I(\bar{s}(t))^2}. \tag{2}$$

Here, $R(\cdot)$ and $I(\cdot)$ denote the real and imaginary parts, respectively. The constraint of the new signal $\bar{s}(t)$ to a band of frequencies yields artificial temporal correlations in the amplitudes and the phases with correlation time $\sigma_T = 1/\sigma_f$. Defining $\sigma_T = N_t/\nu_0$, N_t relates the correlation time to the duration of oscillations of the number N_t at frequency ν_0. By this choice, the phase $\Phi(t)$ corresponds to the Morlet-wavelet phase and the frequency band changes its width σ_f with the centre frequency ν_0 according to $\sigma_f = \nu_0/N_t$. Typically $5 < N_t < 10$ represent a good balance between time and frequency resolution in brain signals.

2.2 Phase Synchronization

In order to quantify the degree of phase synchronization, Kuramoto was one of the first to introduce the circular variance R as an order parameter in ensembles of phase (Kuramoto, 1984). In the context of time series, the circular variance (Mardia and Jupp, 1999) is computed in a sliding window of width ΔT about each time point (Hutt et al., 2003), i.e.

$$\gamma_m(t) = \frac{1}{\tau}\sqrt{\left(\sum_{i=-\tau/2}^{\tau/2} \cos\left(\Delta\phi_m(t + i\Delta T)\right)\right)^2 + \left(\sum_{i=-\tau/2}^{\tau/2} \sin\left(\Delta\phi_m(t + i\Delta T)\right)\right)^2} \tag{3}$$

with $0 \leq \gamma_m \leq 1$ while τ denotes the number of time points in the sliding window of width ΔT. If $\phi_m(t)$ represents the difference between two phases changing in time, $\gamma_m(t)$ defines the bivariate phase synchronization index. The maximum value of $\gamma_m = 1$ reflects maximum bivariate phase synchronization between phases with phase difference ϕ_m, while no bivariate phase synchronization leads to $\gamma_m = 0$.

Essentially the global phase synchronization index may be defined from the bivariate synchronization index (Hutt et al., 2003; Haig et al., 2000; Allefeld and Kurths, 2003) by

$$\gamma_0(t) = \frac{1}{M}\sum_{m=1}^{M} \gamma_m \tag{4}$$

considering M phase pairs. The latter synchronization index (4) is widely used, since it is easy to compute and its interpretation is clear: in case of $\gamma_0 = 1$, all phase pairs exhibit the same phase difference, while $\gamma_0 = 0$ reflects the uniform distribution of phase differences.

However, the width of the time window ΔT is rather arbitrary and has to be chosen a priori. Further $\gamma_0(t)$ takes into account the phase synchronization in space only and neglects temporal structures. In other words, the choice of ΔT focuses on a single timescale only while brain signals are known to exhibit successive time windows of large and low timescales. Consequently we look for another synchronization measure taking into account the space–time structure of the data. Here, the choice of the relevant time window is determined from the data.

Assuming two real time series $s_1(t), s_2(t)$, the generalized phase relation of their corresponding instantaneous phases $\phi_1(t), \phi_2(t)$ is defined by $\Delta\phi_{nm}(t) = n\phi_1(t) - m\phi_2(t)$ for integers n,m. According to Pikovsky et al., (2000), phase synchronization is present if $|\Delta\phi_{nm} - \epsilon| < 2\pi$ for $\epsilon > 0$. Now we extend this definition to

$$|(\Delta\phi_{nm}(t) \mod 2\pi) - \epsilon_{nm}| < \delta_{nm},$$

with $\epsilon_{nm}, \delta_{nm} \in [0, 2\pi), t \in [t, t + \Delta t]$. In words bounded phase differences $\Delta\phi_{nm}(t)$ in a time interval $[t, t + \Delta T]$ reflect the generalized phase synchronization in this time interval. For instance $\delta_{nm} = 2\pi$ tests on no phase synchronization, while $\delta_{nm} \ll 2\pi$ tests on strong phase synchronization. Now we extend this definition to the multivariate case.

Assuming N real time series $s_n(t), n = 1, \ldots, N$, the 1:1 generalized phase relation is defined by $\Delta\phi_{kl} = \phi_k(t) - \phi_l(t)$ and the phase synchronization is present in the time interval $[t, t + \Delta t]$ if

$$|(\Delta\phi_{kl} \mod 2\pi) - \epsilon_{kl}| < \delta \ll 2\pi, \tag{5}$$

holds for $\delta > 0$ and some phase pair k, l. Now, the bounds of the phases $\delta_{nm} = \delta$ are global and independent from the phase pairs.

The set of phase pairs $\mathscr{M}$ for which Eq. (5) holds defines the type of phase synchronization. If $\mathscr{M}$ contains all phase pairs, the signals fulfilling Eq. (5) in a certain time interval exhibit mutual phase synchronization (MPS). Examples are sets of all-to-all pairs, i.e. $\mathscr{M} = \{k, l | k = 1, .., N; l = k+1, .., N\}$, and the set of sufficient pairs $\mathscr{M} = \{k, l | k = 1, l = 2, .., N\}$. Sufficient phase pairs are the pairs which are sufficient to describe all phase relations between the time series. For instance, in case of $N = 3$, ϕ_{12} and ϕ_{13} are sufficient pairs as the missing pair is determined by ϕ_{12} and ϕ_{13}, i.e. $\phi_{23} = \phi_2 - \phi_3 = \phi_2 - \phi_1 + \phi_1 - \phi_3 = \phi_{13} - \phi_{12}$. In mathematical terms, the sufficient pairs build a submanifold in $\Re^M$, which is sufficient to capture the phase dynamics of the system. In cases other than MPS Eq. (5) reflects partial phase synchronization.

This study considers all phase pairs, which is similar to the global synchronization measure introduced in previous studies (Pikovsky et al., 2001; Stam and Dijk, 2002; Breakspear and Terry, 2002). In spite of this similarity, our approach is different from these studies as it does not assume any specific underlying dynamics.

To deal with all phase pairs, we collect all phase differences $\{\Delta\phi_{kl}\}$ in a M-dimensional time series $\Phi(t) = (\Phi_1(t), .., \Phi_M(t))$ with $M = dim(\mathscr{M})$.

2.3 Clustering

According to Eq. (5) MPS is present if all phase differences are bounded. Assuming a high-dimensional torus the trajectory of the signal $\Phi(t)$ casts to a trajectory of phase differences evolving on this torus. If condition (5) is fulfilled, then the data points of the phase time series $\Phi(t)$ aggregate in a bounded region, i.e. they are mutually phase-synchronized. In contrast, unbounded time series of phase differences do not aggregate and thus are not phase-synchronized.

Consequently mutually phase-synchronized time series are bounded distributions of data points and thus represent data agglomerations, i.e. clusters, in the space of phase differences. These clusters may be detected by a clustering algorithm. The present approach aims to detect the time intervals, in which clusters are present. Since clusters are bounded to a specific signal space area and the areas border correspond to time points of the signal, these time windows exhibit MPS (Hutt et al., 2003; Hutt and Munk, 2006). However, some problems have to be solved to achieve this aim. At first, MPS needs to be quantified by an index, which reflects the synchronization degree. Further, the number of clusters is unknown and the clustering algorithm needs to perform on a multidimensional torus, which is a non-standard problem in data analysis. The solutions of these problems are given in the following three sections.

2.3.1 Quantifying MPS

We remark that strong MPS occurs for small bounded regions in the space of phase differences, which contain many data points, i.e. clusters with high data density and $\delta \ll 2\pi$ in Eq. (5). In contrast, weak MPS occurs in rather large regions including few data points only, i.e. low data density. Subsequently, the local data density $P_K(t)$ around each data point at time t gives a reasonable quantity for MPS. Thus let us define the quantity $P_K(t)$ in the space of phase differences by

$$p_n(t) = \frac{T_n}{\Pi_{l=1}^{M}\sigma_l}, \quad n = 1, \dots K \quad , \quad P_K(t) = \frac{p_n(t)}{\sum_{n=1}^{K} p_n}. \tag{6}$$

Here, K gives the total number of clusters, $n = 1, \dots, K$ enumerates the clusters and T_n is the number of data points in the cluster n. The value p_n represents the data density in the hypercuboid of volume $\Pi_{l=1}^{M}\sigma_l$ with side lengths σ_l. Here P_K is normalized to unity. In case of perfect mutual phase synchronization, i.e. all phase differences are equal, it is $P_K = 1/K$ for all time points. In contrast, if there is perfect MPS in a short time interval and no MPS at other time points, it is $P_K \approx 1$ in the short time interval and $P_K \approx 0$ at other time points.

To obtain the borders of the clusters, i.e. T_n and σ_l, the algorithm follows the subsequent idea (Hutt and Riedel, 2003; Hutt et al., 2000): the cluster algorithm computes cluster centres and computes distances from each computed cluster centre to each data point. Then each data point is associated with the nearest cluster centre. Since the data exhibit an order in time, the members of number T_n of the cluster n represent a time window of width $T_n\Delta$ with the sample time Δ. Further, σ_l is the spatial standard deviations of the cluster in dimension l. Further necessary details on the cluster algorithm and the number of clusters K are given below.

2.3.2 The Number of Clusters

To solve the problem of the unknown number of clusters K, the method computes values of $P_K(t)$ for an increasing number of clusters $K = 2, \dots, R$ for a sufficiently large R and sums up all P_K at each time point. If clusters are dominant, they exhibit large values of $P_K(t)$ for most numbers of clusters K. That is, they contribute large values to the sum. In contrast outliers and widespread clusters show low values of P_K and contribute smaller values to the sum. In practice $R = 20$ yields good results for the analysis of typical event-related potentials consisting of $\sim$ 1000 data points and 20 channels.

Considering the previous approach, the final synchronization index for MPS reads

$$P(t) = \frac{1}{R-1} \sum_{K=2}^{R} P_K(t), \tag{7}$$

which shows large values for strong MPS and low values in case of weak MPS (Hutt et al., 2000; Hutt, 2001; Hutt and Riedel, 2003; Hutt, 2004).

Since the cluster results are not unique due to local minima, an average of cluster results obtained from several runs (typically 10) with random initial conditions is computed. This average of cluster results represents a preliminary estimation of the true cluster result in the global minimum and good performance of K-means clustering in phasic data has been observed (Hutt, 2001; Hutt and Riedel, 2003).

2.3.3 The Cluster Algorithm

To detect the clusters of phase differences, we apply a modified K-means cluster algorithm (Hutt et al., 2003) which assumes a priori a fixed number of clusters K. This version of the K-means algorithm is similar to the standard algorithm, see, e.g., Duda and Hart (1973), but has to deal with the fact that the data are periodic, i.e. live on a multidimensional torus.

In some detail, the clustering algorithm aims to find cluster centres in data space, whose average distance to a set of data points is minimized. In our case of phase differences, the distance between two points in data space Φ_i, Φ_j is given by the shortest distance between two points on a torus (Mardia and Jupp, 1999; Hutt et al., 2003)

$$d_{ij} = \sqrt{\pi - |\pi - |(\Phi_i - \Phi_j) \mod 2\pi||}, \tag{8}$$

while the average of data points $\bar{\Phi}$ is given by

$$\bar{\Phi} = (\bar{\Phi}_1, \ldots, \bar{\Phi}_N) \quad , \quad \tan \bar{\Phi}_k = \frac{\sum_{i=1}^{T} \sin \Phi_k(t_i)}{\sum_{i=1}^{T} \cos \Phi_k(t_i)}. \tag{9}$$

As a first step, initial cluster centres are chosen randomly from the set of data points and their distances to each data point are computed. The K-means algorithm defines memberships of data points to a cluster by the smallest distance to its centre. Hence data are segmented into K clusters and new cluster centres are computed as averages of the clustered data points. This procedure is continued until a threshold is reached. This threshold may be a fixed number of iterations or the average change of the cluster centres. In later specific applications, we set the maximum number of iterations to 20 which guaranteed good convergence.

For illustration, Fig. 1 shows an example of a typical trajectory segment. Two cluster centres have been guessed for illustration reasons. Here data between 0 and 90 ms and the two last data points are closer to cluster centre 2 than to cluster centre 1, while the data between 91 and 157 ms belong to cluster 1. In other words, the two cluster centres segment the data into three temporal intervals, whose borders at 90 and 158 ms are determined by the distance from cluster centres. This border defines the time interval mentioned in Sect. 2.3.1. Now, applying the K-means algorithm to the data segment shown in Fig. 1 for $K = 2$, $K = 3$ and $K = 5$ clusters, respectively, we plot the computed distances from cluster centres to data for the

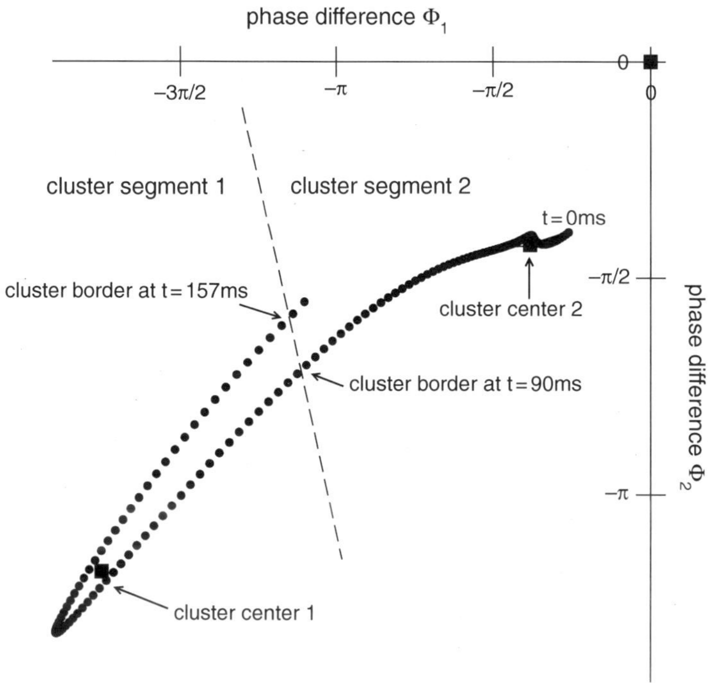

Fig. 1 An example signal segment in the time window $[0\,ms; 159\,ms]$. The signal consists of two time series $\Phi_1(t)$, $\Phi_2(t)$ whose values are plotted on the horizontal and vertical axes, see, e.g. Hutt and Schrauf (2007) for more explanations. The *dashed line* illustrates the border of both clusters

different number of clusters (Fig. 2). The plots exhibit the change of nearest clusters and thus show the temporal segments.

2.4 Summary of the Algorithm

We summarize the procedure steps: first, one chooses a frequency band (here it is $N_t = 10$ if not stated differently) and extracts instantaneous phases according to Eqs. (1) and (2). Then M phase pairs are computed (here it is $M = N(N-1)/2$ if not stated differently) and the K-means cluster algorithm (here with $R = 15$) extracts cluster centres from the new time series of phase differences $\Phi(t)$. Since the cluster results are not unique, e.g. due to local minima, an average of cluster results from 10 runs is computed. Finally, the method computes $P(t)$ for each frequency band.

Further it is necessary to estimate the validity of the clustering results and we compare the cluster results from the original data to the cluster results from data unstructured completely in time. The latter surrogate data set is obtained by randomizing the temporal sequence of the signal corresponding to phase-randomizing of data. In the subsequent sections, the t-values are computed from the R cluster

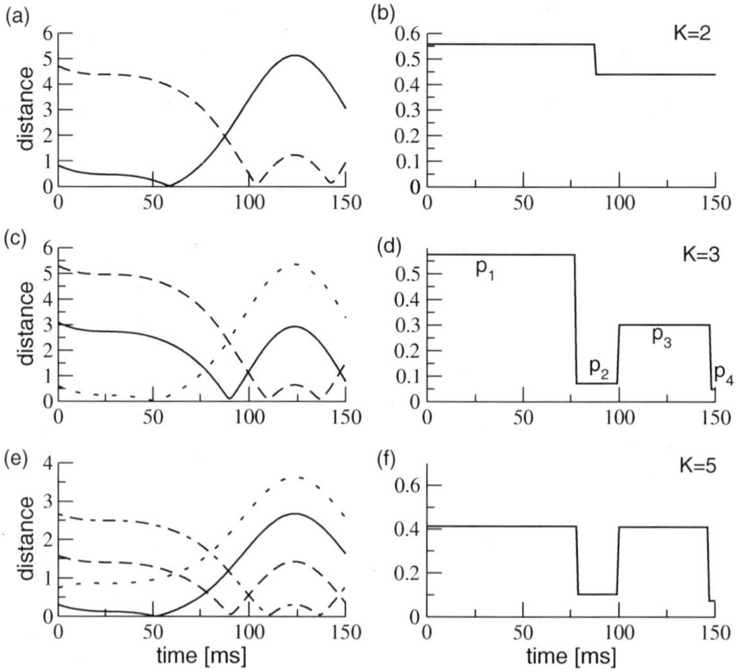

Fig. 2 The basic elements of MPS illustrated for three number of clusters. The panels sketch the distance from cluster centres to each data point with respect to time for $K = 2$ (**a**), $K = 3$ (**c**), $K = 5$ (**e**) and show the corresponding values of P_K and p_k from Eq. (6) for $K = 2$ (**b**), $K = 3$ (**d**) and $K = 5$ (**f**)

results at each time point and each frequency band based on the original data and the surrogate data. The false-positive error for accepting the null hypothesis, i.e. the cluster results obtained from the surrogate data, is set to $\alpha = 10^{-5}$. If the null hypothesis of a cluster result from the original data $P(t)$ is accepted at a time point t and at a frequency band, then we set $P(t) = 0$ for the corresponding frequency band.

3 Applications

This section illustrates the clustering approach by the application to experimental local field potentials. We compare the mutual phase synchronization in randomly chosen single trials with the average of results over single trials.

3.1 Experimental Data

Data were obtained from simultaneous recordings from up to 21 bipolar microelectrodes in awake behaving monkeys. The recordings were performed in multiple visual, parietal, sensorimotor and motor areas of the cortex to determine whether neuronal activity within and across areas might be coordinated in a task-dependent fashion. Electrodes were chronically implanted into the visual areas V1, V2, V4, the parietal area A7, the higher-level somato-sensoric area A5, the post-central somato-sensoric area A1, the primary motor cortex A4, the pre-motor cortex A6 and the frontal eye field A8, see Fig. 3. The electrodes consisted of two wires, a short wire with a tip at 400 μm depth and a long wire with a tip at 1800 μm depth. Signals from the two tips were differentially amplified and subsequently bandpass filtered for 5–150 Hz (3 dB/octave) and notch-filtered for the German power supply frequency of 50 Hz. AD conversion was performed in synchronous mode with 12 bit (1401plus, Cambridge Electronic Design) at 1 kHz and stored on hard disc.

The examples used in this study are taken from a single recording session from one of three monkeys trained to perform a visuo-motor tracking task at high accuracy. In each trial, the monkey was presented with a fixation point at the centre of the screen and 200 ms after the monkey started fixation, a moving sinusoidal grating was presented at a parafoveal position. The monkey had to move a lever against the force of springs along either forwards or backwards depending on the direction of motion of the visual stimulus. If the monkey reached a previously learned position and held the lever at this position for at least 200 ms, the computer presented a new visual instruction by changing the speed of the grating to either twice the drifting velocity or stopping it. This elementary tracking procedure was repeated four times in each trial and if the monkey was successful, it was rewarded with a drop of water at the end of the trial. The markers (stippled lines) presented in the time–frequency plots refer to the first presentation of the instructing visual stimulus (S1), the three subsequent changes of the stimulus (S2, S3, S4), and the four times when the monkey reached the required lever position (R1–R4). The last marker represents the time of an acoustic signal correlated with the reward which was caused by the electromagnetic relais opening the water tube for about 100 ms. In the sample used for this study, trial duration was on average 5.1 s (range 4.6–5.9).

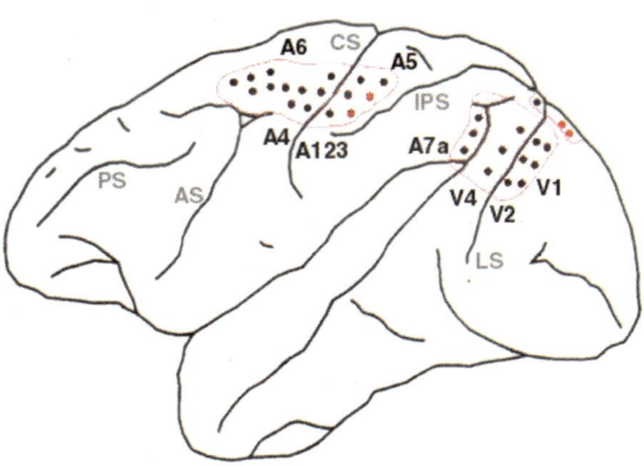

Fig. 3 Spatial locations of implanted electrodes

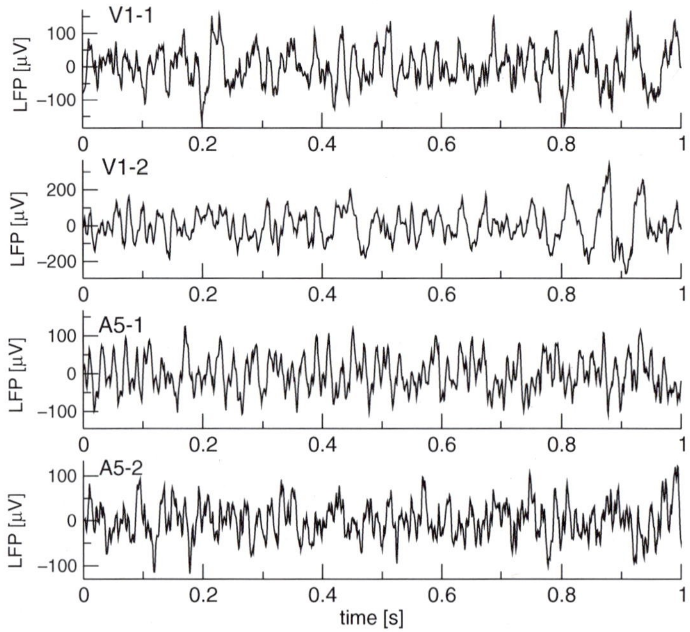

Fig. 4 Examples of local field potentials (single trial no. 17) measured in a monkey in the primary visual cortex V1 (denoted V1-1 and V1-2) and the somato-sensoric area A5 (denoted A5-1 and A5-2). For illustration reasons, the time is limited to the interval [0 ms; 1000 ms], while the original time series is in the range [0 ms; 4987 ms]

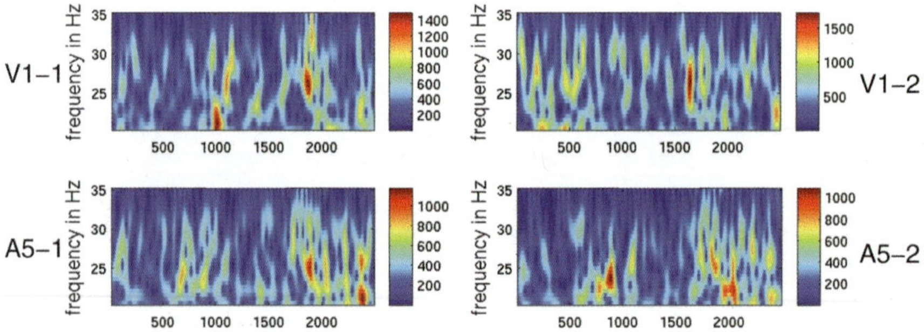

Fig. 5 The spectral power of the time series of trial no. 17 shown in Fig. 4

In the subsequent sections, we examine three trials, which have been chosen arbitrarily. The numbers of these trials indicate their temporal sequence of measurement, though they are not consecutive in time. For illustration reasons, Figs. 4 and 5 show selected time series and their corresponding spectral power distributions in the single trial no. 17, respectively.

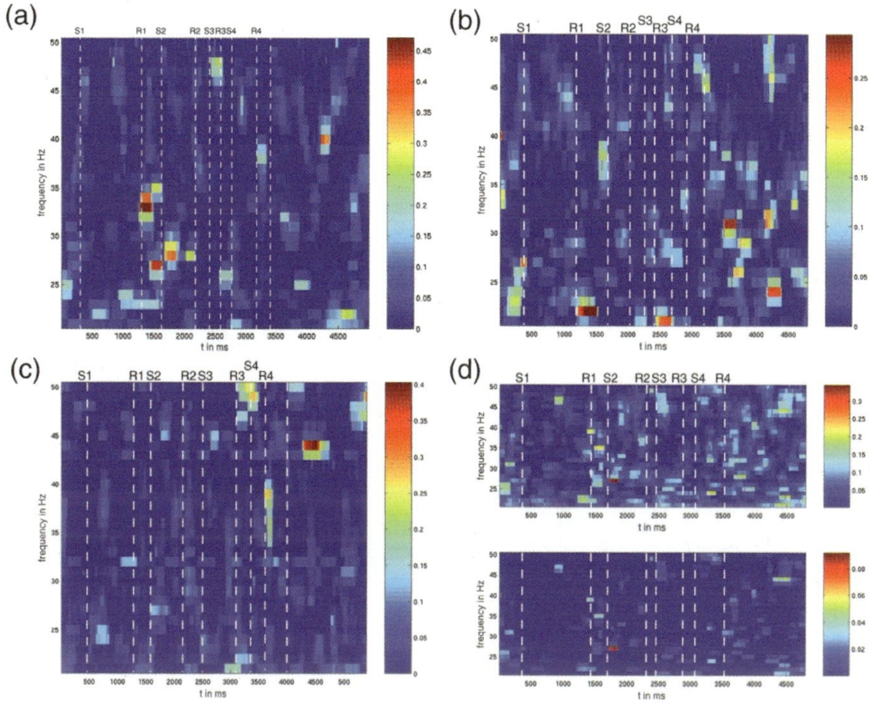

Fig. 6 Time–frequency plots of the MPS of three arbitrarily chosen single trials, their average and standard deviation. (**a**) Trial 17 shown in Figs. 4 and 5, (**b**) trial 28, (**c**) trial 330 and (**d**) average of the trials 17, 28, 330 (*top panel*) and the resulting standard deviation (*bottom panel*). The *color* encodes the index for mutual phase synchronization $P(t)$ from Eq. (7) and the *vertical lines* indicate the occurrence times of events

3.2 Cluster Results

Here and in the following, all phase pairs have been considered for the analysis. Figure 6(a–c) shows clustering results obtained from randomly chosen single trials and the position of the lever with respect to time. The time–frequency plot reveals several time windows with increased MPS. A comparison of these results with their average (Fig. 6(d)) reveals common behaviour in the trials. For instance, at t=1400 ms, all three trials show increased MPS for about 200 ms in the frequency band 21 Hz. Moreover, at t=1000 ms common increased MPS occurs for 200 ms in the frequency band about 47 Hz. Other frequency bands do not show similar MPS in the single trials.

4 Discussion

The proposed cluster algorithm yields an index to quantify mutual phase synchronization at each time point and each frequency band. The statistical evaluation of the synchronization index at each time point is based on time windows, which are extracted from the data set. Moreover, the algorithm takes into account both the temporal and spatial data structures by the high-dimensional treatment of the data.

The application to the local field potentials allows for the extraction of several details of the data. On the one hand, the clustering results elucidate the temporal variability of the occurrences of single trial responses. On the other hand, a comparison of results from single trials and the average of these results yields good accordance in some time windows and frequency bands. The presented results indicate a future approach to single trial data analysis: one extracts synchronization indices from single trials and computes the averages of these results. Hence, this approach allows for a statistical evaluation of single trials even with a small number of trials.

References

Allefeld, C., Kurths, J.: Multivariate phase synchronization analysis of EEG data. IEICE Trans. Fundam. **E86-A**(9), 2218–2221 (2003)

Arieli, A., Shoham, D., Hildesheim, R., Grinvald, A.: Coherent spatio-temporal pattern of ongoing activity revealed by real-time optical imaging coupled with single unit recording in the cat visual cortex. J. Neurophysiol. **73**, 2072–2093 (1995)

beim Graben, P., Saddy, J.D., Schlesewsky, J.D., Kurths, J.: Symbolic dynamics of event-related brain potentials. Phys. Rev. E **62**(4), 5518–5541 (2000)

Boashash, B.: Estimating and interpreting the instantaneous frequency of a signal – part 1: Fundamentals. Proc. IEEE **80**(4), 520–538 (1992)

Breakspear, M., Terry, J.: Detection and description of nonlinear interdependence in normal multichannel human EEG. Clin. Neurophysiol. **113**, 735–753 (2002)

Bressler, S.: Interareal synchronization in the visual cortex. Beh. Brain Res. **76**, 37–49 (1996)

Castelo-Branco, M., Neuenschwander, S., Singer, W.: Synchronization of visual response between the cortex, lateral geniculate nucleus, and retina in the anesthetized cat. J. Neurosci. **18**(16), 6395–6410 (1998)

DeShazer, D., Breban, R., Ott, E., Roy, R.: Detecting phase synchronization in a chaotic laser array. Phys. Rev. Lett. **87**(4), 044101 (2001)

Duda, R., Hart, P.: Pattern Classification and Scene Analysis. Wiley, New York (1973)

Eckhorn, R., Bauer, R., Jordan, W., Brosch, M., Kruse, W., Munk, M., Reitboeck, H.: Coherent oscillations: a mechanism of feature linking in the visual cortex? multiple electrode and correlation analyses in the cat. Biol. Cybern. **60**, 121–130 (1988)

Effern, A., Lehnertz, K., Fernandez, G., Grunwald, T., David P., Elger, C.: Single trial analysis of event related potentials: non-linear de-noising with wavelets. Clin. Neurophysiol. **111**, 2255–2263 (2000)

Engel, A., Koenig, O., Kreiter, A., Singer, W.: Interhemispheric synchronization of oscillatory neuronal response in cat visual cortex. Science **252**, 1177–1179 (1991)

Frien, A., Eckhorn, R., Bauer, R., Woelbern, T., Kehr, H.: Stimulus-specific fast oscillations at zero phase between visual areas v1 and v2 of awake monkey. Neuroreport **5**, 2273–2277 (1994)

Gratton, G., Coles, M., Donchin, E.: A procedure for using multi-electrode information in the analysis of components of the event-related potential: vector filter. Psychophysiology. **26**, 222–232 (1989)

Grindvald, A., Arieli, A., Tsodyks, M., Kenet, T.: Neuronal assemblies: single cortical neurons are obedient members of a huge orchestra. Biopolymers **68**, 422–436 (2002)

Haig, A., Gordon, E., Wright, J., Meares, R., Bahramali, H.: Synchronous cortical gamma-band activity in task-relevant cognition. Neuroreport **11**, 669–675 (2000)

Hutt, A.: Methoden zur Untersuchung der Dynamik raumzeitlicher Signale. MPI Series in Cognitive Neuroscience, vol. 15. PhD thesis. Max Planck-Institute of Cognitive Neuroscience, Leipzig (2001).

Hutt, A.: Spatiotemporal modelling of EEG/MEG. Brain Topogr. **14**(4) (2002)

Hutt, A.: An analytical framework for modeling evoked and event-related potentials. Int. J. Bif. Chaos **14**(2), 653–666 (2004)

Hutt, A., Daffertshofer, A., Steinmetz, U.: Detection of mutual phase synchronization in multivariate signals and application to phase ensembles and chaotic data. Phys. Rev. E **68**, 036219 (2003)

Hutt, A., Munk, M.H.: Mutual phase synchronization in single trial data. Chaos Complex. Lett. **2**(2), 6 (2006)

Hutt, A., Riedel, H.: Analysis and modeling of quasi-stationary multivariate time series and their application to middle latency auditory evoked potentials. Phys. D **177**, 203 (2003)

Hutt, A., Schrauf, M.: Detection of transient synchronization in multivariate brain signals, application to event-related potentials. Chaos Complex. Lett. **3**(1), 1–24 (2007)

Hutt, A., Svensen, M., Kruggel, F., Friedrich, R.: Detection of fixed points in spatiotemporal signals by a clustering method. Phys. Rev. E **61**(5), R4691–R4693 (2000)

Hutt, A., Uhl, C., Friedrich, R.: Analysis of spatio-temporal signals: a method based on perturbation theory. Phys. Rev. E **60**(2), 1350–1358 (1999)

Ioannides, A., Kostopoulos, G., Laskaris, N., Liu, L., Shibata, T., Schellens, M., Poghosyan, V., Khurshudyan, A.: Timing and connectivity in the human somatosensory cortex from single trial mass electrical activity. Hum. Brain Mapp. **15**, 231–246 (2002)

Karjalainen, P.A., Kaipio, J.P.: Subspace regularization method for the single-trial estimation of evoked potentials. IEEE Trans. Biomed. Eng. **46**(7), 849–859 (1999)

Koch, C.: Biophysics of Computation. Oxford University Press, Oxford (1999)

Kuramoto, Y.: Chemical Oscillations, Waves, and Turbulence. Springer, Berlin (1984)

Knig, P., Engel, A., Singer, W.: Relation between oscillatory activity and long-range synchronization in cat visual cortex. Proc Natl Acad Sci USA **92**, 290–294 (1995)

Lachaux, J.P., Rodriguez, E., Martinerie, J., Varela, F.: Measuring phase synchrony in brain signals. Hum. Brain Mapp. **8**, 194–208 (1999)

Laskaris, N., Ioannides, A.: Semantic geodesic maps: a unifying geometrical approach for studying the structure and dynamics of single trial evoked responses. Clin. Neurophysiol. **113**, 1209–1226 (2002)

Le Van Quyen, M., Foucher, J., Lachaux, J., Rodriguez, E., Lutz, A., Martinerie, J., Varela, F.: Comparison of hilbert transform and wavelet methods for the analysis of neuronal synchrony. J. Neurosci. Methods **111**(2), 83–98 (2001)

Lehmann, D., Skrandies, W.: Reference-free identification of components of checkerboard-evoked multichannel potential fields. Electroenceph. Clin. Neurophysiol. **48**, 609 (1980)

Liu, F., Hu, B., Wang, W.: Effects of correlated and independant noise on signal processing in neuronal systems. Phys. Rev. E **63**, 031,907 (2001)

Makeig, S., Jung, T., Bell, A., Sejnowski, T.: Blind separation of auditory event-related brain responses into independent components. Proc. Natl. Acad. Sci. **94**, 10979–10984 (1997)

Makeig, S., Westerfield, M., Jung, T., Enghoff, S., Townsend, J., Courchesne, E., Sejnowski, T.: Dynamic brain sources of visual evoked responses. Science **295**, 690–694 (2002)

Mardia, K., Jupp, P.: Directional Statistics. Wiley, New York (1999)

Nelson, J., Salin, P., Munk, M., Arzi, M., Bullier, J.: Spatial and temporal coherence in cortico-cortical connections: a cross-correlation study in areas 17 and 18 in the cat. Vis. Neurosci. **9**, 21–37 (1992)

Nowak, L., Munk, M., Nelson, J., James, A., Bullier, J.: Structural basis of cortical synchronization. i. three types of interhemispheric coupling. J. Neurophysiol. **74**, 2379–2400 (1995)

Pikovsky, A., Rosenblum, M., Kurths, J.: Phase synchronization in regular and chaotic systems. Int. J. Bif. Chaos **10**(10), 2219 (2000)

Pikovsky, A., Rosenblum, M., Kurths, J.: Synchronization: a universal concept in nonlinear sciences. Cambridge University Press, Cambridge (2001)

Prusseit, J., Lehnertz, K.: Stochastic qualifiers of epileptic brain dynamics. Phys. Rev. Lett. **98**, 138,103 (2007)

Quian Quiroga, R., Garcia, H.: Single-trial event-related potentials with wavelet denoising. Clin. Neurophysiol. **114**, 376–390 (2003)

Salin, P., Bullier, J.: Corticocortical connections in the visual system: structure and function. Physiol. Rev. **75**, 107–154 (1995)

Salinas, E., Sejnowski, T.: Correlated neuronal activity and the flow of neural information. Nat. Rev. Neurosci. **2**, 539–550 (2001)

Shahl, A., Bressler, S., Knuth, K., Ding, M., Mehta, A., Ulbert, I., Schroeder, C.: Neural dynamics and the fundamental mechanisms of event-related brain potentials. Cereb. Cortex **14**, 476–483 (2004)

Singer, W.: Neural synchrony: a versatile code for the definition of relations? Neuron **24**, 49–65 (1999)

Stam, C., Dijk, B.: Synchronization likelihood: an un-biased measure of generalized synchronization in multivariate data sets. Phys. D **163**, 236–251 (2002)

Steinmetz, P., Roy, A., Fitzgerald, P., Hsiao, S., Johnson, K., Niebur, E.: Attention modulates synchronized neuronal firing in primate somatosensory cortex. Nature **404**, 187–190 (2000)

Tallon-Baudry, C., Mandon, S., Freiwald, W., Kreiter, A.: Oscillatory synchrony in the monkey temporal lobe correlates with performance in a visual short-term memory task. Cereb. Cortex **14**, 713–720 (2004)

Tass, P., Rosenblum, M., Weule, J., Kurths, J., Pikovsky, A., Volkmann, J., Schnitzler, A., Freund, H.J.: Detection of n:m phase locking from noisy data: application to magnetoencephalography. Phys. Rev. Lett. **81**(5), 3291–3294 (1998)

Denoising and Averaging Techniques for Electrophysiological Data

Matthias Ihrke, Hecke Schrobsdorff, and J. Michael Herrmann

1 Introduction

Neurophysiological signals are often corrupted by noise that is significantly stronger than the signal itself. In electroencephalographic (EEG) data this may amount to figures of -25 dB (Flexer, 2000), for electromyography (EMG) or functional magnetic resonance imaging (fMRI) the situation is similar. The problem of the recovery of information under noise has been dealt with extensively in the literature of signal and image processing (Whalen, 1971; Castleman, 1996).

The most basic method to improve noisy data is averaging of several measurements of the same signal where the supposedly independent influences of noise are expected to annihilate. Although nowadays more advanced algorithms are available, such pointwise averages over a number of realizations of the same process are still very common in applied research studies. When considering, e.g., the relation of electrophysiological data and cognitive processes by analyzing event-related potentials (ERPs) one relies on the assumption that concomitant ('ongoing') activity is performed at independent time shifts relative to the relevant signal such that averaging progressively reveals the signal from the unwanted concurring processes.

Although mean values often provide robust estimates of underlying parameters even under strong noise, they stand and fall with the validity of the above independence assumption. Of similar importance is the existence and invariance of the moments of the noise distribution, which is far from obvious for real data. To clarify these shortcomings and to propose methods for compensation, we will describe these assumptions in some detail and propose a model that includes temporal variance in the generation of single-trial ERPs.

In order to enable the reader to gain practical experience we refer to an implementation of many of the reviewed algorithms in the software library libeegtools. This library is implemented in C and is available for download, together with an

M. Ihrke (✉)
Bernstein Center for Computational Neuroscience Göttingen, Bunsenstrasse 10, 37073 Göttingen, Germany
e-mail: mihrke@gwdg.de

J.L. Perez Velazquez, R. Wennberg (eds.), *Coordinated Activity in the Brain*, Springer Series in Computational Neuroscience 2, DOI 10.1007/978-0-387-93797-7_9,

extensive documentation, on the book's webpage. Two data sets, one comprising artificial data and the other one consisting of real EEG data, are provided for testing purposes as well. The real data reported in this chapter were obtained in a study featuring a picture-identification task (for details see Ihrke, 2007) where subjects had to compare one of two overlapping pictograms with a written word and responses were given by pressing one of two buttons. The artificial data were generated according to the VCPN model introduced in Sect. 3 (Eq. (7)). The script that generated this data is part of the implementation.

In the remainder of this chapter we will discuss a concept of noise that underlies our work with electrophysiological data. After this, we will discuss descriptive models of the generation of ERPs. Finally, we will outline a method for the integration of external knowledge about the temporal processing of a single trial, i.e., points in time for which the state of the neurophysiological response is known. The methods and algorithms presented in this chapter are motivated by and typical for the application on EEG data. Natural extensions of the range of applicability will become clear in the course of the argument.

2 Noise in Electrophysiological Data

2.1 A Concept of Noise

The extraction of information from an empirical data set requires a model that specifies what information contains a relevant signal and what can be considered as noise. None of the two complementary approaches, namely either to model the signal or to model the noise, is easily defined. This is because both signal and noise arise from a complex interaction of the dynamics of nervous activity and the environment including the measurement devices as well as any task instructions. In order to be able to extract meaningful information at all we have to combine specifications derived from different criteria such as neurophysiological theories and statistical heuristics. It is to be expected that the selection of such background knowledge as well as the construction of a statistical model that is assumed to underlie the data will affect the quality, reliability and effectiveness of the data analysis.

In cases where the signal is known, noise can be considered as a deviation of a measured signal from the original signal (Whalen, 1971). In EEG data this is typically not the case. Moreover, identifiable signals in EEG data often account for only 10% or less of the variance which further justifies the need for sensitive methods in the analysis of these data.

2.2 Event-Related Potentials

A natural approach to relate neural data to behavior is to collect patterns in the data that reoccur under certain conditions, e.g., at the onset of a stimulus or in relation with other events in the course of the experiment. Controllable correlates that repeat significantly more often than chance level are likely to indicate a representation of the external event in the data. This hypothesis is implicit in *event-related potentials* (ERPs). When the single-trial ERPs are averaged, they yield a characteristic pattern (the *averaged event-related potential*; AERP) that is reproducible in comparable experimental setups (Picton et al., 1995). The reliability of the AERP allows for a classification of the curves based on their component structure (i.e., the latency and amplitude of major minima and maxima). Systematic changes in the AERP components between different experimental conditions support the hypothesis that ERP components do reflect stages of brain computation. In this interpretation, the idealized noise-free ERP represents the signal of interest and variability across trials is considered as noise.

The process of signal generation and the insufficient definition of *noise* has been subject to a lively discussion in the methodological literature (Delorme et al., 2007; Wang et al., 2007; Quiroga and Garcia, 2003; Bartnik et al., 1992; de Weerd, 1981); however, no generally acknowledged solution for signal extraction has been proposed so far (see Sect. 4.1 for a review). Hence most studies that try to relate ERPs to brain processes use methodology based on averaging that assumes a very simple noise model. This model can be captured in the following simplifying major assumptions (see also Sect. 3):

(i) The EEG signal contains the relevant components of the brain activity.
(ii) Activations that are specific to the investigated task form a significant though possibly small fraction of the brain activity.
(iii) The brain solves a similar task in a similar way.

The third item does not imply identical spike pattern each time the task is repeated, but excludes the possibility that two variants of the process cancel each other by the average. According to these assumptions the signal of interest can be defined as a minimal variance curve among many repetitions of the same task. The assumptions also imply that variations due to external conditions should be excluded. It is further suggested that the external conditions and even the state of the subject should be kept as constant as possible for all trials.

As the current task is embedded in the so-called *ongoing activity* of the brain (Arieli et al., 1996), there are definitely interactions with the task-specific subprocesses. This circumstance points out that there is no *original signal* to be measured when looking from the outside. Rather ERPs can be thought of as modulations of the spontaneous activity. Every single instance of task execution has its individual neural pattern of processing while producing similar patterns of population activity. Only if we would understand the interplay of the different subprocesses fulfilling the task and their interactions with the ongoing activity, we could try to identify neural correlates of these subprocesses. Such a general insight is improbable in the near future, therefore great care should be taken to select a plausible model that differentiates between signal and noise. The model of choice should be simple, robust and allow to identify as many systematic contributions to noise as possible and cancel stochastic variations.

2.3 Sources of Noise

When proposing a useful definition of noise, there emerges a division on a conceptual level into two types of noise-generating processes that can not directly be influenced: a modification of the original signal by an overlay of activity from other sources uninteresting to the observer on the one hand and intrinsic stochastic behavior of the neural signal source itself on the other hand. Both classes are subsumed as noise. The theoretical border between these two classes of signal disturbance is clear, but on a closer look, it becomes more diffuse as we generally do not know what the original source is that we are observing. The more insight we have into the way the brain computes, the more the above classification boundary is shifted as we identify more sources of irrelevant signals. Therefore a radical view is that noise is any activity from unknown sources while all known sources produce signals (Celka et al., 2006).

Some known sources of disturbing signals are clearly found on the level of recording circumstances like the amplifier and changing conductances of electrodes (Picton et al., 1995). Each of these known sources can be addressed specifically with optimization of the recording environment or special data processing like a 50 Hz notch filter. Furthermore, artifacts from muscular activity (most importantly eye movements) can be diminished (e.g., Delorme et al., 2007; Croft et al., 2005). But when looking at the ongoing activity of the brain, it is hopeless to describe this source of noise precisely enough to derive filtering techniques. Eliminating variance coming from the ongoing activity by improving the recording environment or by application of a simple filter is not possible as their origin is identical to the one of our signal of interest, i.e., the cortex.

The spiking behavior of individual neurons is stochastic (Celka et al., 2006). Only when looking at larger numbers of neurons or recording for longer time intervals a clear pattern becomes visible. But spikes of a single neuron can not be predicted, just like the radioactive spontaneous disintegration of a single atom. There is no determinism in the brain, therefore no filtering technique exists to eliminate stochasticity as a source of noise; we have to apply statistical tools to deal with the intrinsic noise.

Going back to our example of ERPs, investigations with data mining techniques reveal that only a fraction of 60% of the pooled epochs contribute to the AERP waveform while the other 40% just increase the variance (Haig et al., 1995). Thus even very carefully tuned experimental conditions do not necessarily lead to a reliable activation pattern. Whether this nondeterministic nature of ERP epochs has its origin in the stochastic nature of single neuron events, or in the interplay with the overall ongoing activity which is not normalized across trials, or simply in the application of different strategies to solve the task is still unknown. Therefore a careful consideration of the recorded brain activity and the realized experimental situation is needed.

Noise when properly defined, is impossible to eliminate. Nevertheless brain recording techniques like the EEG are an invaluable way to get information about brain processes. Therefore we have to aim at a detailed understanding how the signal we record is generated and to exclude as many unwanted signal sources as we can.

2.4 Strategies to Handle Noise

The majority of studies investigating electrophysiological data still use data analysis techniques like the above mentioned averaging process. The current discussion about EEG correlates is based on results obtained by averaging and thus a terminology biased toward the above assumptions has evolved. However, there are some disadvantages in the averaging approach. For example, the late positive complex is often observed and subject to theoretical speculation on its role, but it is entirely unknown whether it is a brain correlate or just an average of different maxima in the time series that are subject to huge temporal distortions between trials. Especially when considering late components of the ERP, it is possible that averaging *smears out* relevant components (see also Fig. 2).

An existing approach to surmount the shortcomings of data averaging over trials is given by data mining tools. Machine learning algorithms generate a classification of data segments, algebraic methods like the independent component analysis (ICA) reduce the dimensionality of the data by identifying data prototypes that account for most of the variance. In EEG research the ICA has proven to be successful in data cleaning (Dodel et al., 2000; Delorme et al., 2007). Unfortunately, data mining tools are computationally very costly and not well integrated in commonly used software, so that using these techniques requires higher skills in computer programming and access to powerful computing facilities.

In conclusion, applied research focuses on the application of averaging as the main tool because of its simplicity and robustness. By reason of its enormous importance in applied research, we will describe the assumptions underlying averaging schemes more explicitly in the following section. We will show that important information about systematic contributions to the noise can be missed and propose a straightforward extension to incorporate temporal variance. We will then review some of the methodology that is available for data cleaning, clustering, and averaging. Finally, extending the methodology developed in these sections, we will propose a way to incorporate *external* knowledge (i.e., time markers) about the time course of the ERP into the averaging process.

3 Models for Event-Related Potentials

In this section, we will make explicit the assumptions underlying average-based ERP research. Since EEG data are contaminated with strong noise in the sense discussed above, the signal-to-noise ratio (SNR) is typically enhanced by combining data epochs that are supposed to contain a certain signal of interest with a pointwise average

$$\langle s_i(t)\rangle_i = \frac{1}{N}\sum_{i=1}^{N} s_i(t) \qquad i = 1, \ldots, N. \tag{1}$$

Here and in the following $s_i(t)$ is the measured signal in the ith trial at time t. For notational simplicity, the nomenclature chosen here uses functional notation $s(t)$ even though discrete data $s(1), \ldots, s(n)$ is often referred to where n is the number of samples. The implicit model underlying averaging assumes that

(i) signal and noise add linearly together,
(ii) the signal is identical in all trials, and
(iii) noise is a zero-mean random process drawn independently for each trial.

Assuming additive noise of zero mean ($\langle\epsilon(t)\rangle = 0 \forall t$), we can represent the data by the following model:

$$s_i(t) = u(t) + \epsilon_i(t), \tag{2}$$

where $u(t)$ denote the signal that is to be recovered from s. This model is known as *signal-plus-noise* (SPN) model (Truccolo et al., 2002) or *fixed latency* model (de Weerd, 1981). Here simplicity of the ansatz is preferred to realism with respect to the underlying physical situation. Assuming that Eq. (2) applies, the pointwise average is an unbiased and optimal estimate in the mean square error sense. This follows from

$$\langle s_i(t)\rangle_i = \langle u(t) + \epsilon(t)\rangle = u(t) + \langle\epsilon(t)\rangle_i,$$

since $u(t)$ does not depend on i. Thus assuming $\langle\epsilon(t)\rangle = 0$ we have

$$\langle s_i(t)\rangle_i = u(t) \tag{3}$$

for large N, where the error behaves asymptotically as

$$\epsilon(t) \sim \frac{\sigma^2(t)}{N}, \tag{4}$$

with the instantaneous variance $\sigma^2(t)$ being the expectation of $|\langle s(t)\rangle_i - u(t)|^2$. Therefore, averaging over a sufficient number of trials eliminates the noise with the rate from Eq. (4), leaving the constant signal intact. It has been argued on theoretical grounds, that an improvement beyond pointwise averaging is not possible if no a priori knowledge about the characteristics of signal and noise is given (Nagelkerke and Strackee, 1979). The controversy about this assertion led to the insight that an improvement beyond Eq. (4) is possible because a curve is to be found that represents the overall temporal structure of the data (de Weerd, 1981). This can be achieved by either exploiting correlations among the noise processes or by improving the estimator through taking into account neighboring data. While the former requires a complex noise model which can hardly be justified from what is known about the data, the latter is more reasonable if u changes slowly compared to the discretization step of time.

Although the SPN model still guides most ongoing research, its applicability to ERP data has been questioned (Truccolo et al., 2001) on grounds of a number of theoretical and empirical arguments that militate against the assumption of a stationary response: The repetition of a task can be accompanied by different neural activity, either because of setting-dependent (e.g., slightly different displays in the same experimental condition) or subject-dependent variations (e.g., growing tiredness, learning, arousal). Also, variations in the brain states depend on what happened before trial processing and thus can differentially influence the shape of the evoked response. Another, empirical, argument against the stationarity of u comes from the analysis of the residuals ζ_i^{avg} obtained by subtracting the mean from the raw data

$$\zeta_i^{\text{avg}}(t) = s_i(t) - \langle s_i(t) \rangle_i .$$

Given the SPN model, ζ_i^{avg} should not contain any event-related modulation because the noise is assumed to be independent and identically distributed. Therefore statistical coherence measures such as the *auto-correlations* $(\zeta_i^{\text{avg}} \star \zeta_i^{\text{avg}})(\tau) = \int \zeta_i^{\text{avg}}(t)\zeta_i^{\text{avg}}(t+\tau)dt$ and the *power-spectral densities* $\text{PSD}\,(\zeta_i^{\text{avg}}) = \mathscr{F}\{\zeta_i^{\text{avg}} \star \zeta_i^{\text{avg}}\}$ computed on ζ_i^{avg} should not show any event-related modulation (i.e., a flat spectrum and cross-correlations that behave like a delta-function at zero are to be expected). Empirical evidence shows that these assumptions are violated for real data (Truccolo et al., 2001, 2002; see also Fig. 1).

The assumption of nonstationarity of the signal is particularly important for data from tasks where cortical activity is involved. While brainstem potentials show relatively stable characteristics and are thus well described by Eq. (2), *cortical* activity may show considerable trial-to-trial variability (de Weerd, 1981; see also Fig. 5).

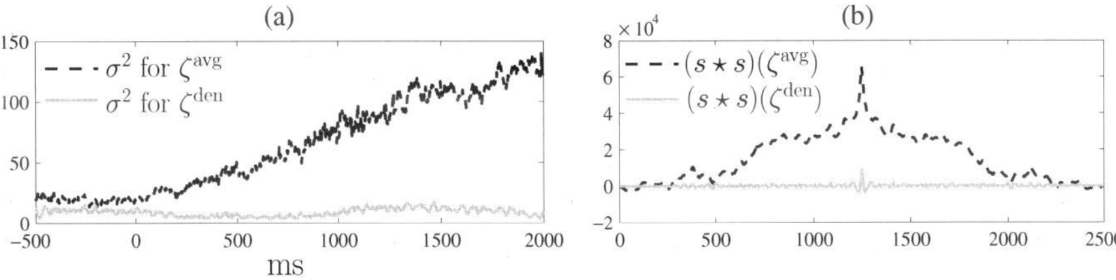

Fig. 1 Coherence measures computed on residuals ζ^{avg}. (**a**) The variance σ^2 over trials shows event-related modulation for the residuals after subtracting the average. Given the SPN model, we expect a flat curve as obtained from computing σ^2 on the single-trial denoised residuals ζ_i^{den} that are obtained as discussed in Sect. 4.1. (**b**) Cross-correlation computed on the residuals for a sample trial. Again, unexpected (from SPN) correlations show up for ζ_i^{avg} whereas the function approximates a delta-function for the single-trial denoised residuals

This makes the applicability of the SPN more problematic as important systematic contributions to the noise may be missed.

A popular extension of the SPN model, the *variable latency model* (VLM) was implicitly assumed by Woody (1967). This model introduces a realization-dependent but constant time lag τ_i by which the evoked potential can be shifted in time as well as a trial-constant scaling factor α_i

$$s_i(t) = \alpha_i u(t + \tau_i) + \epsilon_i(t). \tag{5}$$

It was suggested that a cross-correlation technique was an effective means to estimate the individual τ_i. Using this method, the data can be transformed by shifting the data in individual trials by τ_i that are given as the maximum argument of the *cross-correlation* function

$$\tau_i = \arg\max_t (\upsilon \star s_i)(t) \tag{6}$$

of the trial data and a template $\upsilon(t)$ (e.g., the pointwise average). After that transformation, the data can be interpreted according to the SPN model from Eq. (2). To take care of the scaling factors α_i, the cross-correlation in Eq. (6) is computed on standardized data. The simplicity of this model motivated an analytical treatment to derive predictions for the behavior of statistical measures on the residuals (Truccolo et al., 2002). These authors could show, that the application of the VLM model resulted in a more plausible behavior of some statistics (e.g., behavior of the variance over time) calculated on the residuals. However, they also found patterns that were not consistent with the predictions and it was therefore argued that a more general model would be desirable.

Years earlier Ciganek (1969) showed that the inter-trial variability of the evoked potential can go beyond the simple time shift assumed in Eq. (5) and that individual ERP components (e.g., N100, P300) can be shifted and scaled independently of one another. Therefore, these shifts should be accounted for by designing an appropriate model to base on techniques for their identification. A very general model for an arbitrary shift of the individual data points (although preserving their temporal order) can be formulated as

$$s_i(t) = \alpha_i(t) u(\phi_i^{-1}(t)) + \epsilon(t) \tag{7}$$

where ϕ_i are monotonous functions that map the timescale of the individual trials to that of a template υ (i.e., $||u_\upsilon(t) - u_i(\phi_i(t))||$ is minimal) and α_i are positive functions that indicate the scale of different parts of the curve. It is assumed that a more realistic fit to empirical data from neuropsychological experiments could be achieved by this approach, because an overall latency shift of the complete signal as modeled in the VLM is probably an oversimplification for tasks that involve higher-order cortical activity. The true signal $u(t)$ is thereby assumed to show local variation of the latencies through interaction of noise and signal such that the ERP

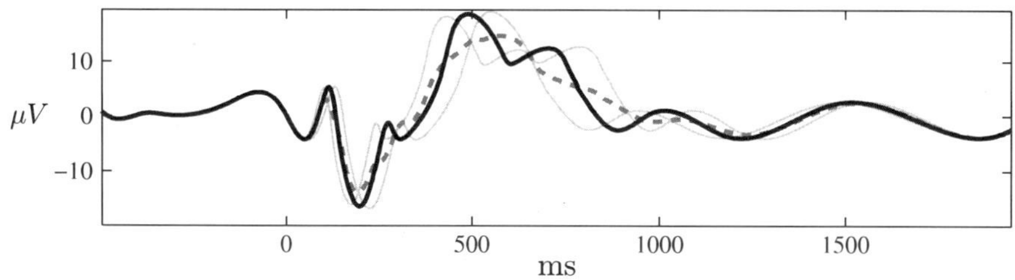

Fig. 2 Smearing of components in the simple average due to temporal variance. Two input signals (*light-gray curves*) were simulated according to the model in Eq. (7). The pointwise average (Eq. (1)) *smears out* some components that are temporally delayed (*dashed curve*). An averaging procedure incorporating temporal variance would produce a curve similar to the *black* one

components of u can be differentially shifted and scaled. Therefore the model will be henceforth referred to as *Variable Components Plus Noise* model (VCPN).

From the standpoint of this model a shortcoming of classical averaging becomes apparent (see Fig. 2). Because identical components may be delayed in time, a pointwise average smears out the components thereby producing a shape of the ERP curve that is distinct from the general features visible in the single-trial ERPs and therefore hampers an optimal interpretation of the curve. An averaging procedure following the VCPN model would average both in scale and in temporal variance to preserve the single-trial ERP's features.

In summary, the VCPN model given in Eq. (7) has the advantage to model temporal variations of the individual signals in addition to differences in scale. The model keeps advantages of the SPN and VLM models as it is a straightforward generalization that keeps both of these models as special cases. This opens up the possibility to find systematic distortions due to temporal fluctuations that otherwise went unnoticed. Of course, the general form of Eq. (7) makes a treatment of the signal estimation more difficult and liable to arbitrariness. Clearly, some theoretically plausible restrictions on the VCPN must be introduced by methods of model selection.

A conceptually inspiring approach is given by minimal description length (MDL) methods (Rissanen, 1978, 2007), where the complexity of the model is compared to the reduction that is achieved when the data are described in terms of the model. Consider as an example the Mandelbrot set that is described by the solubility of a single equation. A set of random numbers on the other hand, cannot be reduced beyond the specification of the underlying distribution. If the data contain deterministic and quasi-random components, MDL provides a scheme that optimizes the complexity of the model. While it is no problem to quantify the data in bits, the specification of the model is often ambiguous, in particular when compared to the data. It might be therefore more convenient to use Akaike's information criterion (AIC; Akaike, 1974) that compares the number of essential model parameters with the likelihood of the model for a given data set.

Although we could vary the number of the parameters in the present model, the calculation of the likelihood function requires further assumptions that we can avoid by using cross-validation which provides a more practical approach to model selection. Cross-validation (Geisser and Eddy, 1979) allows to estimate the *generalization error* of a model by training and testing the model on separate parts of the data. The prediction error with respect to the *test set* (averaged over all possible choices of training and test set, respectively) gives an indicator about how well the model generalizes and thus about its validity.

4 Methods for Signal Estimation

In the following, some methods for a better approximation of the signal will be discussed. First, algorithms to extract the signal from a single realization of an experiment will be presented, followed by data mining methods that try to group clusters of distinct ERPs. Finally, some alternative averaging methods that rely on these algorithms for signal extraction and grouping will be discussed.

4.1 Single-Trial Estimation

Of course the best way to learn about event-related modulation of synchronized brain activity is to extract that information from a single realization of the event. Such an approach has several advantages. First, the discussion about stationarity or nonstationarity of the signal is irrelevant, since no averaging takes place. Second, higher efficiency can be achieved by significantly lowering the number of trials for a single experimental condition. Finally, the extraction of single-trial ERPs allows to investigate ERPs even in situations in which only a single (or very few) realizations are available. Therefore, this methodology bears the potential to be very attractive for a variety of practical applications (e.g., medical applications, brain–computer interfaces).

Due to the enormous potential benefit of a reliable method for signal extraction from single-trial ERP data, various approaches for this task have been proposed in the literature (de Weerd, 1981; Wang et al., 2007; Flexer et al., 2001). Extracting the single-trial ERP from a noisy signal is equivalent to denoising a signal in a way that preserves the relevant information. One simple way, the application of a bandpass filter with fixed cut-off frequencies (i.e., suppressing the frequencies above and below the cut-offs) does not at all guarantee that relevant aspects of the signals are going to be preserved, and therefore requires a very sensitive handling of the cut-off frequencies and a thorough specification of the underlying assumptions. It is generally argued (mostly from empirical experience), that processes that reflect cognitive activity are located in a specific frequency band (e.g., between 0.1 and 30 Hz). However, as discussed above, some noise sources provide disturbances in the same frequency bands, and bandpass filtering can therefore only remove those parts of the noise that are clearly distinct in terms of their frequency distribution. Also, since filtering in the Fourier domain is well covered in a number of books on filter design (Stearns, 1990), we will now focus on other methods that have been proposed more recently.

An approach to improve the SNR of a pointwise average has been the application of *a posteriori Wiener filtering* to the recorded episodes (de Weerd, 1981; de Weerd and Kap, 1981a). Basically, a classical Wiener filter is a time-invariant bandpass filter with a transfer function H that adapts to the power distribution of the input signal

$$H(\omega) = \frac{\Phi_u(\omega)}{\Phi_u(\omega) + \frac{1}{N}\Phi_\epsilon(\omega)}, \tag{8}$$

where Φ_u and Φ_ϵ are the power density spectra of signal and noise, respectively (i.e., $\Phi_u = |\mathscr{F}\{u\}|^2$, where $\mathscr{F}\{u\}$ is the Fourier transform of u). These two spectra can be estimated from the ensemble average and the alternate average $\frac{1}{N}\sum_{i=1}^{N}(-1)^{(i-1)}s_i(t)$ respectively. This estimate is based on the SPN model (for details see de Weerd and Kap, 1981a) and therefore shares the problems associated with it.

The filter from Eq. (8) can be extended to a time-varying version by introducing a time dependency using a spectro-temporal representation of the signal. This approach is superior to choosing a constant filter function due to the complex transient character of the ERP curve. However it poses the fundamental problem in that time and frequency are inversely related quantities, so that a compromise must be found: the higher the frequency resolution (i.e., applying more bandpass filters that are narrower), the lower the temporal resolution (e.g., de Weerd and Kap, 1981b). A possible implementation of this method uses a bank of bandpass filters with different bandwidths and center frequencies (de Weerd and Kap, 1981b) and is thus easily and efficiently implemented.

In practice, the time-varying Wiener filter proves to be superior to the time-invariant version and produces good results for artificial data constructed following the SPN model, even in presence of realistically low SNRs (−25 dB). However, as stated above, a problem inherent to this approach is that it assumes the SPN model or at least the homogeneity of the stochastic processes generating signal and noise. Since the optimal passband is chosen based on the signal and noise spectra derived from averaging, it is not clear whether these are indeed optimal for the single trials. The spectra might differ for single trials, e.g., because of the presence of muscular artifacts.

Robust, detail-preserving signal extraction is typically carried out by applying robust central-tendency statistics (e.g., the median) to the measured signal. Each time point of the estimated signal $\hat{s}(t)$ is approximated by some function of the measured signal in a time window $\omega_t = \{t-w, \ldots, t+w\}$ surrounding t, i.e., $\hat{s}(t) = f(\{s(\tau)|\tau \in \omega_t)\})$. An exemplary function f is the *standard median filter* (Tukey, 1977) approximating the signal as the median of the amplitudes in ω_t $\mathrm{MF}[s(t)] = \mathrm{median}\{s(\tau)|\tau \in \omega_t)\}$. The main advantage of this approach is the good preservation

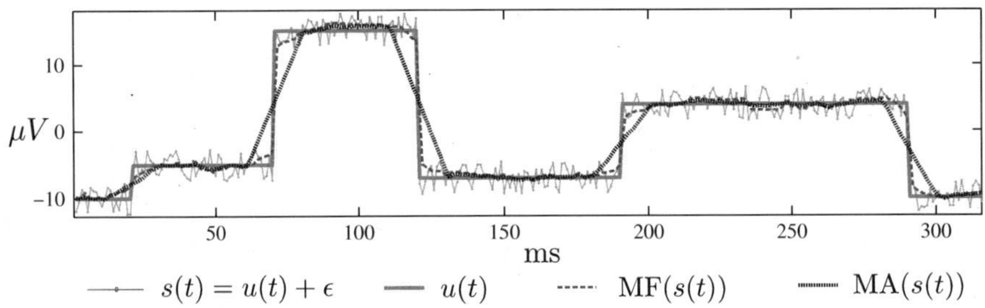

Fig. 3 Median-based filtering techniques. The running median filter (*dashed*) preserves level shifts, while the moving average filter (*fine-dashed*) blurs sharp edges. Preservation of level shifts is important, e.g., for saccade extraction from the EOG

of level shifts in the data as opposed to moving average low-pass-filters that blur sharp edges (see Fig. 3). To preserve shifts with a duration of $k+1$ samples, a running median with window size $w = k$ is appropriate. The assumption underlying this approach is local constancy of the signal (Gather et al., 2006). This assumption is not fully justifiable in the case of cortical EEG data but it can be a useful approach in other scenarios, e.g., for the identification of saccades from electroencephalographic (EOG) data, which are recorded as level shifts (Marple-Horvat et al., 1996).

A remedy for signals that are not locally constant is to design f such that sample points are weighted according to their distance from t. The *weighted median filter* is introduced by defining an arbitrary positive weighting function $w(t, t')$ that assigns a weight to the samples $s(t')$. The weighted median (WM) is the value that minimizes

$$\mathrm{WM}\,(s(t)) = \arg\min_{\mu} \sum_{t' \in \omega_t} w(t, t')|s(t') - \mu|. \tag{9}$$

Replacing each observation $s(t)$ by the weighted median WM($s(t)$) in the time window ω_t gives the weighted median filter (WMF). An efficient algorithm to compute WM ($s(t)$) for each t can be implemented over the sequence $\hat{\omega}_t$ of values from ω_t ordered such that the samples are in ascending order, i.e., $s(\hat{\omega}_t^1) \leq \cdots \leq s(\hat{\omega}_t^N)$ by determining

$$k = \max\left\{ j : \sum_{i=j}^{\#\omega_t} w(t, \hat{\omega}_t^i) \leq \frac{1}{2} \sum_{i=1}^{\#\omega_t} w(t, \hat{\omega}_t^i) \right\} \tag{10}$$
$$\text{then} \quad \mathrm{WMF}\,[s(t)] = s(\hat{\omega}_t^k).$$

In conclusion, robust filtering methods are well suited for the preservation of level shifts as, e.g., present in EOG data. For ERPs their benefit is less clear since a smooth shape is generally considered to be a typical feature of an ERP.

Recently, *wavelet-based techniques* for ERP estimation have been investigated (e.g., Quiroga and Garcia, 2003; Quiroga, 2000; Bartnik et al., 1992). In a comparative study, Wang et al. (2007) directly compared wavelet-based approaches with Wiener filters, least mean squares, and recursive least squares techniques in an animal experiment using direct cortical recordings. Wavelet-based methods showed a general advantage over the competing approaches (see also Fig. 4).

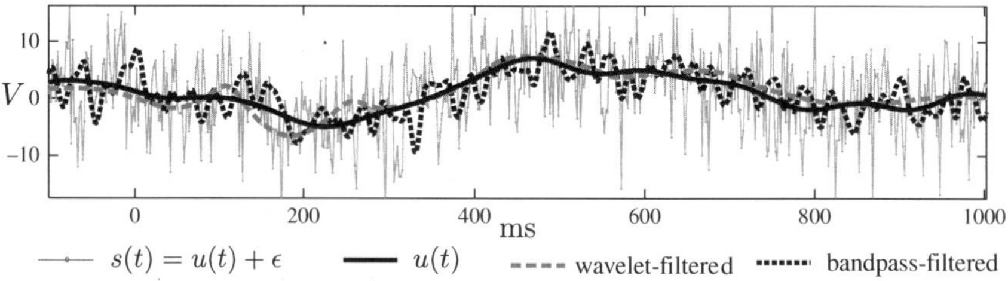

Fig. 4 Comparison of bandpass- and wavelet-filter techniques. The wavelet-filtered curve (*dashed-gray*) approaches the original signal (*solid black*) much more closely than the bandpass filter (*dashed black*) with passband 0.5Hz $< \omega <$ 20 Hz without the need to specify cut-off frequencies because the threshold is estimated from the data

Similar to the Fourier transform, which decomposes a function into bases of sines and cosines, the wavelet transform produces a function representation in a basis of simple functions. Unlike the Fourier transform, however, the wavelet transform $\mathscr{W}$ is not limited to sines and cosines, but decomposes a function s into bases of scaled and shifted functions of a mother function, commonly referred to as the *mother wavelet* ψ (Daubechies, 1992):

$$\mathscr{W}s(a,b) = \int_{-\infty}^{+\infty} s(t)\psi_{a,b}(t)dt \qquad \text{where} \qquad \psi_{a,b}(t) = \frac{1}{\sqrt{a}}\psi\left(\frac{t-b}{a}\right). \quad (11)$$

Restrictions on ψ involve its square and absolute integrability as well as the requirement that the function integrates to zero, thus ensuring a tight localization in time and frequency space. For discrete signals it can be shown that the wavelet transform is equivalent to passing the signal through a *filter bank*, filtering and downsampling the signal in successive steps to obtain wavelet coefficients at different *resolution levels* (Jansen, 2001).

Given an appropriate choice of ψ, it can be assumed that the representation of a signal after the wavelet transform is sparse and a thresholding in the wavelet domain follows naturally. It has been suggested that for EEG data the mother wavelets from the Daubechies family (Daubechies, 1992) are well suited (Quiroga, 2000), because they show a structural similarity to the expected shape of the ERP. Because the wavelet transform is a linear transformation, white noise gets transferred into white noise (i.e., distributed over a wide range of coefficients of relatively small magnitude). Because of the multi-resolution characteristic of the discrete wavelet transform, an *optimal* threshold can be determined separately for the resolution levels, thereby keeping important features on different scales. Several heuristics for finding this threshold have been proposed (Donoho and Johnstone, 1995; Jansen, 2001; Johnstone and Silverman, 1997), most of them based on the distribution of the coefficients on the individual timescales.

When applying wavelet-based filtering techniques to ERP data, care should be taken to set the parameters to reasonable values, since the filtering performance is very sensitive to changes in its main parameters (for a study investigating different parameter settings with randomized input signals, see Taswell, 2001). Our experiments on simulated ERP data involving randomized signals (Ihrke, 2008) showed that for low SNRs as encountered in cortical ERPs a translation invariant filter (Wang et al., 2007) proved to be most successful. Generally, single-trial estimation produced good results in a number of studies and bears strong potential in terms of applicability in various settings. However, given the extraordinarily low signal-to-noise ratios commonly encountered in EEG research even elaborate signal-extraction methods meet their limits. Single-trial estimation in the context of ERP research should therefore be seen as an important preprocessing step to increase single-trial SNRs. However, it should not be used for interpretation as brain correlates because of their sensitivity to disturbing influences.

4.2 Classification Methods

As discussed above, it is relatively unlikely that the ERP remains constant over different trials, i.e., $u_i(t) \approx u_j(t)$ for $i \neq j$. The variability of the signal can be induced by different causes. There are ERPs that differ slightly in general shape because of the stochastic nature of spike patterns underlying the measured electrophysiological response. Another cause of differences is the way how the subject processed the stimulus (e.g., due to a different cognitive strategy or situational variable). Figure 5 illustrates this situation. While most of the trials look relatively similar, some clearly diverging trials (e.g., missing N2) are present. Also, a general shift in the pattern with growing number of trial is observable (P2/N4 amplitude). This could be due to increased tiredness (e.g., Boksem et al,. 2005, 2006) at later stages of the experiment or habituation processes.

Therefore, correlating neurophysiological data to cognitive processes is the quest for the possibility of distinguishing or grouping the trials that reflect different cognitive processes from those that merely differ in a stochastic manner. Machine learning algorithms seem to best provide such a classification. The application of unsupervised classification strategies is a useful approach to that problem, whereas supervised learning algorithms are less applicable, since generally nothing is known about the real ERP to give the teacher in supervised learning strategies an objective criterion for correctness of classification. Supervised algorithms, however, can be applied in other situations involving ERP data, for example brain–machine interfaces, where it is known which effect a given brain–electrical pattern should cause (e.g., movement of a computer cursor to the left or to the right).

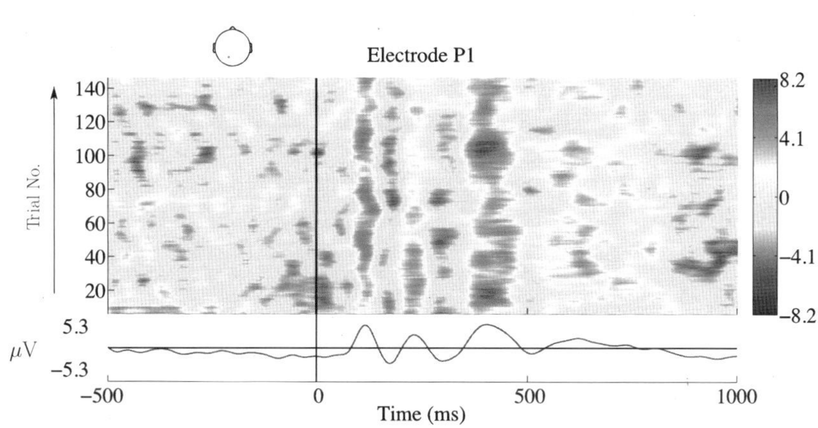

Fig. 5 Real data ERPs for 150 trials (color-coded) and their average (lower part). With growing number of trials (i.e., time spent in the experiment) the shape of the ERP is subject to changes. For example, while the P2 is very pronounced for the first 50 trials, its amplitude is decreased later. Also N2 amplitude decreases over time

Results from unsupervised learning approaches like data mining studies using neural-network classifiers (Masic & Pfurtscheller, 1993; Lange et al., 2000), fuzzy-clustering algorithms (Zouridakis et al., 1997), cross-correlation clustering (Dodel et al., 2002; Voultsidou et al., 2005), or a combined PCA/vector-quantization approach (Haig et al., 1995) support the above notion of how ERPs are generated. For example, Haig et al. (1995) found distinct clusters of ERPs. The shape of the ordinary average was mainly determined by the single-trial ERPs from the largest cluster ($\approx 60\%$). Within each of the clusters, the distances among single-trial ERPs were relatively small. Thus, the authors concluded, that clusters represent different strategies or states of the subject while the ERPs within a cluster are only subject to minor stochastic variation. Given this potential of unsupervised classification techniques to reveal a hidden structure in ERP ensembles, it has been suggested that more researchers should apply such methodology (Handy, 2005).

In the following paragraph, a short overview over the methodology of *cluster analysis* will be given to point to accompanying problems. We will outline the conceptually relatively simple K-means and K-medoids cluster-analysis technique and direct the reader to pertinent literature on methods for unsupervised learning and classification (e.g., Hastie et al., 2001) for a discussion of other approaches.

In order to apply any cluster-analytic algorithm to data, a metric (or at least pairwise distances) must be applied to the objects that are to be grouped (in our scenario the N ERP curves) such that dissimilarities $\Delta(s_i, s_j)$ are provided. The metric to be used is crucial for the performance of the algorithm and should therefore be carefully chosen on the grounds of theoretical considerations. Often an Euclidean metric based on squared distances cumulated over *object features* (here individual points of the ERP) is used

$$\Delta(s_i, s_j) = ||s_i - s_j||^2 = \sum_t d_t(s_i(t), s_j(t)) = \sum_t (s_i(t) - s_j(t))^2. \quad (12)$$

The choice of an Euclidean metric might not be very useful for ERP research, since individual time points can not be assumed to be independent features (except in the SPN model).

After choosing a dissimilarity measure, the clustering algorithm attempts to find K clusters of objects such that the distance according to Δ is minimal within and maximal between clusters. The number K of clusters to be formed is given as input. Generally, clustering algorithms start from K random initial cluster centers and repeat the following steps until convergence:

(i) each trial is assigned to the closest cluster center according to Δ, and
(ii) new cluster centers are determined based on the current clustering (e.g., average of the trials of a specific cluster).

Depending on how Δ is chosen, either K-means clustering (in case of Euclidean distances) or K-medoids (independent of chosen metric) can be used to cluster the data. For K-means clustering, the second step consists of choosing the average of the within-cluster trials as new cluster center (hence the assumption of Euclidean distances). The more robust K-medoids procedure would take as new cluster center the trial $i_C^\star$ that minimizes the sum of distances to all other trials in cluster C

$$i_C^\star = \arg\min_{i \in C} \sum_{j \in C} \Delta(s_i, s_j). \quad (13)$$

Both algorithm variants depend heavily on the starting conditions since they converge relatively quickly and can therefore run into a local optimum instead of the global one. Output quality can thus be improved by running the algorithm several times with randomized initial conditions.

High dimensionality can cause problems when applying clustering algorithms to ERP data. As should be clear from the arguments in Sect. 3 it also becomes clear, that the dimensions that do correspond to sampling points are not necessarily independent across trials. Using a dimensionality-reduction scheme in advance is thus highly advised. For example in Haig et al. (1995) a principal component analysis was applied to the data, with only the first few principal components being used for the representation that underwent the clustering. Despite the robustness and computational efficiency of this approach, the components generated by the PCA do not necessarily represent meaningful dimensions that can accurately distinguish between different processing strategies. Instead, a distance measure that directly implements assumptions on the distinguishing features would be desirable. Such a measure can be obtained from *dynamic time warping* introduced in Sect. 4.3 (see also Fig. 6).

A second problem refers to the number K of distinct clusters that are to be extracted from the data. Since K enters the clustering algorithms as a parameter,

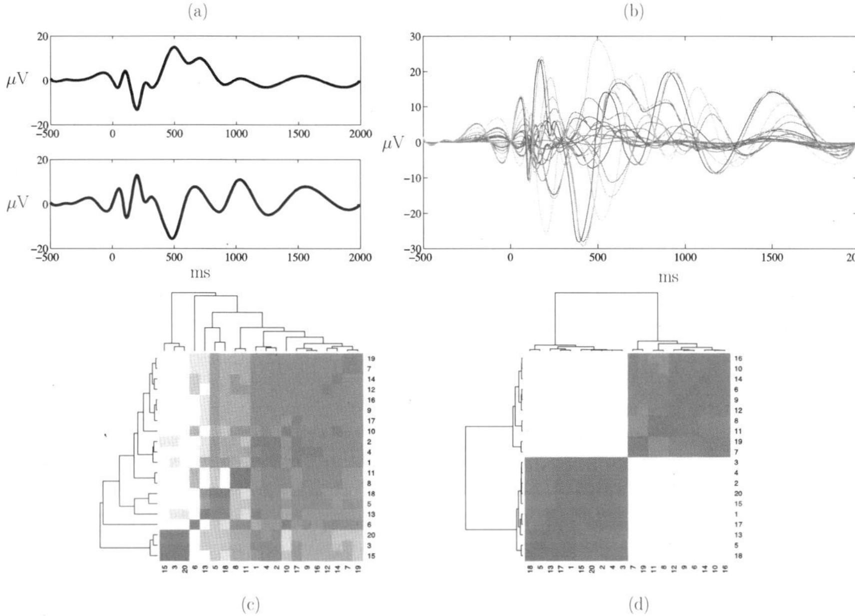

Fig. 6 Cluster analysis on denoised single-trial ERP data: (**a**) shows two template trials which were used to derive the single-trial instances in (**b**) according to Eq. (7); (**c**) shows the resulting heat map applying Euclidean distances (Eq. (12)) while (**d**) is based on the *dynamic time-warping algorithm* (DTW) explained in Sect. 4.3. While the DTW metric correctly classifies all trials to be generated by one of the templates in (**a**), the Euclidean fails to do so in several instances

its choice is arbitrary. There are however strategies to choose K appropriately based on the within-cluster similarities (e.g., the Gap-statistic, Tibshirani et al., 2001). Another convenient way is to use a hierarchical clustering approach that yields so-called *dendrograms* which list the results in terms of a root-tree where each node represents a cluster on a given level along with the separating distances (e.g., see Fig. 6c and d).

In summary, unsupervised classification techniques can help to identify ERPs generated by distinct brain processing. It does not help, however, to resolve the issue of temporal variance introduced in the VCPN model (Eq. (7)). Averaging techniques, therefore should be applied selectively to trials within distinct clusters. In the following section, we will furthermore outline some averaging procedures that try to dispose of the temporal variance difficulty.

4.3 Averaging Procedures

A number of alternative averaging schemes have been proposed besides the simple pointwise mean to address some of the issues discussed in Sect. 3. Basar et al. (1975) proposed a *selective averaging* scheme, where only specific episodes are included in the average. Since data is occasionally subject to very strong artifacts (e.g., from muscle activity), this can greatly improve the average. However, the choice of the trials to be used for the average has to be reached by manual investigation of the data, which is of course far too subjective. Following the methodology developed in the previous section, we propose to limit selective averaging to within-cluster trials, an approach that has been successfully applied in some studies (e.g., Lange et al., 2000).

Selective averaging, however, does not solve the problem of temporal variance presented in the VCPN model. In the following paragraphs, we will briefly review some methods that try to integrate model assumptions into the averaging process. We will discuss at more length dynamic time warping because it constitutes the foundation of the time-marker-based averaging method outlined in the next section.

A very simple means of improving the signal-to-noise ratio is to use *multi-electrode averaging*, i.e., to combine the signals of neighboring electrodes. Of course some spatial resolution is lost, but that might be not so crucial because it is already very low. The poor spatial resolution of the EEG is due to spatial sampling, reference electrode distortion, smearing of cortical potentials because of skull, and separation of sensors from sources (Nunez et al., 1994). Even modern setups of 64 or 128 electrodes do not increase spatial resolution as measured signals of adjacent electrodes do hardly differ. Downsampling, however, can only cancel out noise sources that are specific to a single electrode (e.g., impedance fluctuations due to insufficient preparation of the skin).

As mentioned above, Woody (1967) proposed a method for accounting for ERP-onset latencies, *variable latency averaging*, as assumed in the VLM (Eq. (5)). In this approach, cross-correlations between (cleaned) single-trial data and a *template* ERP (starting with the pointwise average) are computed and the single-trial ERPs shifted according to the time lag that produced the largest value in the cross-correlation curve. The shifted trials are averaged and the whole process repeated with the new average as template until the process converges. Assuming the VLM, this technique does not account for differences in reaction times. That these differences are produced exclusively by delays shortly after stimulus onset is probably an unrealistic assumption.

Another related approach, by Woldorff (1993), termed *adjacent response overlap averaging* (ADJAR), focuses on removing temporal jitter coming from the response that precedes the current trial. The reasoning is that in the presence of short response stimulus intervals (RSI), the interval between the subject's response to trial n and onset of trial $n + 1$, the cognitive processes (and thus the EEG) occurring after trial n are shifted by the response-lag and thus have a differential impact on trial $n + 1$. This assumption is meaningful only for the special case of very short RSIs, as non-random cognitive processes (that are caused by the previous trial) can be assumed to be finished after a reasonably long interval. Also, the technique does not correct the trial's EEG using the associated response latency but does so based on distribution assumptions about the preceding responses.

The *dynamic time-warping algorithm* was first applied to the problem of computer-based speech analysis (e.g., Myers and Rabiner, 1981). Researchers in this domain deal with similar segments of one-dimensional data that can vary in amplitude and timing but not in general shape (i.e., digitized realizations of the same phoneme). This is a similar situation as the one depicted in the VCPN model: the two curves are very similar in basic shape but may be shifted and scaled in time. Noting these similarities, Picton et al. (1988, 1995) proposed an averaging method for the evaluation of brain-stem ERPs.

Generally, DTW is used to compare or transform two curves such that the difference in temporal characteristics is minimized. It was developed for cases in which there is a general shape in the curves, but they are differently aligned on the x-axis (see Fig. 7b). First, a pointwise dissimilarity measure between two signals s_1, s_2 is defined, such as

$$d(s_2(t_1), s_2(t_2)) := |\tilde{s_1}(t_1) - \tilde{s_2}(t_2)| + |\tilde{s_1}'(t_1) - \tilde{s_2}'(t_2)| \tag{14}$$

where $\tilde{s}(t) := \frac{s(t)-\langle s(t)\rangle_t}{\sqrt{\langle s(t)^2\rangle_t}}$ denotes the *normalized* signal and s' is the first derivative of s. The data is normalized to find the best match regardless of absolute amplitude, since only the temporal difference between the signals is of interest. The distance used here is referred to as *derivative DTW* (Keogh and Pazzani, 2001) because it considers both amplitude and slope of the signals.

This measure constitutes the dissimilarity matrix $\mathbf{d}_{jk} = d(s_1(j), s_2(k))$. An optimal time mapping according to the metric chosen above is produced by finding a path p_i that is described recursively by

$$\text{if } p_i = (j, k) \text{ then } p_{i+1} \in \{(j+1, k), (j, k+1), (j+1, k+1)\} \tag{15}$$

through $\mathbf{d}_{jk}$ from the top left to the bottom right corner that minimizes the sum of the d_{jk} (for proof, see Picton et al., 1988).

This path can be found by a dynamic programming strategy, computing the *cumulated cost matrix*

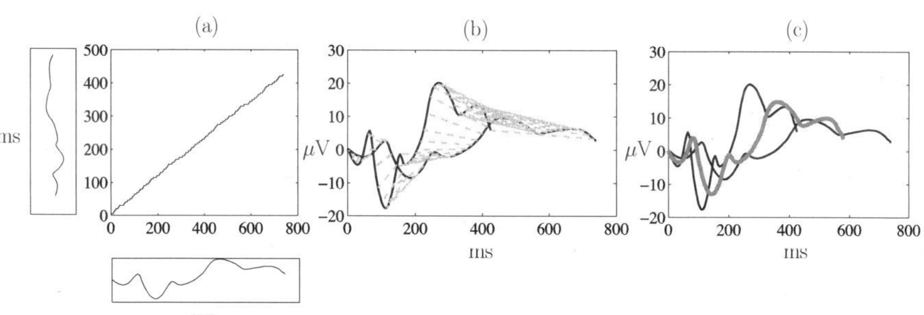

Fig. 7 Dynamic Time Warping. (**a**) The optimal path p_i through the cost matrix $\mathbf{D}_{jk}$ for the two signals (*black curves*) from (**b**) and (**c**) is shown. (**b**) How DTW matches corresponding points in s_1 is illustrated and s_2 is illustrated and (**c**) The average produced by ADTW (*gray*) is shown

$$\mathbf{D}_{jk} = \mathbf{d}_{jk} + \min\{\mathbf{D}_{j,k-1}, \mathbf{D}_{j-1,k}, \mathbf{D}_{j-1,k-1}\} \tag{16}$$

and backtracking via the minimum of the three neighboring entries (down, down-right, right) from $\mathbf{D}_{J,K}$ to $\mathbf{D}_{1,1}$. The final element $\mathbf{D}_{J,K}$ constitutes a measure for the (dis-)similarity of the two curves based on their overall shape. Once this path is available, it is easy to average the curves (called *averaging dynamic time warping*, ADTW) to reduce both temporal and scale variances (see Fig. 7) by setting

$$\text{ADTW}\{s_1, s_2\}(t) = \frac{s_1(j) + s_2(k)}{2}, \tag{17}$$

where $(j, k) = p_t$ as introduced in Eq. (15) and $t = 1, \ldots, J + K$.

For N trials, a straightforward solution proposed in Picton et al. (1988) is to simply combine pairs of single-trial ERPs using ADTW. In a next step, pairs of the results from this combination can be averaged again and the entire process iterated until only one average is left. Recursively,

$$\text{ADTW}\{s_1, \ldots, s_{2N}\}(t) = \text{ADTW}\{\text{ADTW}\{s_1, \ldots, s_N\}, \text{ADTW}\{s_{N+1}, \ldots, s_{2N}\}\}(t) \tag{18}$$

with base case from Eq. (17).

It has been argued that DTW is not suitably applicable to single-trial data because it requires the curves to be smooth. This is because the algorithm is relatively sensitive to fitting noise in the data since it was designed for nearly identical curves. Counteracting this problem, all methods for single-trial ERP estimation discussed above can of course be applied prior to the DTW algorithm. Furthermore, it is possible to introduce constraints on the DTW method that penalize path deviations (Picton et al., 1995) from the main diagonal, thereby lowering the bias at the cost of an increased variance. In their original study, Picton et al. (1988) applied their algorithm to auditory evoked brain-stem potentials, which show a very reliable shape over trials. For studies investigating cortically evoked ERPs this reliability is inexistent. Therefore, before applying the DTW algorithm, one should ensure that trials are sufficiently similar to each other, e.g., by applying time warping only within distinct clusters of ERPs. However, because of the large variability of cortical ERPs, it is not always clear which criteria should be used for clustering. In the next section, we propose to apply external time markers that can act as an objective reference for trial matching and advanced averaging.

5 Enhancing Averaging by Integrating Time Markers

In neuropsychological research it is a frequent problem that both neurophysiological processes that are occurring pre- and post-stimulus as well as those that occur pre- and post-*response* are of interest. A straightforward solution to that problem is to realign the data based on the response markers, an approach which results in the distinction between stimulus-locked (sERP) and response-locked (rERP) ERPs. In many cases, when late aspects of the ERP are of interest the curve becomes easier to interpret if the rERP is considered. Several shortcomings of this approach are obvious. First, a dual analysis is required, which requires extra computation. Second, it is not at all clear which parts of the resulting curves can be interpreted reliably. This is particularly true when there is a large variance of reaction times, which is a common observation in psychological experiments that require higher cognitive functions.

This problem does not seem to have played a big role in methodological research on ERPs so far. To our knowledge only one research paper, by Gibbons and Stahl (2007), has dealt explicitly with that problem. These authors proposed to stretch or compress the single-trial signals in order to match the average reaction time by simply moving the sampling points in time according to a fixed family of functions (linear, quadratic, cubic, or to-the-power-of-four functions). This approach can be seen as an initial attempt to integrate knowledge about the external time-marker *reaction time* into the computation of the average. Extending this idea to integrate arbitrarily many time markers and generalizing it to a larger family of functions that are theoretically more plausible could obtain better estimates for the underlying brain activity.

To account for possible different classes of ERPs as discussed in Sect. 4.2 a cluster analysis should be carried out as a first step. We propose to use the dissimilarity measure $\mathbf{D}_{JK}$ computed by the time-warping algorithm (Eq. (16)). This measure is particularly useful for distinguishing between clusters, because it gives an indication of how far from each other the curves are in terms of their *general shape* rather than a pointwise measure (the cumulated path coefficient indicates how much warping was needed to fit the curves). This comes much closer to the problem posed in Sect. 3 and makes it possible to separate clusters of ERPs based on their general waveform. After this clustering, the model in Eq. (7) can be assumed to be valid *within* the clusters, and the temporal variance given by the ϕ_i can be reduced by application of a dynamic time-warping strategy. Of course, a separate analysis of the emerging clusters is necessary (where the number of clusters must be determined by some heuristic, e.g., the above mentioned Gap-statistic by Tibshirani et al., (2001). However it is probable that in practical situations only a relatively small number of clusters emerges since visual inspection (see, e.g., Fig. 5) shows most trials to be relatively similar in early parts of the ERP.

We propose an extension to the dynamic time-warping algorithm discussed above to integrate knowledge about latency variability during the processing of a single trial into the formation of an ERP estimate. We extend the DTW scheme by (i) selectively choosing pairs of trials before averaging based on their *similarity* and (ii) artificially inserting corresponding points in the two curves based on external knowledge about trial processing. Dynamic time-warping works best and requires fewest transformation of the curves if the underlying signals are similar. It is therefore proposed to group the trials into pairs of most similar trials before averaging. As a measure for the dissimilarity of two curves, again the cumulated path coefficient obtained by application of DTW can be used. Selecting the minimum element from the resulting difference matrix $\Delta_{ij} = \text{DTW}\{s_i, s_j\}$ by $(i, j) = \arg\min_{(j,k)} \Delta_{jk}$ gives the indices of the two most similar trials. These trials are combined in the first step after which row i and column j are removed from Δ_{jk} and the process repeated until the matrix is empty. After one iteration of this process, only on the order of $N_c/2$ trials are left, where N_c is the number of trials in the current cluster. The entire process is repeated until only one average is left ($N_c/2 = 1$). We refer to this method as *pyramidal ADTW* (PADTW) because of the successive division of N_c in the main iteration step.

This extension of the classical DTW algorithm has the merit of providing a simple tool for including knowledge about arbitrarily many known time points during trial processing in the temporal averaging procedure. This can be achieved by combining selected parts of the single-trial ERPs, namely, by concatenating those that correspond to each other with respect to additional time markers. For example, in a neuropsychological setting with known time-marker stimulus onset and response, the first segment from stimulus onset to response could be combined and concatenated with the curve resulting from the combination of the segments from response to the onset of the next trial. This approach is equivalent to calculating the pointwise dissimilarity matrix $\mathbf{d}_{jk}$ from Eq. (14) on the complete time series of n samples and manipulating this matrix before continuing with the steps outlined in the PADTW algorithm. In the manipulated matrix $\hat{\mathbf{d}}_{jk}$, the fields corresponding to an event in both trials are set to zero (minimal dissimilarity). This instructs the algorithm that the points of the two signals match and force the DTW path to lead through these fixed points.

From this formulation, it follows naturally that not only two but arbitrarily many time markers that are known to correspond to each other in the two trials can be integrated. This is relevant for some experimental setups where more than two such markers are given. For example, Ihrke (2007) considered the onset of an eye movement as an additional marker. In similar experiments it would be possible to add a large number of time markers by simultaneously acquiring the exact location of the eye movement in situations in which a strict sequence of task processing is provided.

The advantage of the presented approach is that known shortcomings of the DTW algorithm in case of very different curves can be compensated. The warp algorithm has less degrees of freedom to find an optimal path (in terms of the specified distance measure) and thus avoids paths that can be considered unlikely when taking into account the external events that accompanied the trial processing, thereby decreasing the bias of the model.

6 Discussion and Conclusion

Averaging of electrophysiological data from similar trials is distinguished by robustness and computational simplicity. There are, however, indications that more complex procedures of "data cleaning" and signal estimation may be well justified. This is clearly the case for data sets with many similar trials or a large number of electrodes with crosstalk noise. We have emphasized the point that differences in the processing of identical stimuli may be covered by a model in particular if the variability can be expressed in terms of processing speed only.

While we concentrated on the electroencephalogram it is obvious that most aspects apply to other types of data as well. Starting from considerations on the interpretation of the electroencephalographic data and the analysis of event-related potentials, we presented several approaches to improve the estimation of the signal supposedly underlying the measured activity. We have provided evidence that the assumption of a stationary signal with an additive noise component can introduce a blurring of the data which can be compensated by a statistical model that is only moderately more complex. This model assumes an underlying signal that is distorted in time and amplitude by the interaction with ongoing activity together with an additive noise component.

After cleaning of single-trial data, a statistical analysis is performed to reveal reoccurring patterns in the data based on this scheme. Data mining studies showed the existence of different classes of ERPs for the same experimental condition (Haig et al., 1995). Cluster analysis appeared to be necessary to partition the data into clusters of comparable trials, i.e., trials where similar brain processes can be assumed. Within these clusters averaging can be applied in order to reduce noise. We reviewed established approaches to improve the signal obtained by simple averaging as well as a new time-warping approach that we have developed for dealing with latency variations. This algorithm does allow for differences due to stimulus-locked and response-locked averaging. Time-warped averaging is a flexible tool to integrate behaviorally produced time series into the analysis of electrophysiological data. In the case of ERPs we have shown how the averaging can simultaneously align both stimulus onset and the subject's reaction. When collecting multiple time markers the presented algorithm can be extended to map them in a straightforward manner.

References

Akaike, H. (1974). A new look at the statistical model identification. *Automatic Control, IEEE Transactions on*, *19*(6), 716–723.

Arieli, A., Sterkin, A., Grinvald, A., and Aertsen, A. (1996). Dynamics of ongoing activity: Explanation of the large variability in evoked cortical responses. *Science*, *273*(5283), 1868.

Bartnik, E., Blinowska, K., and Durka, P. (1992). Single evoked potential reconstruction by means of wavelet transform. *Biological Cybernetics*, *67*(2), 175–181.

Basar, E., Gonder, A., Ozesmi, C., and Ungan, P. (1975). Dynamics of brain rhythmic and evoked potentials. I. Some computational methods for the analysis of electrical signals from the brain. *Biological Cybernetics*, *20*(3–4), 137–143.

Boksem, M., Meijman, T., and Lorist, M. (2005). Effects of mental fatigue on attention: An ERP study. *Cognitive Brain Research*, *25*(1), 107–116.

Boksem, M., Meijman, T., and Lorist, M. (2006). Mental fatigue, motivation and action monitoring. *Biological Psychology*, *72*(2), 123–132.

Castleman, K. (1996). *Digital Image Processing*. Prentice Hall Press, Upper Saddle River, NJ.

Celka, P., Vetter, R., Gysels, E., and Hine, T. J. (2006). Dealing with randomness in biosignals. In B. Schelter, M. Winterhalder, and J. Timmer (Eds.), *Handbook of Time Series Analysis*. Wiley-Vch, Berlin.

Ciganek, L. (1969). Variability of the human visual evoked potential: Normative data. *Electroencephalogr Clin Neurophysiol*, *27*(1), 35–42.

Croft, R. J., Chandler, J. S., Barry, R. J., Cooper, N. R., and Clarke, A. R. (2005). Eog correction: A comparison of four methods. *Psychophysiology*, *42*(1), 16–24.

Daubechies, I. (1992). *Ten Lectures on Wavelets*. Society for Industrial and Applied Mathematics, Philadelphia, PA.

de Weerd, J. (1981). A posteriori time-varying filtering of averaged evoked potentials. I. Introduction and conceptual basis. *Biological Cybernetics*, *41*(3), 211–222.

de Weerd, J. and Kap, J. (1981a). A posteriori time-varying filtering of averaged evoked potentials. II. Mathematical and computational aspects. *Biological Cybernetics*, *41*(3), 223–234.

de Weerd, J. and Kap, J. (1981b). Spectro-temporal representations and time-varying spectra of evoked potentials. *Biological Cybernetics*, *41*(2), 101–117.

Delorme, A., Sejnowski, T., and Makeig, S. (2007). Enhanced detection of artifacts in eeg data using higher-order statistics and independent component analysis. *Neuroimage*, *34*(4), 1443–1449.

Dodel, S., Herrmann, J. M., and Geisel, T. (2000). Localization of brain activity: Blind separation for fMRI data. *Neurocomputing*, *32*(33), 701–708.

Dodel, S., Herrmann, J. M., and Geisel, T. (2002). Functional connectivity by cross-correlation clustering. *Neurocomputing*, (44–46), 1065–1070.

Donoho, D. L. and Johnstone, I. M. (1995). Adapting to unknown smoothness via wavelet shrinkage. *Journal of the American Statistical Association*, *90*(432), 1200–1224.

Flexer, A. (2000). Data mining and electroencephalography. *Statistical Methods in Medical Research*, *9*(4), 395.

Flexer, A., Bauer, H., Lamm, C., and Dorffner, G. (2001). Single trial estimation of evoked potentials using Gaussian mixture models with integrated noise component. *Proceedings of the International Conference on Artificial Neural Networks*, *3*, 609–616.

Gather, U., Fried, R., and Lanius, V. (2006). Robust detail-preserving signal extraction. In J. T. B. Schelter and M. Winterhalder (Ed.), *Handbook of Time Series Analysis*, chapter 6 (pp. 131–157). Wiley-Vch Verlag, Weinheim.

Geisser, S. and Eddy, W. (1979). A predictive approach to model selection. *Journal of the American Statistical Association*, *74*(365), 153–160.

Gibbons, H. and Stahl, J. (2007). Response-time corrected averaging of event-related potentials. *Clinical Neurophysiology*, (118), 197–208.

Haig, A., Gordon, E., Rogers, G., and Anderson, J. (1995). Classification of single-trial ERP subtypes: Application of globally optimal vector quantization using simulated annealing. *Electroencephalogry Clinical Neurophysiology*, *94*(4), 288–297.

Handy, T. C. (2005). *Event-Related Potentials: A Methods Handbook*. MIT Press, Cambridge, Mass.

Hastie, T., Tibshirani, R., and Friedman, J. (2001). *The Elements of Statistical Learning: Data Mining, Inference, and Prediction*. Springer, New York.

Ihrke, M. (2007). Negative priming and response-relation: Behavioural and electroencephalographic correlates. Master's thesis, University of Göttingen. Available from: http://www.psych.uni-goettingen.de/home/ihrke

Ihrke, M. (2008). Single trial estimation and timewarped averaging of event-related potentials. B.Sc. thesis, Bernstein Center for Computational Neuroscience. Available from: http://www.psych.uni-goettingen.de/home/ihrke

Jansen, M. (2001). *Noise Reduction by Wavelet Thresholding*. Springer, New York.

Johnstone, I. and Silverman, B. (1997). Wavelet threshold estimators for data with correlated noise. *Journal of the Royal Statistical Society: Series B (Statistical Methodology)*, *59*(2), 319–351.

Keogh, E. and Pazzani, M. (2001). Derivative dynamic time warping. In *First SIAM International Conference on Data Mining (SDM2001)*, Chicago, IL.

Lange, D., Siegelmann, H., Pratt, H., and Inbar, G. (2000). Overcoming selective ensemble averaging: Unsupervised identification of event-related brain potentials. *Biomedical Engineering, IEEE Transactions on*, *47*(6), 822–826.

Marple-Horvat, D., Gilbey, S., and Hollands, M. (1996). A method for automatic identification of saccades from eye movement recordings. *Journal of Neuroscience Methods*, *67*(2), 191–195.

Masic, N. and Pfurtscheller, G. (1993). Neural network based classification of single-trial EEG data. *Artificial Intelligence in Medicine*, *5*(6), 503–513.

Myers, C. and Rabiner, L. (1981). A level building dynamic time warping algorithm for connected word recognition. *IEEE Transactions on Acoustics, Speech, and Signal Processing*, *29*(2), 284–297.

Nagelkerke, N. and Strackee, J. (1979). Some notes on the statistical properties of a posteriori Wiener filtering. *Biological Cybernetics*, *33*(2), 121–123.

Nunez, P., Silberstein, R., Cadusch, P., Wijesinghe, R., Westdorp, A., and Srinivasan, R. (1994). A theoretical and experimental study of high resolution EEG based on surface Laplacians and cortical imaging. *Electroencephalography Clinical Neurophysiology*, *90*(1), 40–57.

Picton, T., Hunt, M., Mowrey, R., Rodriguez, R., and Maru, J. (1988). Evaluation of brain-stem auditory evoked potentials using dynamic time warping. *Electroencephalogry and Clinical Neurophysiology*, *71*(3), 212–225.

Picton, T. W., Lins, O. G., and Scherg, M. (1995). The recording and analysis of event-related potentials. In F. Boller and J. Grafman (Eds.), *Handbook of Neuropsychology* (pp. 3–73). Elsevier Science B.V., Amsterdam

Quiroga, R. Q. (2000). Obtaining single stimulus evoked potentials with wavelet denoising. *PHYSICA D*, *145*, 278.

Quiroga, R. Q. and Garcia, H. (2003). Single-trial event-related potentials with wavelet denoising. *Clinical Neurophysiology*, *114*(2), 376–390.

Rissanen, J. (1978). Modeling by shortest data description. *Automatica*, *14*(5), 465–471.

Rissanen, J. (2007). *Information and Complexity in Statistical Modeling*. Springer, New York, London.

Stearns, S. (1990). *Digital Signal Processing*. Prentice Hall International, Englewood Cliffs, NJ.

Taswell, C. (2001). Experiments in wavelet shrinkage denoising. *Journal of Computational Methods in Sciences and Engineering*, *1*, 315–326.

Tibshirani, R., Walther, G., and Hastie, T. (2001). Estimating the number of clusters in a data set via the gap statistic. *Journal of the Royal Statistical Society: Series B (Statistical Methodology)*, *63*(2), 411–423.

Truccolo, W. A., Ding, M., and Bressler, S. L. (2001). Variability and interdependence of local field potentials: Effects of gain modulation and nonstationarity. *Neurocomputing*, *38–40*, 983–992.

Truccolo, W. A., Ding, M., Knuth, K. H., Nakamura, R., and Bressler, S. L. (2002). Trial-to-trial variability of cortical evoked responses: implications for the analysis of functional connectivity. *Clinical Neurophysiology*, *113*(2), 206–226.

Tukey, J. (1977). *Exploratory Data Analysis*. Addison-Wesley, Reading, Mass.

Voultsidou, M., Dodel, S., and Herrmann, J. M. (2005). Neural networks approach to clustering of activity in fmri data. *IEEE Transactions on Medical Imaging*, *24*(8), 987–996.

Wang, Z., Maier, A., Leopold, D. A., Logothetis, N. K., Liang, H. (2007). Single-trial evoked potential estimation using wavelets. *Computers in Biology and Medicine*, *37*(4), 463–473.

Whalen, A. (1971). *Detection of Signals in Noise*. Academic Press, New York.

Woldorff, M. G. (1993). Distortion of erp averages due to overlap from temporally adjacent erps: Analysis and correction. *Psychophysiology*, *30*, 98–119.

Woody, C. D. (1967). Characterization of an adaptive filter for the analysis of variable latency neuroelectric signals. *Medical and Biological Engineering and Computing*, *5*, 539–553.

Zouridakis, G., Jansen, B., and Boutros, N. (1997). A fuzzy clustering approach to EP estimation. *Biomedical Engineering, IEEE Transactions on*, *44*(8), 673–680.

Dissection of Synchronous Population Discharges In Vitro

Ivan Cohen and Liset Menendez de la Prida

1 Introduction

Synchronous population activities are generated by neuronal circuits of most brain regions. Distinct behavioural states are associated with the emergence of oscillations such as the theta and gamma rhythms of the hippocampus (Bragin et al., 1995; Buzsaki et al., 1983; Leung, 1992). Sleep phases are recognized by distinct rhythmic synchronous patterns (Steriade and Timofeev, 2003). Neuronal synchrony is observed during perinatal development, in many brain structures including the hippocampus (Leinekugel et al., 1998), retina (Meister et al., 1991) and spinal cord (O'Donovan, 1999). Pathological neuronal synchronization is associated with several disease states including Parkinson (Levy et al., 2000) and the epilepsies (Jefferys, 1994). This list shows that synchrony is ubiquitous in the nervous system, observed in many areas with different spatial and temporal properties and physiological or pathological correlates. In this review, we will focus on epileptiform and allied activities generated in the hippocampal formation attempting to show how different techniques and methods of analysis have propelled recent progress.

Work on epileptiform synchronous discharges has shown that multiple original mechanisms must be considered to link different spatial levels from cells to larger brain areas, and a timescale that ranges from millisecond to minutes. Epileptic discharges detected from the surface EEG of patients consist of oscillations of millivolt amplitude in the frequency range of a few hertz to a few hundred hertz. This activity takes place for periods of seconds to minutes, either focally or spread across brain areas. Epileptiform discharges were first assumed to be highly synchronous events. It was thought that such large voltage deflections could not be expected to emerge from weak independent somatic or dendritic generators (Penfield and Jasper, 1954). Applying this idea to spatially distributed EEG recordings tended, however, to avoid questions on mechanisms operating at a local network level.

I. Cohen (✉)
INSERM UMRS975/CNRS UMR7225, CHU Pitié-Salpêtrière, 105 boulevard de l'hôpital, 7501 Paris, France
e-mail: coheni@upmc.fr

J.L. Perez Velazquez, R. Wennberg (eds.), *Coordinated Activity in the Brain*, Springer Series in Computational Neuroscience 2, DOI 10.1007/978-0-387-93797-7_10,

A paroxysmal discharge is the cellular hallmark of an epileptic focus (Johnston and Brown, 1984). Recorded with extracellular macroelectrodes and wide-band filters, these discharges are seen as sharp waves associated with fast oscillations representing very synchronous population activities (Goldensohn et al., 1977). Surface EEG records, typically band-pass filtered over the range 0.1–1 to 70–100 Hz, reveal distorted sharp waves usually termed interictal spikes (Lueders et al., 1981). These two different spatiotemporal scales are traditionally intermixed in the synchronization literature.

The principles underlying the generation of these activities have been studied in epileptic animal models and slice preparations. Here we focus mostly on acute in vitro preparations that permit pharmacological studies and the association of intracellular, field and multiunit recordings. These tools offer a panel of space and time resolutions that have been used to explore the facets of the transition from baseline to synchronous activity modes, mechanisms that control burst discharge cycles and to distinguish how distinct cell types contribute to these synchronous discharges.

2 In Vitro Approaches to Studying Population Activities

The use of brain slices to model synchronous neuronal discharges has a rich history. Such in vitro approaches certainly reduce the complexity of the systems studied, but they permit precise work on synaptic interactions that can be linked to population activities measured by field or multiunit records. They allow clean pharmacological studies as well as precise changes in and measurement of extracellular variables, such as ionic concentration or pH.

The acute brain slice preparation maintains tissue samples of 300–500 μm thickness with enough oxygen, glucose and a realistic extracellular ionic composition for many hours. Slices can be prepared from different brain regions of diverse laboratory animals and also from postsurgical human brain (Schwartzkroin and Knowles, 1984). Specific preparations have been developed to examine long-range effects as in thalamocortical (Agmon and Connors, 1991) or septo-hippocampal slices (Tóth et al., 1997). Long-term effects have been studied over several weeks in organotypic slices prepared aseptically and maintained in culture (Gähwiler et al., 1997).

Cellular mechanisms of population activities have usually been studied in experiments where brain slices are exposed to convulsant drugs or to an altered artificial cerebrospinal fluid (ACSF), including $GABA_a$ receptor blockers, such as picrotoxin, potassium channel blockers or media with low Mg^{2+}, high K^+ and low Ca^{2+}. Under these conditions, slices from rodents typically produce epileptiform discharges that resemble interictal epileptic events (Fig. 1A). As we will describe, the canonical circuit responsible for the generation of these activities varies between models.

Disinhibition has been a popular approach to induce epileptiform activity in hippocampal slices (Fig. 1B). This technique was introduced by P. Schwartzkroin and D. Prince, on the assumption that a reduced inhibition was a critical component of human epileptic foci (Schwartzkroin and Prince, 1977). While that assumption might not be justified in the same way today, this model has revealed some aspects

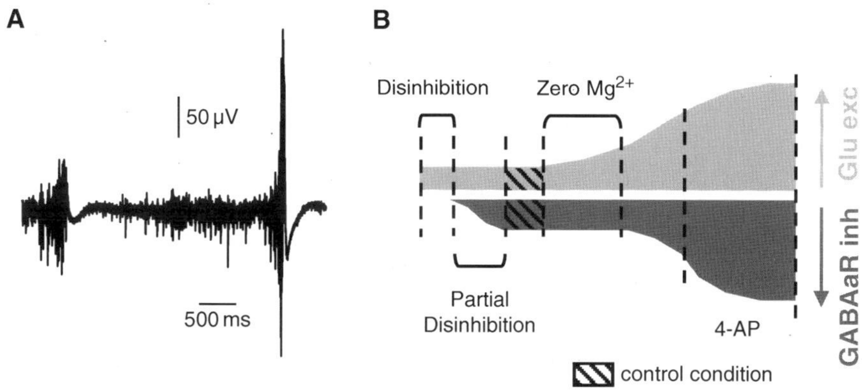

Fig. 1 (**A**) Representative example of population epileptiform activities induced by disinhibition in hippocampal slices. (**B**) Scheme illustrating the different putative mechanisms underlying synchronous population activity in different in vitro models. The curves represent the relative strength of $GABA_aR$-mediated inhibition and glutamatergic excitation. Curve changes represent how different manipulations differentially affect these two forces. Disinhibition model operates under glutamatergic excitation and eventually zero phasic $GABA_aR$ transmission. Partial disinhibition affects only the strength of $GABA_aR$ neurotransmission. Zero Mg^{2+} conditions directly potentiate glutamatergic excitation, whereas the potassium blocker 4-AP is known to enhance both types of neurotransmission. The shaded region represents balanced inhibition and excitation

of how disinhibited circuits generate population synchrony (Hablitz, 1984; Gutnick et al., 1982; Traub and Miles, 1991). In brief, antagonists at $GABA_a$ receptors, such as picrotoxin or bicuculline, lead to the expression of spontaneous interictal-like discharges. This synchrony may emerge from recurrent excitatory synaptic interactions that synchronize neuronal firing into repeated population bursts (Miles and Wong, 1987; Traub and Wong, 1982). We will review recent work that is revealing interesting aspects of interactions between fast glutamatergic and delayed $GABA_bR$-mediated neurotransmission in this preparation.

Now we know that GABAergic inhibition is anatomically and functionally present in focal tissue from epileptic human brain (DeFelipe, 1999; Menendez de la Prida et al., 2002). This has prompted researchers to examine the effects of a partial disinhibition using lower concentrations of $GABA_aR$ blockers to explore activity generated with a moderate reduced fast GABAergic signalling (Fig. 1B; Chagnac-Amitai and Connors, 1989; Traub and Miles, 1991). Other in vitro approaches include reducing extracellular [Mg^{2+}] or increasing [K^+] (Mody et al., 1987; Traynelis and Dingledine, 1988). Both conditions increase population excitability by enhancing NMDA transmission and intrinsic firing, respectively. Similarly the convulsant 4-aminopyridine (4-AP), which both increases cellular activity and enhances release at glutamatergic and GABAergic synapses by actions on specific potassium channels (Perreault and Avoli, 1991), has been used widely as a model of epileptiform synchrony (Fig. 1B).

Oscillations that may be more relevant to physiological activities can also be induced in vitro, often by manipulations that tend to enhance excitability. Thus, reducing extracellular [Ca^{2+}] and [Mg^{2+}] to 1 mM, closer to the brain interstitial fluid in situ, results in the emergence of spontaneous UP and DOWN states in cortical slices (Sanchez-Vives and McCormick, 2000) and sharp wave-like events in hippocampal slices (Foffani et al., 2007). With a different focus, acetylcholine, carbachol or kainate superfusion, which may have specific effects on interneurones as well as on glutamatergic cells, has been used to initiate gamma- or theta-like rhythmic activities in vitro (Bartos et al., 2007; Mann et al., 2005).

Thus, multiple experimental approaches have been used to induce spontaneous population activity in vitro. We will discuss recent data on epileptiform activities in the context of the cellular and network mechanisms underlying the generation of synchrony. Since these spontaneous activities reflect the functional organization of the underlying circuits, dissecting their mechanisms at several spatiotemporal scales may provide general insights into the dynamics of neuronal ensembles and macroscopic measures of synchrony as well as specific information on epileptiform synchrony.

3 Spatiotemporal Resolution of Extracellular Recordings During Population Activities

Extracellular recording is a major tool used to explore neuronal population activities induced in hippocampal slices using the approaches described above. Here we discuss conceptual and methodological questions arising from their use that must be considered in assessments of neuronal population activities and synchrony.

Extracellular recordings use tungsten wires or glass pipettes with impedances in the range of 300–1500 kΩ. Tungsten wires with tips etched to a point of diameter of about 10–15 μm (uninsulated tip resistance ~500 kΩ) permit records of local field potentials generated by hundreds of neurons in their vicinity. The term *field potential* generally refers to a change in the extracellular voltage of larger amplitude and longer duration than single unit action potential (1–10 mV versus 50–200 μV, 10–100 ms versus 1 ms). These are summed potentials generated by synchronous action potentials in groups of neurons or by simultaneous synaptic currents which, in the hippocampus, often impinge on aligned dendrites of many pyramidal cells (Anderson et al., 1971). Two types of signals can then be extracted from records made with these electrodes. Signals in the same frequency band as a traditional human EEG (0.1–100 Hz) largely reflect synchronous synaptic events. In a higher frequency band, 100–5000 Hz, extracellular electrodes detect action potentials which generate an extracellular potential due to the loops of current flowing to and from transmembrane sources and sinks. These two signals can be obtained from the same electrode by recording with a wide-band amplifier (0.1–5000 Hz), and then separating them with two different band-pass filters (Fig. 2A). They represent separate features of population activities providing insight into both neuronal firing and the generation of synaptic potentials at different timescales.

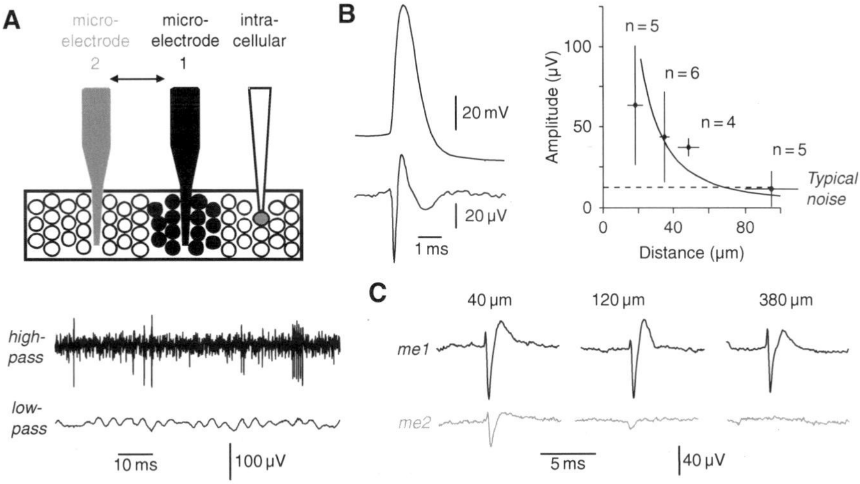

Fig. 2 (**A**) Scheme of multiunit extracellular recordings (microelectrodes 1 and 2) versus intracellular recording from a single cell. The extracellular electrode detects spikes from neighbouring neurons (*black*) intermixed with field potential fluctuations. These two signals can be separated using appropriate filters. (**B**) *Left*. Simultaneous intracellular and extracellular recordings of a single action potential. *Right*. The amplitude of extracellular action potentials, measured with the electrode at different distances from the recorded cell diminishes with increasing separation. Modified from Cohen and Miles (2000) with permission. (**C**) Extracellular detection of action potentials decreases with distance so that spikes can be detected up to 50–100 μm away from the unit. Modified from Menendez de la Prida et al. (2006)

Synaptic signals may be generated locally or originate from neurons with distant soma, but extracellular action potentials originate locally. Action potentials produce transmembrane current loops that are sensed when an electrode is in close proximity to a neuron (Hubel, 1957). We have estimated the limits of this proximity by combining intra- and extracellular recordings from the same neuronal unit showed (Fig. 2B). We showed that moving the extracellular electrode away in the stratum pyramidale reduced the amplitude of the detected spike to a value less than typical noise levels, of about 10 μV, at a distance of 60–80 μm (Cohen and Miles, 2000). Thus, given typical neuron packing density in stratum pyramidale, such an extracellular electrode might detect multiunit firing from 400 to 500 neurons in a 400 μm thick hippocampal slice. Therefore, to avoid oversampling from the same neuronal population, two extracellular electrodes must be separated by more than 200 μm in the stratum pyramidale (Fig. 2C). This limited detection range necessitates care to minimize mechanical damage during electrode placement. In long-term in vivo recordings, microelectrodes will likely cause bleeding and gliosis on a layer thicker than the detection radius.

Another important point is a possible variation in the waveform of extracellular action potentials according to the relative position of the recording electrode and the current sources involved (Gold et al., 2006). This issue may be especially important in laminated structures like the hippocampus, where action potentials backpropagate into dendrites expressing voltage-dependent channels (Spruston et al., 1995). Consequently, detection of extracellular action potentials may vary according to the cellular compartment where it was generated and its position relative to the fixed electrode. These differences may be especially important for spike-sorting analyses in conditions of elevated excitability when the somato-dendritic origin of action potentials may vary unpredictably (Helmchen et al., 1999).

Another problem, exacerbated in conditions of high excitability, occurs when action potentials originating from different sources occur at such high frequency that they overlap in time. Thus, coincident firing of several neurons during population activities can lead to underestimate firing rate. Simulations of independent events showed that our detection algorithm could resolve two spikes separated by more than 0.4–0.8 ms and that firing frequency derived from detection deviated from expected value above 700 spikes per second (Menendez de la Prida et al., 2006). Beyond this cutoff, the distortion may be corrected numerically if the following condition of convergence is fulfilled (Ankri and Korn, 1999):

$$freq \cdot \Delta T < \exp(-1)$$

Thus, for an overlapping window $\Delta T = 0.4$ ms correction is possible up to 900 Hz. However, possible synaptic interaction implies that the limit may be actually lower.

One last comment should be made on relations between electrode characteristics and those of the signals that they detect best. Microelectrodes, with tips of 10–15 μm diameter, sample activity from local neuronal populations better than do needle-type macroelectrodes, with diameters of 100–125 μm. This type of electrode provides only poor records of action potentials generated by single cells but faithfully records field deflections. Because they record from regions larger than 200 μm wide, they are biased to sample local synchronous firing or synaptic activity. Other electrode configurations include planar electrodes, as used in electrocorticography, which enable to record average activity from an area immediately beneath the contact. It is estimated that the minimal focus necessary to produce a measurable event in the surface EEG is about 2 mm wide, corresponding to an effective recording area of about 12 mm^2 (Lueders et al., 1981). Thus the spatiotemporal features of recorded activity depend strongly on electrode characteristics (Niedermeyer and Rocca, 1972).

4 Analytical Approach

Analytical tools for use in studies on spontaneous population activities generated by interactions within a neuronal circuit must be adapted to handle irregular and unpredictable events. A major part of the analysis of spontaneous firing rests on comparing various indicators of activity in association with the spontaneous events.

This approach is similar to a peri-stimulus statistics (Gerstein and Perkel, 1969), using a spontaneous field event, rather than an experimental trigger, as the time reference point. Activity indicator changes may also be usefully studied during transitions between different states of a neuronal system like those induced by progressive actions of neurotransmitter antagonists.

The extraction of robust and informative activity indicators is the starting point of this analysis. Processing steps for extracellular signals are detailed in Fig. 3. Bold and plain grey text indicates, respectively, that some data may have a time-continuous form while event data may consists of a series of event timings which is a point process (Daley and Vere-Jones, 2003). Basic steps help to separate field and multiunit data as spikewaves and action potentials. To this purpose, extracellular recordings can be processed to extract distinct components of the extracellular signal. As described, typical band-pass filter settings of 0.1–1 Hz to 30–100 Hz have been used to separate field potentials, while multiunit action potentials are detected with a band pass that ranges typically from 50–500 Hz to 2500–4000 Hz. Field and action potentials are detected with a thresholding procedure. Events including extracellular action potentials and spike-wave sequences are then extracted using a reliable and computationally efficient algorithm that generates data on event timing (Cohen and Miles, 2000).

It may be useful to transform point-like data on event timing into a continuous parameter so that methods available for continuous signals can be applied. We have used smoothing procedures to derive such continuous measures of event frequency (Abeles, 1982). Frequency curves are computed with a smoothing kernel, typically a Gaussian curve. In practice, Gaussian curves of fixed width, equivalent to a smoothing time constant, are centered on each detected event, and all the elementary curves are summed together. For spike-wave complexes, time constants of 5–30 s have been used to estimate frequency transients and 5–10 min to produce a more heavily smoothed frequency curve. For action potentials, the time constants are typically 2–10 s and 30 s–3 min. Using multiple time frames provides a useful vision of the same data at different timescales (Cohen et al., 2006).

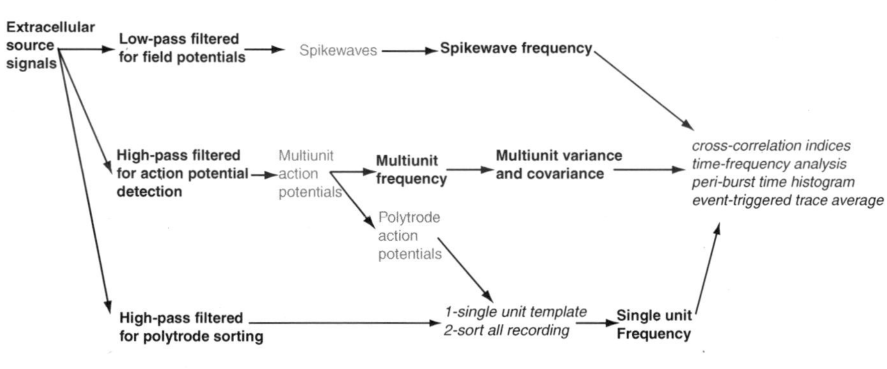

Fig. 3 Processing steps for extracellular signals

Ultimately, trace and event data can be compared from distinct extracellular channels or between extracellular and intracellular signals. Different analytical tools can be applied to extract measures of field to field synchrony or to compare the dynamics of field events and single-cell behaviour (Fig. 3, in italic). Recent reviews describe methodological approaches applied to neuroscience data (Le Van Quyen and Bragin, 2007; Brown et al., 2004). In this chapter, we will focus on our work on the mechanisms underlying the emergence of population activities in the hippocampus in vitro. We follow the rationale summarized in Fig. 3, which is implemented in the package Spikoscope, available through a dedicated website (http://spikoscope.sourceforge.net/Spikoscope/Home.html). Additional tools developed to handle specific features of data analysis written in MATLAB (http://www.mathworks.com/) or Labview (http://www.ni.com/labview/) as well as all analytical derivations are also available on request.

5 Intermediate Levels of Firing Rate Fluctuations

Spontaneous multiunit action potentials are always detected in extracellular records from the CA3 stratum pyramidale in an interface recording chamber bathed with standard ACSF. The frequency of basal firing detected by all cells recorded by an electrode is typically about 50–100 action potential spikes per second. Frequency fluctuates but recurring temporal patterns are rarely evident. Transition to a disinhibited state uncovers polysynaptic circuits normally under strong GABAergic control (Miles and Wong, 1987). During perfusion of $GABA_aR$ blockers, recurring patterns corresponding first to partial and then to full synchrony, become visible in both field and multiunit activity (Traub and Miles, 1991; Cohen et al., 2006). As a stable bath concentration of GABAergic receptor antagonists is achieved, fully synchronous population bursts are generated in recurring fashion at intervals of 3–10 s. Each burst is followed by a silent period and then multiunit activity increases before the next burst.

In the transition period, multiunit activity typically shows a progressive increase and patterning. We quantified this process by deriving a measure of multiunit frequency from event timings using a smoothing procedure (Cohen et al., 2006). As described, detected events were each replaced by a Gaussian curve of fixed standard width ranging from 5 to 5000 ms, which were summed together to build a firing rate index (Fig. 4A). With an elementary curve of normalized surface, the integral of the summed elementary curves represents multiunit firing rate as a continuous signal. A firing rate curve computed with a long time constant for the Gaussian (>1 s) gives a slow trend of multiunit activity as perfusion progresses, $F(t)$ (Fig. 4B, black curve). Use of a short time constant (5–50 ms) highlights changes in the collective firing pattern, $f(t)$ (Fig. 4B, grey curve). Comparing the two curves reveals periods of local synchronous firing, synchronous silence or asynchronous activity (Cohen et al., 2006).

This approach allows us to derive a continuous measure of the synchrony in discharges of a local group of a few hundred neurons. This local synchrony index,

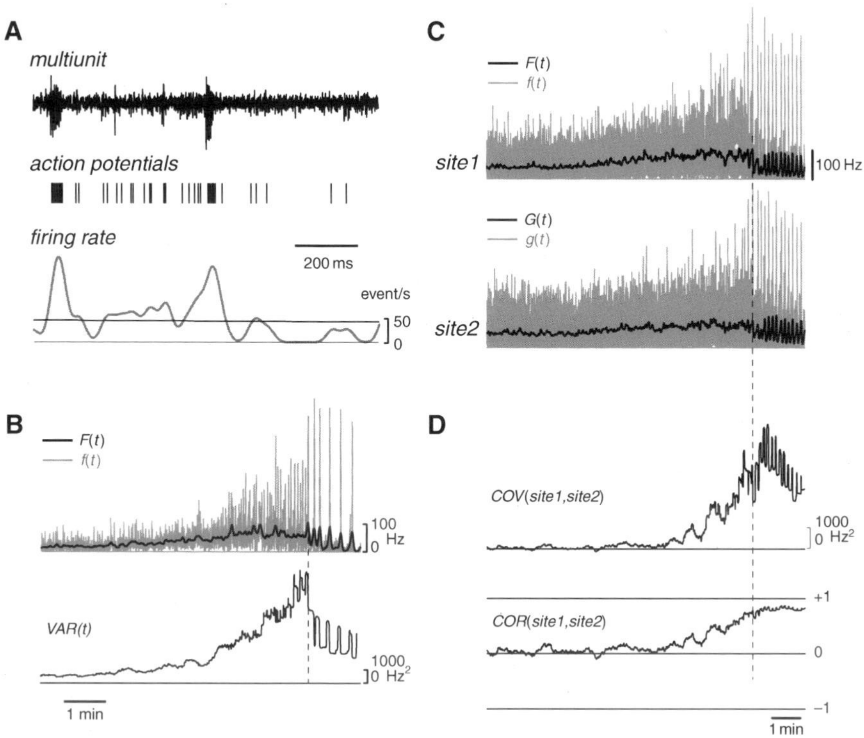

Fig. 4 (**A**) Action potentials (AP) are detected in the extracellular multiunit recording and turned into firing rate curves by convolution with smoothing kernels. (**B**) Using short (*grey line*, 20 ms) and long (*black line*, 5 s) time constants reveal different temporal scales of multiunit firing rate. A variance index (*lower trace*) can be derived from the two frequency curves (*upper black* and *grey traces*) to follow the transition to synchronous bursting induced by disinhibition. (**C**) Same representation from two separate microelectrodes sampling activity at two different sites. (**D**) Covariance and correlation indices can similarly track progressive multiunit synchronization between two recording sites. All panels are modified from Cohen et al., (2006) with permission

equivalent to an instantaneous variance of multiunit activity, is computed from the two firing rate curves with short and long time constants. The squared difference is averaged over a sliding window of length 5–30 s depending on the desired temporal scale:

$$VAR(t) = mean((f - F)^2)$$

As disinhibition progresses, mean firing rate increases. Then patterned fluctuations in population firing emerge with little further increase in frequency (Fig. 4B, bottom curve). They consist of transient sequences of increases followed by decreases in firing that recur at intervals of 0.5–2 s. The amplitude and variability of these frequency fluctuations increases with time, and is associated with the emergence and growth of periodic field potentials. We used this index to estimate the variability of multiunit firing as disinhibition proceeds in minislices of the CA3 region (Cohen et al., 2006). In records from single sites, the variance index increased from about 900 Hz^2 in control conditions to above 5000 Hz^2 just before the onset of epileptiform discharges (Cohen et al., 2006). This more than fivefold increase in variance was much larger than that expected (~70%) for the ~30% increase in frequency if firing of independent neurones were scaled. Thus epileptiform synchrony is preceded by increasing variation and effectively partial synchronous firing at the level of a local neuronal population. This initial increase in CA3 population activity presumably results from the reduction in the inhibitory control of pyramidal cell firing (Miles and Wong, 1987; Hausser and Clark, 1997; Cohen et al., 2000).

6 Transition to Fully Synchronous Network States

The spatial extent of multiunit synchrony during the transition to epileptiform synchrony was estimated by comparing frequency fluctuations in extracellular records from distant cell populations in the CA3 region of the same slice. These further synchrony indices reflect a covariance and a correlation coefficient. They are based on multiunit frequencies computed with two time constants at each site (*f*, *F* for site 1 and *g*, *G* for site 2), with the same sliding window average (Fig. 4C):

$$COV(site1, site2, t) = mean((f - F) \times (g - G))$$

$$COR(site1, site2, t) = \frac{COV(site1, site2, t)}{\sqrt{VAR(site1, t) \times VAR(site2, t)}}$$

The correlation index can vary between +1, indicating a perfect coherence of the two signals, and –1, corresponding to changes in frequency in opposing directions for the two recordings. In our experiments using CA3 minislices, these indices fluctuated by about 0.07 units in normal ACSF (Cohen et al., 2006). The mean correlation index increased more than 10 times as disinhibition increased towards the onset of full synchronous epileptiform discharges (Fig. 4D). Thus, the coherence of population activities recorded from separate sites increased together with the increase in fluctuations of activity at a single site.

Larger synchronous epileptiform bursts emerge, usually abruptly, after $GABA_aR$ blockers reach a maximal concentration in the bath. These events were associated with a maximal multiunit firing and cross-correlation between two electrodes (Fig. 4D). At the transition from partial synchrony to interictal-like events, there was a large increase in field potential amplitude, suggesting a discontinuity may exist as fully synchronous firing emerges. Thus, the amplitude of the largest field potential associated with partial synchrony was about 100 μV, whereas that associated with the first fully synchronous burst was larger than 2 mV (Cohen et al., 2006). Therefore, as disinhibition proceeds polysynaptic glutamatergic circuits mediate a progressive spread of excitation throughout the circuit, evident as small partial synchronous events. Presumably, the gradual increase of indices of synchrony preceding the onset of the first interictal-like burst reflects the growth of excitation.

As the first full burst emerges, synchronization indices reach a maximum suggesting that most, or all, cells are recruited, in contrast with partial events. Even so, partial events and full discharges exhibit some common features (Fig. 5A). Both are preceded by accelerating multiunit firing and a slow field prepotential (Cohen et al., 2006; Menendez de la Prida et al., 2006). Field potentials of similar form, but different amplitudes, are generated in both stratum pyramidale and radiatum for both types of event. Those corresponding to partial events typically consist of a fast negative deflection of duration 20–50 ms followed by a slower positive wave of duration 150–300 ms (Fig. 5B). The fast negative deflection is typically correlated with the peak increase in extracellular firing. Simultaneous intracellular recordings show an associated depolarizing event with some isolated spikes. Similarly, the first full event

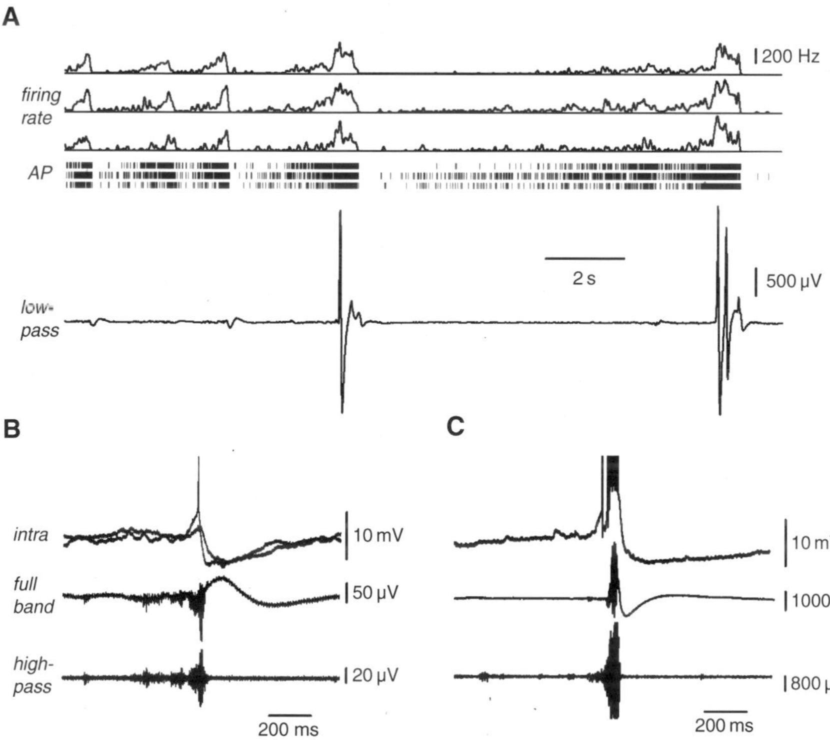

Fig. 5 (**A**) Multiunit activity recorded from three extracellular microelectrodes is represented as firing rate (*upper traces*) and action potentials (AP). The low-pass field potential signal from one electrode is shown below. During transition to full synchronous event there are episodes of partial synchronization that are associated with firing rate increases and small field potential deflections. (**B,C**) Simultaneous intracellular and extracellular recordings during a partial and a full synchronous event. Panels **A** and **B** are modified from Cohen et al. (2006) with permission

is associated with multiunit firing increase and an initial fast negative deflection of field recordings that precedes the discharge onset. A depolarizing ramp is evident in local intracellular records (Fig. 5C). These observations converge to suggest that full, epileptiform synchrony may be an extreme of a range of potential partially synchronous activities (Traub and Miles, 1991).

Similar partial behaviour has been described in the initiation of population activities in the developing hippocampus (Menendez de la Prida and Sanchez-Andres, 1999) and spinal cord (Wenner and O'Donovan, 2001). They have been also observed in hippocampal slices under stable regimes of partial disinhibition (Schneiderman, 1986; Scanziani et al., 1991). Probably, partial events are similar to sharp wave bursts recorded in vivo in the CA3 region during specific behavioural states (Suzuki and Smith, 1985; Buzsaki, 1986). Sharp waves in vivo are thought to emerge in the CA3 region by correlated activity from a fraction of CA3 pyramidal cells and interneurones (Csicsvari et al., 2000), but the basic mechanisms are still under investigation.

7 Threshold Behaviour in the Initiation of Population Activity

In epilepsy, population activities generated at one or more focal sites can recruit distant regions to interictal discharges or seizures (Graf et al., 1984; Alarcon et al., 1997; Pastor et al., 2006). In records made with macroelectrodes that integrate EEG fields from a large spatial scale it may be difficult to resolve the microdynamics of cellular activities within a focus. However, depth recordings with microelectrodes implanted in the temporal lobe of epilepsy patients are revealing a wide range of activities within the parahippocampal region including multifocal contributions to interictal activities of human temporal lobe (Engel et al., 2003; Bourien et al., 2005).

The existence of a focal region and the factors responsible has been examined in the CA3 region. The spatial identity of the triggering zone differs according to the method used to induce synchrony and can also vary between species (Colom and Saggau, 1994). For disinhibition of the guinea pig hippocampus, the initiation zone is located within CA3a, in the last 800–1000 μm of the mossy fibre plexus (Colom and Saggau, 1994; Knowles et al., 1987; Wong and Traub, 1983). Extracellular records from this region reveal a slow field potential correlated with a cascade of multiunit firing before synchronous burst onset (Fig. 6). These events start at 50–150 ms before a population event and are virtually absent in other areas (Menendez de la Prida et al., 2006; Wittner and Miles, 2007). Processes preceding fully synchronous population events are reminiscent of those associated with partial synchrony. As in partial events, intracellular records from cells in the CA3a driver region show that synaptic activity increases just before population bursts. Our previous data show that an increase of EPSP frequency was correlated with the build-up of multiunit firing and a negative field potential deflection (Menendez de la Prida et al., 2006). EPSPs occurring during the build-up period in the driver region sum into a slow depolarization with similar time course to the extracellular field

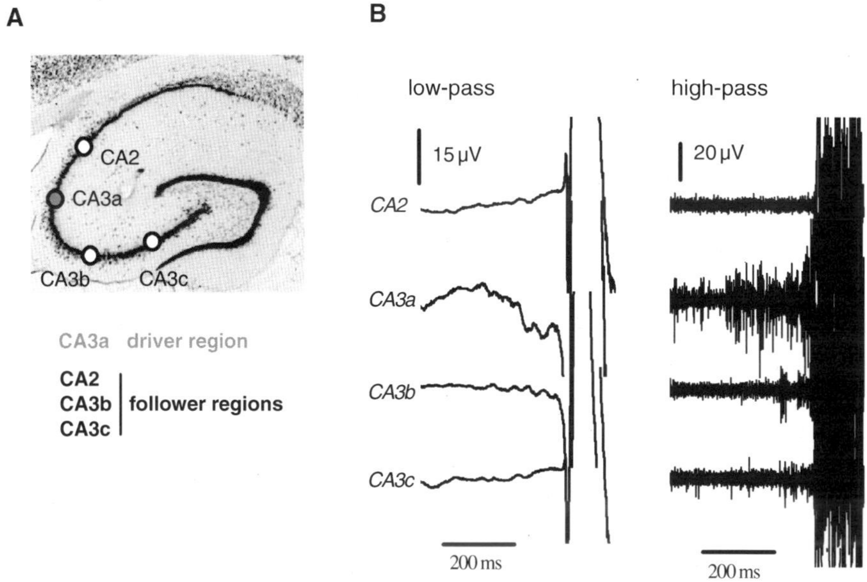

Fig. 6 (**A**) Schematic representation of the CA3 subregions from where simultaneous recordings were obtained. (**B**) Field potentials (*left*) and multiunit activity (*right*) recorded before the onset of a full synchronous event. Note the negative field potential deflection and the multiunit firing increase confined to the CA3a-triggering region

deflection. Therefore the increased synaptic activity recorded just before the onset of a full synchronous event is likely to reflect a kind of partial synchronization of the recurrently connected CA3 cell population (Traub and Miles, 1991).

Distinctive dynamical features characterize partial events. Field potential deflections preceding full events are larger than those associated with partial events (Menendez de la Prida et al., 2006). Multiunit firing rate at the onset of the full discharge is higher than at any other point during the period between bursts (Fig. 7), as is EPSP frequency. This has led us to suggest that population bursts are initiated when partial synchronization exceeds a threshold firing frequency (Menendez de la Prida et al., 2006). Such a threshold condition should be associated with an unstable state of the system, because after this critical point explosive excitation spreads throughout the circuit (Tabak et al., 2001; Latham et al., 2000). What is the cellular basis of such threshold behaviour?

We have shown that small changes in the efficacy of either fast glutamatergic or $GABA_bR$-mediated signalling alter the threshold firing frequency for full synchronization in the disinhibition model. In low concentration of NBQX, which partially blocks AMPA/kainate receptors, the frequency of multiunit firing preceding full events is increased and most partial synchronous events remain subthreshold, being unable to elicit population discharges (Fig. 8A, B). Full synchrony can be rescued by decreasing $GABA_bR$-mediated inhibition with low concentrations of

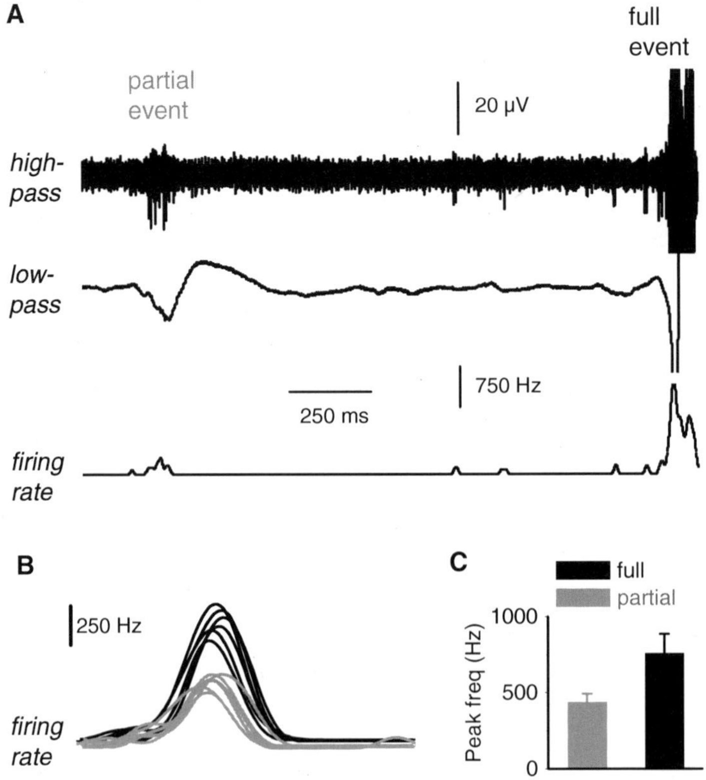

Fig. 7 (**A**) Multiunit activity and field recordings associated with partial and full synchronous events. Multiunit activity is transformed in firing rate data to evaluate differences of firing between partial and full events. (**B**, **C**) Comparison of firing rate values during several episodes of partial and full events reveals that firing at the onset of the full synchronization is higher than any partial event recorded

CGP52432 (Fig. 8B, C). This manipulation then reduces the firing threshold for the initiation of population discharges. Thus, a dynamic balance between glutamatergic and GABAergic networks is involved in shaping these events (Menendez de la Prida et al., 2006).

8 Synaptic Processes Underlying Network Synchronization

With both excitatory and inhibitory forces involved in setting a network threshold for full synchrony, we can seek to separate these synaptic conductances during subthreshold partial events. Changes of membrane conductance, ΔG, and the composite reversal potential, *Vrev*, associated with partial events can be estimated from intracellular recordings following the method of Borg-Graham et al. (1998). In this case

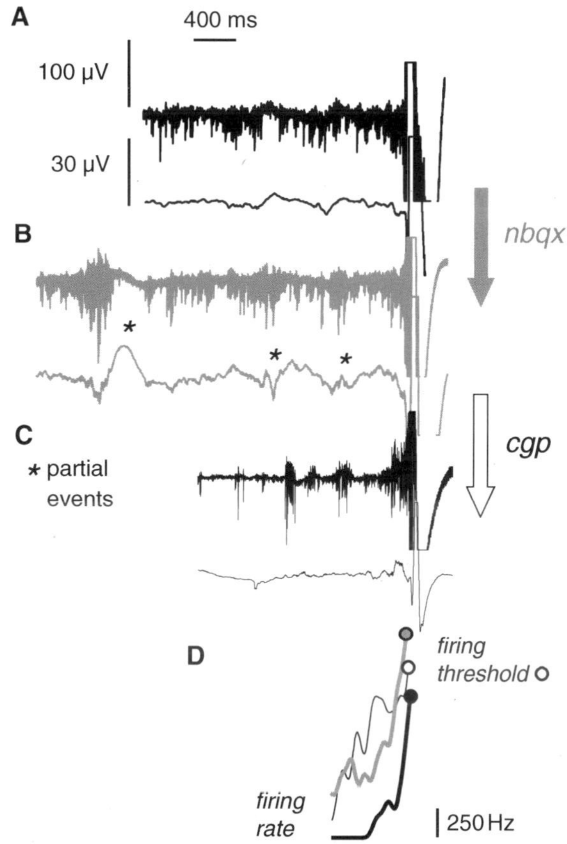

Fig. 8 (**A**, **B**) Modulation of the network frequency threshold by mild reduction of AMPA/kainate receptors. In low concentration of NBQX (0.2–0.7 μM) most partial events remain subthreshold to initiate full synchronous discharges. (**B**, **C**) Full synchronization can be rescued by decreasing GABAbR-mediated inhibition with CGP52432. (**D**) Comparison of firing rate values at the population discharge onset reveals modulation of frequency threshold with these manipulations. Note that NBQX (*grey trace*) increases the firing threshold and that CGP (*thin trace*) further reduces it

with contributions from just two synaptic processes, we could dissect an excitatory and an inhibitory component of the composite synaptic event (Menendez de la Prida et al., 2006):

$$\Delta Ge = \Delta G \times (Vrev - Vi)/(Ve - Vi)$$
$$\Delta Gi = \Delta G \times (Vrev - Ve)/(Vi - Ve)$$

Here *Ve* and *Vi* are the reversal potential of glutamatergic (+10 mV) and $GABA_bR$ (–85 mV)-mediated synaptic currents.

This approach showed that partial events recorded under disinhibition were associated with a sequence of glutamatergic and $GABA_bR$-mediated signals (Menendez de la Prida et al., 2006). Similar sequences of summed EPSPs followed by summed $GABA_bR$-mediated IPSPs were described by Scanziani et al. (1991). We note that the initial multiunit firing characteristic of partial events grows in parallel with a net synaptic depolarization. A delayed inhibitory conductance then emerges as firing is suppressed (Fig. 9A). Thus, the threshold condition may reflect a need for activity

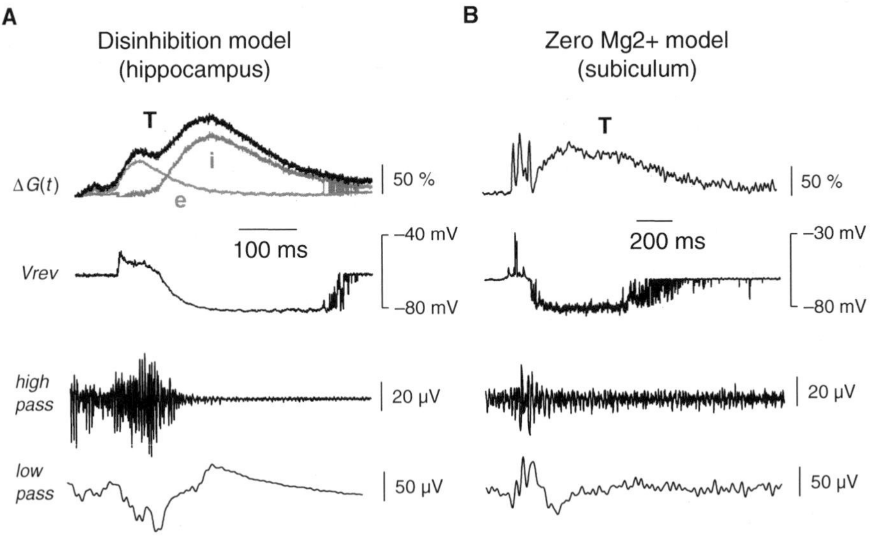

Fig. 9 (**A**) Evolution of the synaptic conductances and the composite reversal potential associated with partial events in the disinhibition model. Simultaneous extracellular recordings are shown at the bottom to separate multiunit firing (high-pass) and field potentials (low-pass). In this model, the contribution of glutamatergic (*e*)- and GABAbR (*i*)-mediated transmission can be separated. (**B**) Same analysis in the subiculum under zero Mg^{2+}. Here the excitatory and inhibitory components cannot be separated. Panel **A** is modified from Menendez de la Prida et al. (2006). Panel **B** is modified from Menendez de la Prida and Gal (2004) with permission

to spread quickly through the network via divergent glutamatergic signalling before $GABA_bR$-mediated inhibition can suppress cell firing. In the CA3 circuit, a presynaptic cell induces firing in a postsynaptic cell with probability less than 1 (Miles and Wong, 1986). However, the divergence of recurrent synapses ensures that sufficient cells are recruited for a build-up in population activity to proceed (Traub and Miles, 1991). The G-protein-dependent delay of several tens of milliseconds in $GABA_bR$-mediated inhibition provides a short time window for a partial event to reach threshold. Cells that are excited late will be more likely to receive EPSPs from several presynaptic cells. This determines a critical point that a partial event must cross to initiate a full synchronous discharge (Menendez de la Prida et al., 2006).

However, population activities with fast $GABA_aR$-mediated inhibition blocked or partially reduced are one extreme of the different regimes for circuit function (Fig. 1B). In the disinhibition model, the dynamics of partial events are tightly controlled by delayed $GABA_bR$-mediated inhibition providing a large window for synaptic integration. Experiments using partial disinhibition suggest that beyond a certain reduction in efficacy of fast inhibitory signalling, further reduction cannot account for changes in the ability to initiate population discharges. Once full events are generated neither their amplitude nor their propagation speed change with further disinhibition (Menendez de la Prida and Pozo, 2002; Pinto et al., 2005). The only clear effect is a consistent reduction in the fraction of partial synchronization at the expense of full events (Cohen et al., 2006). Therefore, fast $GABA_aR$-mediated transmission determines a critical point for the spatiotemporal dynamics of partial synchronization.

The zero Mg^{2+} seizure-like model in which inhibition is intact displays both partial and full events in the isolated subiculum (Menendez de la Prida and Gal, 2004). This approach let us examine both glutamatergic and GABAergic components of partial synchrony in a more natural framework (Fig. 1B). Our data suggest that both $GABA_aR$- and $GABA_bR$-mediated inhibition are critical in shaping partial events (Menendez de la Prida and Gal, 2004). The study of synaptic conductances and *Vrev* evolution during partial synchronization reveals that fast $GABA_aR$-mediated inhibition is probably shunting glutamatergic-driven forces involved during the initial phases of a partial event (Fig. 9B). Although we could not separate the excitatory conductance and the two distinct inhibitory processes (Borg-Graham et al., 1998), this delayed effect is clearly depicted by the concurrent evolution of ΔG and *Vrev* (Fig. 9B). Therefore, in regimes where both fast and slow inhibitions are operative, the time available for fast glutamatergic activity to spread is severely constrained by $GABA_aR$.

This concept of a threshold for network synchronization extends from other biological threshold behaviours such as the generation of an action potential by a neuron. For a neuron to fire an action potential, Na^+ channels must open fast enough to depolarize the cell to threshold before K^+ channels with slower kinetics can prevent further depolarization. Similarly, in a neuronal circuit glutamatergic synaptic events may be analogous to Na^+ channel openings in that they act to enhance firing in a positive feedback fashion. In contrast, GABAergic synaptic events may be analogous to K^+ channel openings in that they depress population firing with a delay. Single cells are recruited to bursts when depolarized to their action potential threshold. This recruitment depends on the successful integration of net depolarizing conductances before inhibitory forces repolarize cells and stop firing. Summing this process across the population implies the existence of a network firing threshold.

9 Refractory Periods for the Initiation of Full Population Synchrony

One fundamental property of biologically excitable systems, such as those involved in the generation of an action potential, is that a period of recovery may be needed before the process can be repeated. Fully synchronous population discharges seem to follow similar dynamics. Indeed, because full synchronous discharges recur at intervals of 3–10 s, it has been proposed that the recovery of cellular or synaptic processes determines the timing of periodic events. Both cellular excitability and synaptic efficacy must recover from a previous burst. Hyperpolarizing synaptic events (Hablitz, 1984) as well as potentials dependent on Ca^{2+}-dependent K^+ channels and the Ih current (Agmon and Wells, 2003) probably terminate too quickly to control the timing of a succeeding event. Instead, it has been suggested that the recovery of synaptic vesicles available for release may control the interval between fully synchronous events in the hippocampus (Staley et al., 1998) and the developing spinal cord (Fedirchuk et al., 1999).

The ability of partial synchronization to initiate a new population discharge after a previous full event totally reflects such a refractory effect (Fig. 10A). During an absolute period after a full event, even partial synchronization cannot be induced by focal stimuli. This ability progressively recovers during the activity cycle (Fig. 10B), as cellular excitability, synaptic traffic and EPSP-spike coupling are reestablished (Menendez de la Prida et al., 2006). Thus, population events seem to be initiated by active processes involving synaptic transmission of activity within the population after recovery from a previous full event is complete. Such mechanisms may imply that a full burst is initiated when a threshold frequency of synaptic or cellular activity is exceeded (Menendez de la Prida et al., 2006). How to reconcile the recovery and the threshold hypothesis?

In systems where recovery processes dominate, the duration of an event is correlated with the preceding interevent interval (Grzywacz and Sernagor, 2000;

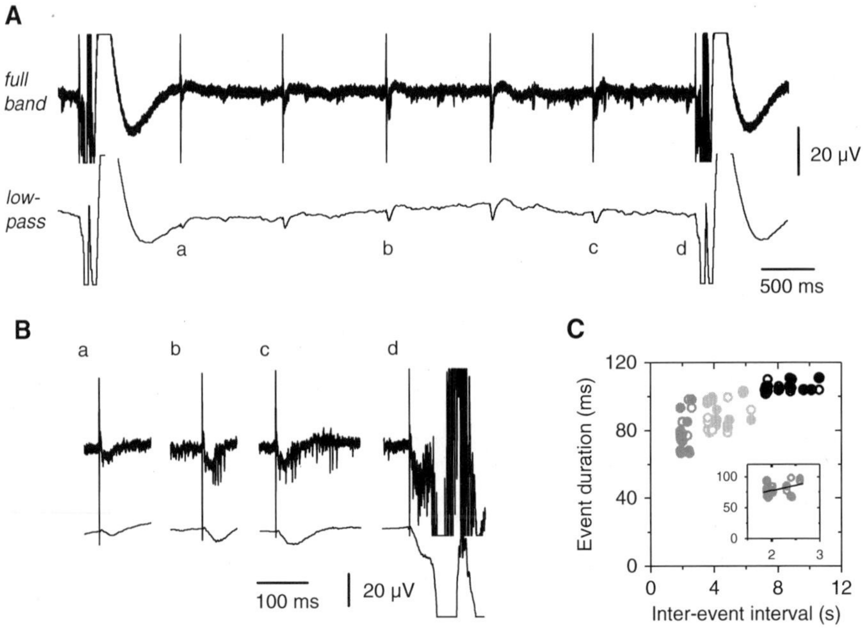

Fig. 10 (**A**) Recovery dynamics of partial events induced by mild stimulation after a previous full population discharge. Letters denote partial events shown in **B**. (**B**) Details of partial events at different times of the activity cycle between two full events. (**C**) Full event duration versus the preceding (*filled dots*) and the following (*open dots*) interevent intervals for different concentrations of extracellular potassium: 4 mM (*black*), 6 mM (*light grey*) and 8 mM (*dark grey*). Inset highlights the case of 8 mM K^+ (Pearson correlation $r^2 = 0.6$, at $p = 0.05$). Panel **C** is modified from Menendez de la Prida et al. (2006) with permission

Staley et al., 2001; Tabak et al., 2001). However, for spontaneous population discharges induced by disinhibition at 4 mM K^+ these parameters are poorly correlated (Menendez de la Prida et al., 2006). Increasing extracellular K^+in the range of 6–8 mM reduces the interval between full synchronous discharges (Fig. 10C). At 8 mM K^+discharge duration is better correlated with the preceding interevent interval suggesting that recovery processes dominate. Therefore, the probability that partial events become supra-threshold for full synchronization varies during the activity cycle. If cellular excitability increases then the interevent interval is reduced so that recovery determines event timing. At lower excitability levels, network threshold processes govern the exact timing of full event initiation.

10 Spatial Patterns of Activity During Partial and Full Synchronization

One important issue to characterize the relationship between partial synchronization and full population events is to understand their spatial patterns of activity. These patterns would reflect both the local initiation dynamics and the subsequent fast propagation of local synchrony.

Multiunit records from several electrodes show that partial synchronization can remain spatially segregated. In the disinhibited hippocampus, partial synchrony may be restricted to its initiating site in CA3a region (Fig. 11A) while multiple independently active cell clusters can be detected in the isolated subiculum under zero Mg^{2+} conditions (Fig. 11B). The amplitude, number, and sequence of cell firing during consecutive events vary substantially, in line with a variable recruitment. This is also reflected in the large variability of the associated field events and in the irregular dynamics of their initiation (Menendez de la Prida and Gal, 2004).

In contrast, full population bursts are never spatially constrained. They are typically initiated in the driver region following the build-up processes we have described and then spread rapidly throughout the tissue (Miles et al., 1988). One exception to this rule is the seizure-like activity induced in the absence of synaptic transmission (low Ca^{2+}). It has been shown that seizure patterns at a large spatiotemporal scale (1–2 mm and several seconds) proceed after clustering of full synchronous events within spatially segregated modules (Bikson et al., 2003; Li et al., 2007). The mechanism leading to this spatial aggregation seems to depend on non-synaptic interactions (gap junctions and field effects) and intrinsic properties of a critical mass of neurons (Fox et al., 2007).

In the disinhibited hippocampus, latencies of 1–2 ms are detected between the peak of field activity in CA3a and CA2 regions and the lag is about 3 ms between CA3a and CA3b during full event discharges (Wittner and Miles, 2007). This latency difference between distant sites suggests that synchronous firing propagates in a polysynaptic wave-like fashion (Colom and Saggau, 1994; Knowles et al., 1987; Miles and Wong, 1987). Propagation speed in longitudinal hippocampal slices has been estimated at about 0.1 mm/ms, considerably slower than axonal conduction

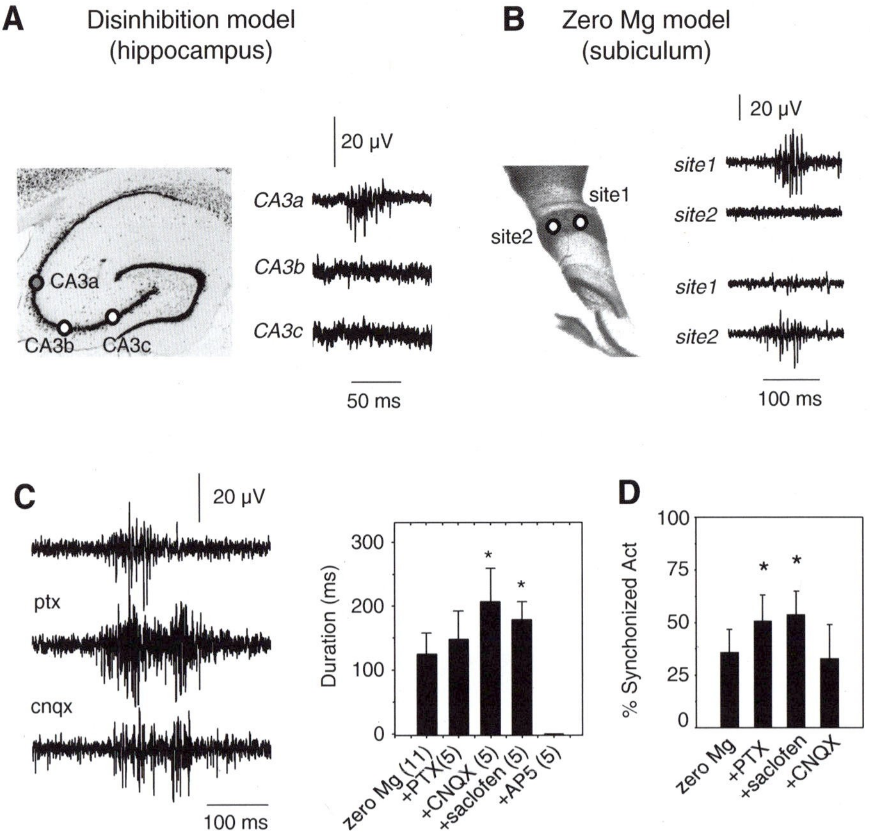

Fig. 11 (**A**) Multiunit recordings from the hippocampal CA3 region under disinhibition show that partial synchronization is spatially restricted to the CA3a-triggering area. (**B**) In the subiculum under zero Mg^{2+} partial synchronization can emerge from multiple clusters. Note that both sites 1 and 2 can initiate partial events that remain spatially constrained. (**C**) Modulation of the duraxtion of partial events in the isolated subiculum by drugs that affect synaptic transmission. $GABA_aR$: picrotoxin (PTX); AMPA/kainate receptors: CNQX; $GABA_bR$: saclofen. NMDA receptors: AP5. (**D**) Changes of synchronization between originally segregated spatial events after superfusion with drugs that affect synaptic transmission. Panels **B**, **C** and **D** are modified from Menendez de la Prida and Gal (2004) with permission

velocity (Miles et al., 1988; Traub and Miles, 1991). The slower conduction presumably involves time taken for the processes of synaptic transmission and initiation of firing as well as axonal conduction.

Synaptic processes controlling the spatial behaviour of full and partial events can be examined using pharmacological tools. Following the initiation of a fully synchronous event in the triggering region, its amplitude and propagation speed depend on the excitatory glutamatergic network but not on GABAergic inhibition for distances smaller than 1.5 mm (Pinto et al., 2005). In longitudinal slices of the CA3 region, of length several millimeters, full events may attenuate and slow if $GABA_aR$-mediated inhibition is only partially impaired (Miles et al., 1988). If full events propagate into different structures, such as from the hippocampus to the subiculum, regional differences in the inhibitory control of polysynaptic excitation can affect propagation dynamics (Menendez de la Prida and Pozo, 2002).

On the contrary, partial events are strongly controlled by the GABAergic circuits due to their local nature. Suppressing either $GABA_aR$- or $GABA_bR$-mediated inhibition increases the duration and the spatial spread of partial events (Fig. 11C; Menendez de la Prida and Gal, 2004; Cohen et al., 2007). In the isolated subiculum under zero Mg^{2+}, activity clusters remain mostly segregated and few episodes involve synchronous firing (Menendez de la Prida and Gal, 2004). A correlation index between two sites increases abruptly following exposure to $GABA_aR$- and $GABA_bR$-mediated antagonists (Fig. 11D). Nevertheless, $GABA_aR$-mediated inhibition may play a more critical role in controlling the spatial extent of cells involved in synchronous events (Menendez de la Prida and Gal, 2004). Therefore, the triggering region remains spatially constrained as fast inhibition fractionates active cell clusters into multiple modules.

This critical role of fast GABAergic inhibition is evident intracellularly as a disruption of the glutamatergic-inhibitory sequence described in the triggering area (Fig. 12). Instead of a clear excitatory activation followed by inhibitory repolarization and hyperpolarization, some neurons in the border between partial synchrony clusters exhibit an opposite synaptic sequence (Menendez de la Prida and Gal, 2004). Conductance analysis in these cells revealed the presence of inhibitory synaptic currents at the onset of partial events (Fig. 12). The composite *Vrev* showed a drop toward negative potentials near the reversal potential of $GABA_aR$-mediated

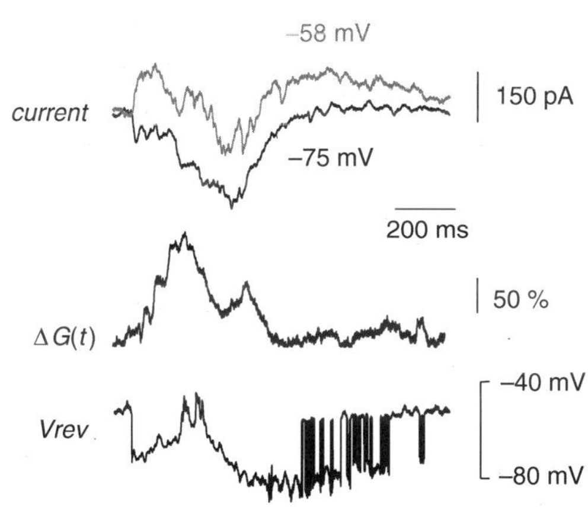

Fig. 12 Evolution of the synaptic conductance and the composite reversal potential recordings from a cell in the border between partial synchrony clusters in the isolated subiculum under zero Mg^{2+}. Upper traces show the associated synaptic currents at two different holding potentials. Modified from Menendez de la Prida and Gal (2004) with permission

currents. A second barrage of excitatory synaptic activity was evident, as a shift of *Vrev* towards depolarizing values. This synaptic activity pattern is consistent with the presence of an inhibitory surround acting to define the spatial limits of active cell clusters, as previously suggested in acute artificial epilepsy focus (Dichter and Spencer, 1969; Prince and Wilder, 1967). It imposes an inhibitory veto to the build-up processes and determines the spatial segregation of partial synchrony (Trevelyan et al., 2006).

11 Cell-Type Dissection of Full Population Discharges: Leader Cells

We must also consider how single cells contribute to partial and full synchronous activities. Cell-type specific firing patterns impose a structure on interactions between different neurons during the construction of synchronous discharges. Examination of the cell-type firing patterns during synchronous discharges focuses on the cellular relationships between different neuronal types and population synchrony.

The participation of single cells of a given type during population events is clearly heterogeneous. In disinhibited hippocampal slices from guinea pigs, some but not all cells at the triggering region initiate firing during the 50–100 ms before the onset of population bursting (Wittner and Miles, 2007). Similarly, in the isolated rat subiculum under zero Mg^{2+}, a group of pyramidal cells is firing before the emergence of network synchrony (Menendez de la Prida and Gal, 2004). Since these cells fire during the build-up or initiation of population events, they presumably represent leader cells.

We have examined temporal correlations between single-cell firing and population activity using field-triggered data analysis. The timing of cellular firing with respect to a population event is calculated to provide a peri-event firing histogram (Gerstein and Perkel, 1969). Since single cell firing can vary significantly between successive population discharges, the contribution of a cell must be averaged over several events. To this purpose, we introduced the driving index, σ, as a measure of the probability of a given cell to fire before the field potential (Menendez de la Prida and Gal, 2004):

$$\sigma = \frac{\#action\_potential(t < 0)}{\#action\_potential(total)}$$

Leader cells that always fire before the field possess a driving index of 1, while follower cells that consistently fire after the field have a driving index of 0. Cells with an intermediate index of 0.5 cannot be defined as either follower or leader cells. We use this index to define the electrophysiological and morphological properties of leader cells.

In the isolated subiculum, cells that typically fire within the 100 ms preceding the initiation of full events are a subpopulation of bursters (Menendez de la Prida

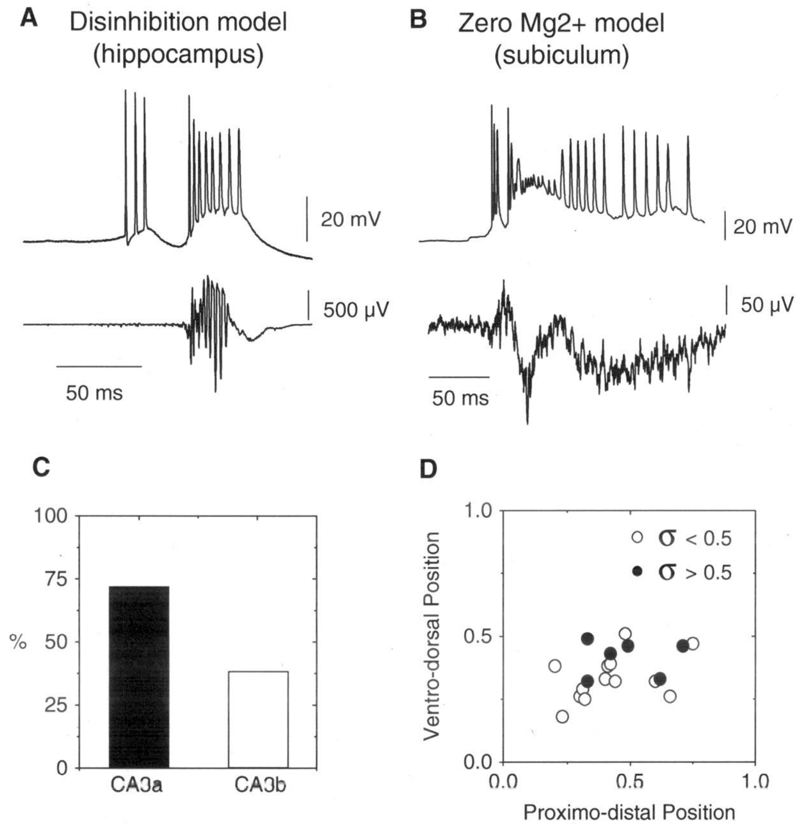

Fig. 13 (**A**, **B**) Leader cells of the hippocampus and the subiculum typically fire during the build-up period preceding the onset of full population events. (**C**) Distribution of leader cells in the disinhibited hippocampus. (**D**) Spatial distribution of subicular cells exhibiting lower ($\sigma < 0.5$) and higher ($\sigma > 0.5$) driving index. Panels **B** and **D** modified from Menendez de la Prida and Gal (2004) with permission. Panel **C** is modified from Wittner and Miles (2007) with permission

and Gal, 2004; Fig. 13B). Most subicular regular-spiking cells are activated at the onset of the field activity associated with full events. Similarly, cells that fired consistently during the build-up of full discharges in the disinhibited CA3 region are burst-firing cells (Wittner and Miles, 2007; Fig. 13A). In the subiculum, a clear spatial organization of leader cells was not evident (Fig. 13D), but most leader cells in the hippocampus are located in the CA3a region (Fig. 13C).

By recruiting other neurons to fire, leader cells can contribute to the initiation of population events. While it is not clear whether leader cells establish more effective synapses than follower cells, single leader cells have been shown to induce both field potentials and multiunit activity with latencies on the order of 5–50 ms (Fig. 14A, B; Cohen and Miles, 2000; Menendez de la Prida et al., 2006; Wittner and Miles, 2007). This type of field activity is similar to spontaneous partial events. First they are both associated with negative–positive field deflections which correspond intracellularly with sequences of excitation and inhibition. Second, when multiunit firing is induced it is usually followed by a transient suppression of firing. Furthermore, discharging repetitively single leader cells in the disinhibited hippocampus accelerates full synchronous bursting after the induction of supra-threshold partial events (Fig. 14D). Therefore, leader cells can truly initiate both partial and fully

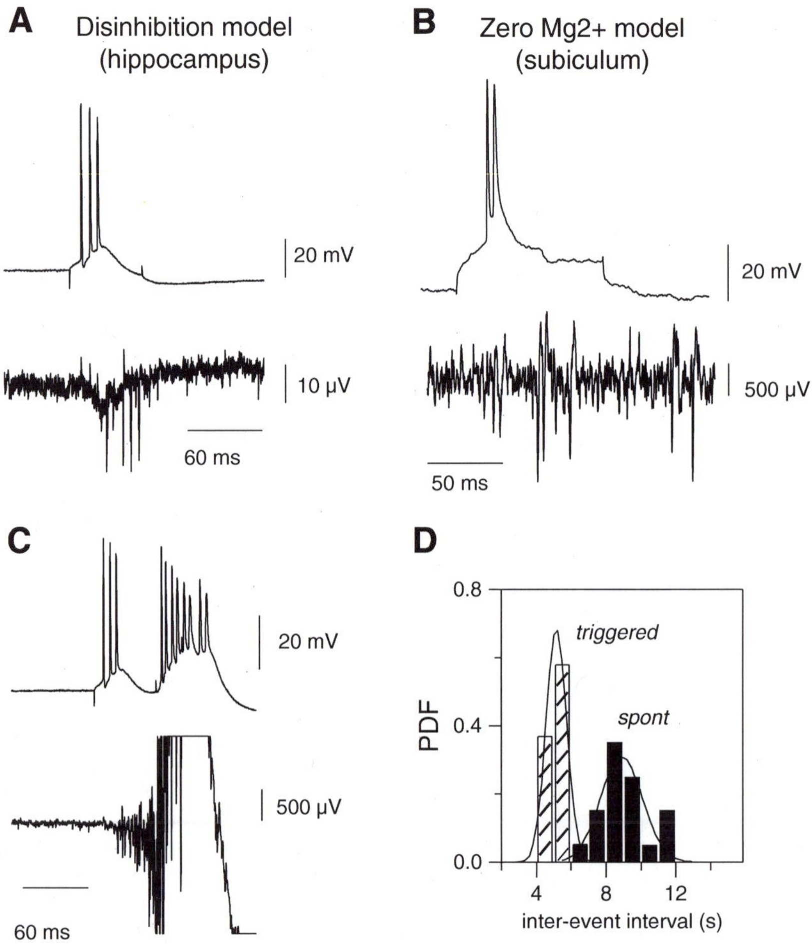

Fig. 14 (**A**, **B**) Influences of single hippocampal and subicular leader cells in population activities. Firing from a single cell can initiate multiunit activity and small field deflection indicating the activation of polysynaptic circuits. (**C**) In the disinhibited hippocampus, discharging leader cells can trigger full population events after partial synchronization. (**D**) Probability density function (PDF) showing how full population events accelerate by discharging a single leader cell in the hippocampus. Panel **B** is modified from Menendez de la Prida (2006) with permission. Panels **C** and **D** modified from Menendez de la Prida et al. (2006) with permission

synchronous population events (Miles and Wong, 1983; Menendez de la Prida et al., 2006).

A leading role of bursting cells has been recognized in different in vitro models of interictal activity (Miles and Wong, 1983; Chagnac-Amitai and Connors, 1989; Jensen and Yaari, 1997) and in chronic epileptic tissue (Sanabria et al., 2001; Cohen et al., 2002). Burst firing from a group of pyramidal cells act to initiate population firing in the developing hippocampus (Sipilä et al., 2005). It has been shown that a permanent up-regulation of intrinsic bursting contributes to hyperexcitability in chronic epilepsy (Wellmer et al., 2002). Firing from single cortical cells can even initiate whisker movements with latencies of 50–100 ms (Brecht et al., 2004). Excitable initiating cells have been implicated in the genesis of sharp waves (Buzsáki, 1989)

What is the nature of these leader cells? While both subicular and CA3a leader cells fire in bursts, they seem to possess other electrophysiological and morphological specificities. In the subiculum, there is a negative correlation between action potential-firing threshold and the cell-driving index (Fig. 15B), which is not present in the hippocampus (Fig. 15A). CA3a leader cells may possess small differences in firing threshold and depolarizing after-potentials, which facilitate burst firing (Wittner and Miles, 2007). Furthermore, dendritic arbors of both subicular and CA3a leader cells are longer and more elaborated than those of follower neurons (Fig. 15C, D). A strong positive correlation between the mean dendritic crossings and the driving index in the subiculum further support this connection (Fig. 15E).

Recurrent excitatory synapses drive synchronous firing (Miles and Wong, 1983). Differences in the connectivity of recurrent synapses might then contribute to the leader role of these cells. Synaptic connectivity is dependent on contacts between presynaptic terminals and postsynaptic dendrites. More complex, longer dendritic arborization of leader cells, especially in zones receiving recurrent synapses, might underlie a numerically stronger connectivity. Indeed, by convolving the number of axon terminals with the estimated dendritic length of reconstructed CA3 pyramidal cells, it was suggested that hippocampal leader cells in the CA3a region receive a higher density of recurrent inputs (Wittner and Miles, 2007). Also, in contrast to single spikes, bursts may secure spike transmission better, depending on the release probability and the facilitating nature of synapses (Miles and Wong, 1986; Deuchars et al., 1994). Therefore, leader bursting cells are better connected into the circuit and recruit target cells to fire thereby mediating polysynaptic activation.

12 Cell-Type Firing Dynamics of Pyramidal Cells and GABAergic Interneurons

If firing induced by current injection into some leader cells can initiate partial-like events, we must ask how these cells can generate spontaneous cascades of firing. Understanding the function of different cell types during partial activities would

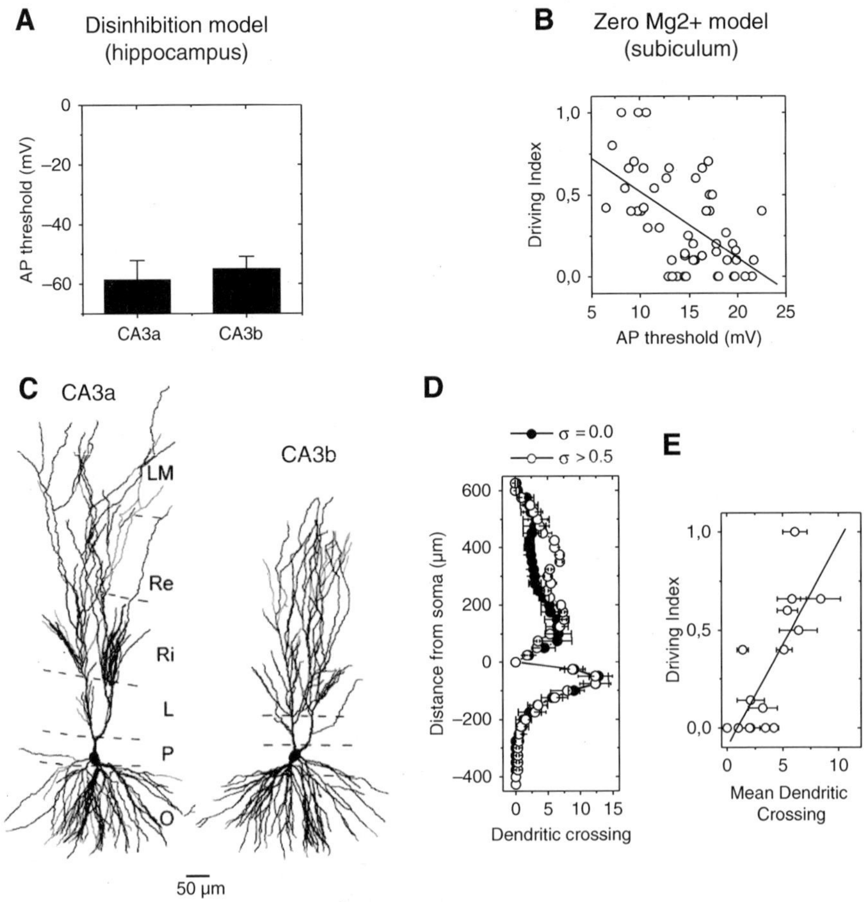

Fig. 15 (**A**, **B**) Relationship between action potential threshold and the leader capacity of a cell in the disinhibited hippocampus and the isolated subiculum. (**C**) Leader cells from the hippocampus are typically located at the CA3a region and exhibit more dendritic arborization compared with CA3b cells. (**D**) Quantification of dendritic arborization of leader ($\sigma > 0.5$) and follower ($\sigma = 0$) cells of the subiculum. (**E**) Correlation between the mean number of dendritic crossing at 250–450 μm from the soma and the driving index of subicular cells. Panels **B**, **D** and **E** are modified from Menendez de la Prida and Gal (2004) with permission. Panels **A** and **C** are from Wittner and Miles (2007) with permission

help define mechanistic links between cellular firing, synaptic events and activities of neuronal populations (Menendez de la Prida, 2006).

In the subiculum, superfused with zero Mg^{2+} ACSF, leader bursting cells initiate a local recruitment process associated with partial synchrony (Menendez de la Prida and Gal, 2004). Typically, regular-spiking cells fire after the local field potential. The timing of firing is often very variable and a given cell may remain silent over several successive episodes (Fig. 16). Therefore, the initial glutamatergic conductance

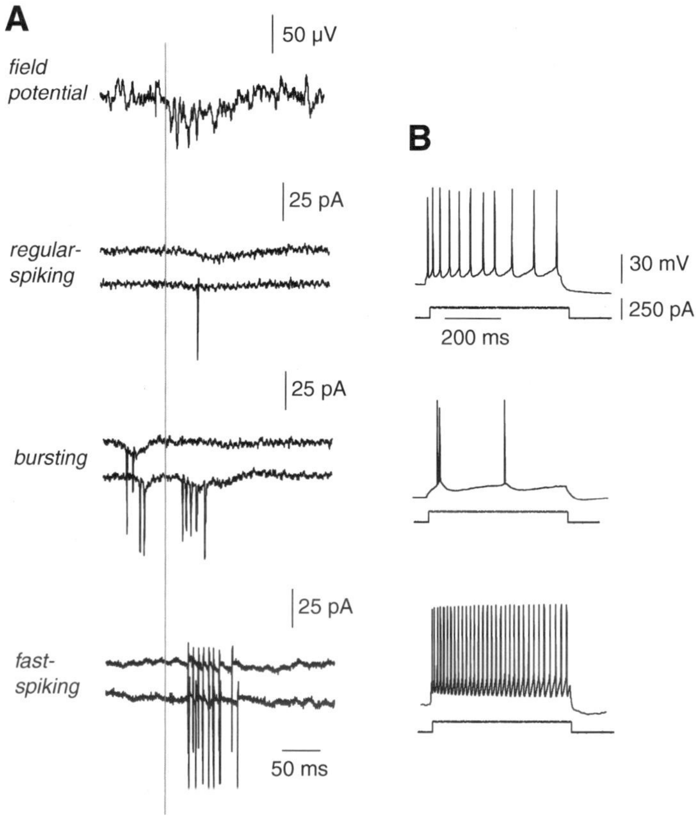

Fig. 16 (**A**) Cell-type firing dynamics during partial events in the isolated subiculum under zero Mg^{2+}. (**B**) Electrophysiological responses to depolarizing current injection of the three different cell types: regular-spiking, bursting and fast-spiking. Modified from Menenedez de la Prida and Gal (2004) and Menendez de la Prida (2006) with permission

increase coupled to partial events is mainly driven by the activity of a group of bursting cells. Possibly, other postsynaptic pyramidal cells help in setting the excitatory wave front, recorded as a negative field potential or detected from multiunit firing. Remarkably, presumed GABAergic cells consistently fire 10–15 ms after the start of partial events (Fig. 16). This late activity of fast-spiking interneurons is responsible for a delayed onset of inhibitory conductances, which curtail pyramidal cell firing and generate the positive field deflections that terminate partial events.

Similar cell firing sequences are detected in the isolated subiculum during full events: (1) a group of bursting cells tends to fire during the build-up period preceding burst onset; (2) most regular-spiking and other bursting cells start firing at the onset of the field event, together with fast-spiking interneurons (Menendez de la Prida and Gal, 2004). This type of circuit in which leader cells activate the feedback of a group of GABAergic interneurons generates a sequence of excitation and inhibition that

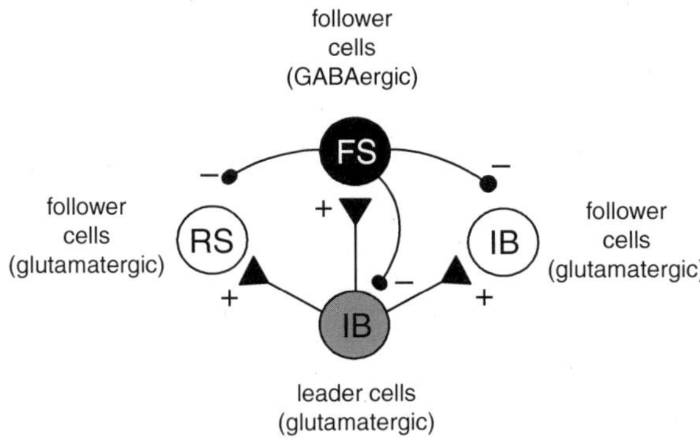

Fig. 17 Schematic representation of a neuronal circuit organized in a leader–follower fashion. Leader bursting cells (IB) recruit firing from follower glutamatergic (either regular-spiking, RS and bursting cells) and GABAergic (FS) neurons. GABAergic follower cells feedback onto leader and follower glutamatergic neurons to attenuate their activity

efficiently controls runaway excitation (Fig. 17). With this design, inhibition can be adjusted by excitation since GABAergic interneurons firing is time-locked to that of leader cells.

However, the control system sometimes 'fails' and full events are initiated. This 'failure' of inhibition seems to depend on firing rate during the synchronization process corresponding to a threshold-like behaviour (Menendez de la Prida et al., 2006). Possible factors underlying this transition include secondary activation driven by follower glutamatergic cells. Short-term changes in synaptic efficacy during higher frequency firing may differ at glutamatergic and GABAergic synapses (Galarreta and Hestrin, 1998), or frequency-dependent changes may occur in the reversal potential of $GABA_a$R-mediated currents (Thomson and Gähwiler, 1989; Staley and Proctor, 1999).

Other network interactions will also complicate this simple circuit design. In the hippocampus under high K^+, firing from horizontal *stratum oriens* (SO) interneurons precedes and outlasts action potential firing in most pyramidal cells (Aradi and Maccaferri, 2004). A similar behaviour of SO cells was described during seizure-like activity induced by 4-AP (Ziburkus et al., 2006) or high-frequency stimulation (Perez Velazquez and Carlen, 1999). Seizure-like discharges may also involve additional or distinct network processes (Avoli et al., 2002), so the behaviour and roles of leader–follower cells identified in the subiculum under zero Mg^{2+}, may be different in vivo. Even so, the broad class of interneurons includes cells with different axonal patterns targeting distinct postsynaptic domains and displaying specific firing behaviour during population activities in vivo (Somogyi and Klausberger, 2005). Inhibition from a group of SO 'leader' interneurons at the distal dendritic branches can differentially control the pyramidal cell output (Pouille and Scanziani, 2004). So, probably, different types of interneurons exhibit specific firing patterns during partial events and the build-up period preceding full synchronization, resulting in more subtle and complex circuit interactions than our present description.

13 Conclusions

> There are two reasons that a population of neurons might fire together: in response to an afferent input that is itself already synchronized; as a result of forces that couple the behaviours of the neurons together. For a synchronized burst in an isolated population of neurons, especially a burst in which the triggering stimulus excites only a small number of neurons, synchrony throughout the population can arise only because of cellular interactions.

Perhaps this paragraph from the book by Traub and Miles (1991) clearly poses the central problem that we have reexamined in this chapter. How do interactions of a small number of neurones lead to synchrony? We have shown that this behaviour can be examined at levels from macroscopic field measures of synchrony via records of multiunit activity to interactions and even the identity of single cells that can affect population activity. Now it seems that synchrony arises from a situation where two types of synaptic signals with different time course are opposed. Excitatory glutamatergic signals are transmitted from a subpopulation of pyramidal cells that discharge earlier perhaps due to an enhanced excitability and higher connectivity. These leader cells appear to be heavily arborizing intrinsic busters that discharge some of their synaptic targets with a delay of several milliseconds. Delayed and slower inhibitory signals transmitted by the activation of $GABA_a$ and $GABA_b$ receptors can limit synchrony in time and in space. Or, if glutamatergically transmitted firing spreads fast enough to escape inhibition and exceed a population firing threshold, a fully synchronous population burst will result with a delay of several tens of milliseconds. The frequency condition emerges from the need for activity to spread quickly, since a critical number of cells must fire synchronously before inhibitory feedback forces suppress firing.

Neuronal synchronization then emerges from the postsynaptic integration across a neuronal population of synaptic events initiated by many single cells over a timescale of tens or hundreds of milliseconds. This summed effect induces a coherent extracellular current flow that rescales as the local field reflecting a macroscopic activity pattern correlated with the synchronous output of 400–500 neurons. These are the elementary processes underlying interictal spikes or sharp-wave events recorded as the EEG. The microdynamics of partial synchrony, spatially restricted to a distance of perhaps a few hundred micrometers, contrasts with the spread of fully synchronous discharges over a few millimetres or even further before attenuation. EEG spikes of different amplitude, and perhaps duration thus represent different degrees of local neuronal coordination with different spatial and temporal extents. At this level, the understanding of longer-range neuronal interactions involving much larger numbers of neurons, circuits and regions represents the challenge of studies on the whole brain.

References

Abeles M (1982) Quantification, smoothing, and confidence limits for single-unit histograms. J Neurosci Methods 5:317–325.

Agmon A, Connors BW (1991) Thalamocortical responses of mouse somatosensory (barrel) cortex in vitro. Neuroscience 41: 365–379.

Agmon A, Wells JE (2003) The role of the hyperpolarization-activated cationic current I(h) in the timing of interictal bursts in the neonatal hippocampus. J Neurosci 23: 3658–3668.

Alarcon G, Garcia Seoane JJ, Binnie CD, Martin Miguel MC, Juler J, Polkey CE, Elwes RDC, Ortiz Blasco JM (1997) Origin and propagation of interictal discharges in the acute electrocorticogram. Implications for pathophysiology and surgical treatment of temporal lobe epilepsy. Brain 120: 2259–2282.

Anderson P, Bliss VP, Skrede KK (1971) Lamellar organization of hippocampal excitatory pathways. Exp Brain Res 13:222–238.

Ankri N, Korn H (1999) A statistical method for correcting distortions of amplitude distribution histograms due to collisions of synaptic events. J Neurosci Methods 91:83–99.

Aradi I, Maccaferri G (2004) Cell type-specific synaptic dynamics of synchronized bursting in the juvenile CA3 rat hippocampus. J Neurosci 24:9681–9692.

Avoli M, D'Antuono M, Louvel J, Köhling R, Biagini G, Pumain R, D'Arcangelo G, Tancredi V (2002) Network and pharmacological mechanisms leading to epileptiform synchronization in the limbic system in vitro. Prog Neurobiol 68:167–207.

Bartos M, Vida I, Jonas P (2007) Synaptic mechanisms of synchronized gamma oscillations in inhibitory interneuron networks. Nat Rev Neurosci 8:45–56.

Bikson M, Fox JE, Jefferys JG (2003) Neuronal aggregate formation underlies spatiotemporal dynamics of nonsynaptic seizure initiation. J Neurophysiol 89:2330–2333.

Borg-Graham LJ, Monier C, Fregnac Y (1998) Visual input evokes transient and strong shunting inhibition in visual cortical neurons. Nature 393:369–373.

Bourien J, Bartolomei F, Bellanger JJ, Gavaret M, Chauvel P, Wendling F (2005) A method to identify reproducible subsets of co-activated structures during interictal spikes. Application to intracerebral EEG in temporal lobe epilepsy, Clin Neurophysiol 116:443–455.

Bragin A, Jandó G, Nádasdy Z, Hetke J, Wise K, Buzsáki G (1995) Gamma (40–100 Hz) oscillation in the hippocampus of the behaving rat. J Neurosci 15:47–60.

Brecht M, Schneider M, Sakmann B, Margrie TW (2004) Whisker movements evoked by stimulation of single pyramidal cells in rat motor cortex. Nature 427:704–710.

Brown EN, Kass RE, Mitra PP (2004) Multiple neural spike train data analysis: state-of-the-art and future challenges. Nat Neurosci 7:456–461.

Buzsaki G (1986) Hippocampal sharp waves: their origin and significance. Brain Res 398:242–252.

Buzsáki G (1989) Two-stage model of memory trace formation: a role for 'noisy' brain states. Neuroscience 31:551–570.

Buzsaki G, Eidelberg E (1983) Phase relations of hippocampal projection cells and interneurons to theta activity in the anesthetized rat. Brain Res 266:334–339.

Chagnac-Amitai Y, Connors BW (1989) Synchronized excitation and inhibition driven by intrinsically bursting neurons in neocortex. J Neurophysiol 62:1149–1162.

Cohen I, Huberfeld G, Miles R (2006, Feb 1) Emergence of disinhibition-induced synchrony in the CA3 region of the guinea pig hippocampus in vitro. J Physiol 570(Pt 3):583–594.

Cohen I, Miles R (2000) Contributions of intrinsic and synaptic activities to the generation of neuronal discharges in in vitro hippocampus. J Physiol 524:485–502.

Cohen I, Navarro V, Clemenceau S, Baulac M, Miles R (2002). On the origin of interictal activity in human temporal lobe epilepsy in vitro. Science 298:1418–1421.

Colom LV, Saggau P (1994) Spontaneous interictal-like activity originates in multiple areas of the CA2–CA3 region of hippocampal slices. J Neurophysiol 71:1574–1585.

Csicsvari J, Hirase H, Mamiya A, Buzsaki G (2000) Ensemble patterns of hippocampal CA3–CA1 neurons during sharp wave-associated population events. Neuron 28:585–594.

Daley D, Vere-Jones D (2003) An Introduction to the Theory of Point Process, 2nd ed. Springer-Verlag, New York.

DeFelipe J (1999) Chandelier cells and epilepsy. Brain 122:1807–1822.

Deuchars J, West DC, Thomson AM (1994 Aug 1) Relationships between morphology and physiology of pyramid-pyramid single axon connections in rat neocortex in vitro. J Physiol 478(Pt 3):423–435.

Dichter M, Spencer WA. (1969) Penicillin-induced interictal discharges from the cat hippocampus. II. Mechanisms underlying origin and restriction. J Neurophysiol 32:663–687.

Engel J Jr, Wilson C, Bragin A (2003) Advances in understanding the process of epileptogenesis based on patient material: what can the patient tell us? Epilepsia 44 (Suppl 12):60–71.

Fedirchuk B, Wenner P, Whelan PJ, Ho S, Tabak J, O'Donovan MJ (1999) Spontaneous network activity transiently depresses synaptic transmission in the embryonic chick spinal cord. J Neurosci 19:2102–2112.

Foffani G, Uzcategui YG, Gal B, Menendez de la Prida L (2007) Reduced spike-timing reliability correlates with the emergence of fast ripples in the rat epileptic hippocampus. Neuron 55: 930–941.

Fox JE, Bikson M, Jefferys JG (2007) The effect of neuronal population size on the development of epileptiform discharges in the low calcium model of epilepsy. Neurosci Lett 411: 158–161.

Gähwiler BH, Capogna M, Debanne D, McKinney RA, Thompson SM (1997) Organotypic slice cultures: a technique has come of age. Trends Neurosci 20:471–477.

Galarreta M, Hestrin S (1998) Frequency-dependent synaptic depression and the balance of excitation and inhibition in the neocortex. Nat Neurosci 1:587–594.

Gerstein GL, Perkel DH (1969) Simultaneously recorded trains of action potentials: analysis and functional interpretation. Science 164:828–830.

Gold C, Henze DA, Koch C, Buzsáki G (2006) On the origin of the extracellular action potential waveform: a modeling study. J Neurophysiol 95:3113–3128.

Goldensohn ES, Zablow L, Salazar A (1977) The penicillin focus. I. Distribution of potential at the cortical surface. Electroencephalogr Clin Neurophysiol 42:480–492.

Graf M, Niedermeyer E, Schiemann J, Uematsu S, Long DM (1984) Electrocorticography: information derived from intraoperative recordings during seizure surgery. Clin Electroencephalogr 15:83–91.

Grzywacz NM, Sernagor E (2000) Spontaneous activity in developing turtle retinal ganglion cells: statistical analysis. Vis Neurosci 17:229–241.

Gutnick MJ, Connors BW, Prince DA (1982) Mechanisms of neocortical epileptogenesis in vitro. J Neurophysiol 48:1321–1335.

Hablitz JJ (1984) Picrotoxin-induced epileptiform activity in hippocampus: role of endogenous versus synaptic factors. J Neurophysiol 51:1011–1027.

Hausser M, Clark BA (1997). Tonic synaptic inhibition modulates neuronal output pattern and spatiotemporal synaptic integration. Neuron 19:665–678.

Helmchen F, Svoboda K, Denk W, Tank DW (1999) In vivo dendritic calcium dynamics in deep-layer cortical pyramidal neurons. Nat Neurosci 2:989–996.

Hubel DH (1957) Tungsten microelectrodes for recording single units. Science 125:549–555.

Jefferys JG (1994) Experimental neurobiology of epilepsies. Curr Opin Neurol 7:113–122.

Jensen MS, Yaari Y (1997) Role of intrinsic burst firing, potassium accumulation, and electrical coupling in the elevated potassium model of hippocampal epilepsy. J Neurophysiol 77: 1224–1233.

Johnston D, Brown TH. (1984) The synaptic nature of the paroxysmal depolarizing shift in hippocampal neurons. Ann Neurol 16 Suppl:S65–S71.

Knowles WD, Traub RD, Strowbridge BW (1987) The initiation and spread of epileptiform bursts in the in vitro hippocampal slice. Neuroscience 21:441–455.

Latham PE, Richmond BJ, Nelson PG, Nirenberg S (2000) Intrinsic dynamics in neuronal networks. I. Theory. J Neurophysiol 83:808–827.

Le Van Quyen M, Bragin A (2007) Analysis of dynamic brain oscillations: methodological advances. Trends Neurosci 30:365–367.

Leinekugel X, Khalilov I, Ben-Ari Y, Khazipov R (1998) Giant depolarizing potentials: the septal pole of the hippocampus paces the activity of the developing intact septohippocampal complex in vitro. J Neurosci 18:6349–6357.

Leung LS (1992) Fast (beta) rhythms in the hippocampus: a review. Hippocampus 2:93–98.

Levy R, Hutchison WD, Lozano AM, Dostrovsky JO (2000) High-frequency synchronization of neuronal activity in the subthalamic nucleus of parkinsonian patients with limb tremor. J Neurosci 20:7766–7775.

Li X, Cui D, Jiruska P, Fox JE, Yao X, Jefferys JG (2007) Synchronization measurement of multiple neuronal populations. J Neurophysiol 98:3341–3348.

Lueders H, Bustamante LA, Zablow L, Goldensohn ES (1981) The independence of closely spaced discrete experimental spike foci. Neurology 31:846–851.

Mann EO, Suckling JM, Hajos N, Greenfield SA, Paulsen O (2005) Perisomatic feedback inhibition underlies cholinergically induced fast network oscillations in the rat hippocampus in vitro. Neuron 45:105–117.

Meister M, Wong RO, Baylor DA, Shatz CJ (1991) Synchronous bursts of action potentials in ganglion cells of the developing mammalian retina. Science 252:939–943.

Menendez de la Prida L (2006) Functional features of the rat subicular microcircuits studied in vitro. Behav Brain Res 174:198–205.

Menendez de la Prida L, Benavides-Piccione R, Sola R, Pozo MA (2002) Electrophysiological properties of interneurons from intraoperative spiking areas of epileptic human temporal neocortex. Neuroreport 13:1421–1425.

Menendez de la Prida L, Gal B (2004) Synaptic contributions to focal and widespread spatiotemporal dynamics in the isolated rat subiculum in vitro. J Neurosci 24:5525–5536.

Menendez de la Prida LM, Huberfeld G, Cohen I, Miles R. (2006) Threshold behavior in the initiation of hippocampal population bursts. Neuron 49:131–142.

Menendez de la Prida L, Pozo MA (2002) Excitatory and inhibitory control of epileptiform discharges in combined hippocampal/entorhinal cortical slices. Brain Res 940:27–35.

Menendez de la Prida LM, Sanchez-Andres JV (1999) Nonlinear frequency-dependent synchronization in the developing hippocampus. J Neurophysiol 82:202–208.

Miles R, Traub RD, Wong RK (1988) Spread of synchronous firing in longitudinal slices from the CA3 region of the hippocampus. J Neurophysiol 60:1481–1496.

Miles R, Wong RK (1983) Single neurones can initiate synchronized population discharge in the hippocampus. Nature 306:371–373.

Miles R, Wong RK (1986) Excitatory synaptic interactions between CA3 neurones in the guinea-pig hippocampus. J Physiol 373:397–418.

Miles R, Wong RK (1987) Inhibitory control of local excitatory circuits in the guinea-pig hippocampus. J Physiol 388:611–629.

Mody I, Lambert JD, Heinemann U (1987) Low extracellular magnesium induces epileptiform activity and spreading depression in rat hippocampal slices. J urophysiol 57:869–888; Nature 429:717–723.

Niedermeyer E, Rocca U (1972) The diagnostic significance of sleep electroencephalograms in temporal lobe epilepsy. A comparison of scalp and depth tracings. Eur Neurol 7:119–129.

O'Donovan MJ. (1999) The origin of spontaneous activity in developing networks of the vertebrate nervous system. Curr Opin Neurobiol 9:94–104.

Pastor J, Menendez de la Prida LM, Hernando V, Sola RG (2006) Voltage sources in mesial temporal lobe epilepsy recorded with foramen ovale electrodes. Clin Neurophysiol 117: 2604–2614.

Penfield W, Jasper H (1954) Epilepsy and the Functional Anatomy of the Human Brain. Boston, MA: Little, Brown & Co.

Perez Velazquez JL, Carlen PL (1999) Synchronization of GABAergic interneuronal networks during seizure-like activity in the rat horizontal hippocampal slice. Eur J Neurosci 11: 4110–4118.

Perreault P, Avoli M (1991) Physiology and pharmacology of epileptiform activity induced by 4-aminopyridine in rat hippocampal slices. J Neurophysiol 65:771–785.

Pinto DJ, Patrick SL, Huang WC, Connors BW (2005) Initiation, propagation, and termination of epileptiform activity in rodent neocortex in vitro involve distinct mechanisms. J Neurosci 25:8131–8140.

Pouille F, Scanziani M (2004) Routing of spike series by dynamic circuits in the hippocampus. Nature 429:717–723.

Prince DA, Wilder BJ. (1967) Control mechanisms in cortical epileptogenic foci. "Surround" inhibition. Arch Neurol 16:194–202.

Sanabria ER, Su H, Yaari Y (2001) Initiation of network bursts by Ca2+-dependent intrinsic bursting in the rat pilocarpine model of temporal lobe epilepsy. J Physiol 532:205–216.

Sanchez-Vives MV, McCormick DA (2000) Cellular and network mechanisms of rhythmic recurrent activity in neocortex. Nat Neurosci 3:1027–1034.

Scanziani M, Gahwiler BH, Thompson SM (1991) Paroxysmal inhibitory potentials mediated by GABAB receptors in partially disinhibited rat hippocampal slice cultures. J Physiol 444: 375–396.

Schneiderman JH (1986) Low concentrations of penicillin reveal rhythmic, synchronous synaptic potentials in hippocampal slice. Brain Res 398:231–241.

Schwartzkroin PA, Knowles WD (1984) Intracellular study of human epileptic cortex: in vitro maintenance of epileptiform activity? Science 223:709–712.

Schwartzkroin PA, Prince DA (1977) Penicillin-induced epileptiform activity in the hippocampal in vitro prepatation. Ann Neurol 1:463–469.

Sipilä ST, Huttu K, Soltesz I, Voipio J, Kaila K. (2005) Depolarizing GABA acts on intrinsically bursting pyramidal neurons to drive giant depolarizing potentials in the immature hippocampus. J Neurosci 25:5280–5289.

Somogyi P, Klausberger T (2005, Jan 1) Defined types of cortical interneurone structure space and spike timing in the hippocampus. J Physiol 562(Pt 1):9–26 (Epub 2004 Nov 11, Review).

Spruston N, Schiller Y, Stuart G, Sakmann B (1995) Activity-dependent action potential invasion and calcium influx into hippocampal CA1 dendrites. Science 268:297–300.

Staley KJ, Bains JS, Yee A, Hellier J, Longacher JM (2001) Statistical model relating CA3 burst probability to recovery from burst-induced depression at recurrent collateral synapses. J Neurophysiol 86:2736–2747.

Staley KJ, Longacher M, Bains JS, Yee A (1998). Presynaptic modulation of CA3 network activity. Nat Neurosci 1:201–209.

Staley KJ, Proctor WR (1999) Modulation of mammalian dendritic GABA(A) receptor function by the kinetics of Cl- and HCO3-transport. J Physiol 519:693–712.

Steriade M, Timofeev I (2003) Neuronal plasticity in thalamocortical networks during sleep and waking oscillations. Neuron 37:563–576.

Suzuki SS, Smith GK (1985) Single-cell activity and synchronous bursting in the rat hippocampus during waking behavior and sleep. Exp Neurol 89:71–89.

Tabak J, Rinzel J, O'Donovan MJ (2001) The role of activity-dependent network depression in the expression and self-regulation of spontaneous activity in the developing spinal cord. J Neurosci 21:8966–8978.

Thomson SM, Gähwiler BH (1989) Activity-dependent disinhibition. I. Repetitive stimulation reduces IPSP driving force and conductance in the hippocampus in vitro. J Neurophysiol 61:501–511.

Tóth K, Freund TF, Miles R. (1997) Disinhibition of rat hippocampal pyramidal cells by GABAergic afferents from the septum. J Physiol 500:463–474.

Traub RD, Miles R (1991) Neuronal Networks of the Hippocampus. Cambridge University Press, New York.

Traub RD, Wong RK (1982, May 14) Cellular mechanism of neuronal synchronization in epilepsy. Science 216(4547):745–747.

Traynelis SF, Dingledine R (1988) Potassium-induced spontaneous electrographic seizures in the rat hippocampal slice. J Neurophysiol 59:259–276.

Trevelyan AJ, Sussillo D, Watson BO, Yuste R. (2006) Modular propagation of epileptiform activity: evidence for an inhibitory veto in neocortex. J Neurosci 26:12447–12455.

Wellmer J, Su H, Beck H, Yaari Y (2002) Long-lasting modification of intrinsic discharge properties in subicular neurons following status epilepticus. Eur J Neurosci 16:259–266.

Wenner P, O'Donovan MJ (2001) Mechanisms that initiate spontaneous network activity in the developing chick spinal cord. J Neurophysiol 86:1481–1498.

Wittner L, Miles R (2007) Factors defining a pacemaker region for synchrony in the hippocampus. J Physiol 584:867–883.

Wong RK, Traub RD (1983) Synchronized burst discharge in disinhibited hippocampal slice. I. Initiation in CA2–CA3 region. J Neurophysiol 49:442–458.

Ziburkus J, Cressman JR, Barreto E, Schiff SJ (2006) Interneuron and pyramidal cell interplay during in vitro seizure-like events. J Neurophysiol 95:3948–3954.

Time–Frequency Methods and Brain Rhythm Signal Processing

Jesse Gillis

1 Introduction

Physiological signal analysis could encompass any of the methods of signal analysis being brought to bear upon a problem in physiology. In practice, physiological signal analysis is a narrower field than this because physiological signals share common features which limit the applicability of different methods. An important property of physiological signals is that they typically possess features determined by their historical contribution to organism fitness. This gives rise to the problem of identifying and classifying features present in a physiological signal. The primary statistical feature of physiological signals is that they are nonstationary, meaning their statistical characteristics change with time, and a variety of different techniques exist to analyze such data. The suitability of such techniques is data dependent, and particularly upon the degree and manner in which the signal changes in time and frequency. This is a well-known problem and specialized techniques have been developed for particular data sets (e.g., Achermann and Borbély, 1998). Developing data-specific methodologies presents its own difficulties in terms of hypothesis and verifiability (e.g., separation into training and testing sets for the methodology). In this article, I focus on introducing standard and well-defined techniques relevant to signal processing of brain rhythms as well as the organization of time–frequency activities in a hippocampal rhythm. In general, analyses consist of three stages: (1) Transformation into the time–frequency domain; (2) Quantification of properties in the time–frequency domain; and (3) Grouping together of these properties. For example, wavelet transform (time–frequency transform) coefficients (quantification) might be extracted and clustered (grouping) (Quiroga et al., 2002). Filtering of a neurophysiological signal might be performed (an implicit time–frequency step), a threshold chosen (an implicit quantification), and the remaining data averaged (grouping). Below, I suggest a generally useful way to incorporate

J. Gillis (✉)
The Centre for High-Throughput Biology (CHiBi), 177, Michael Smith Laboratories, 2185 East Mall, University of British Columbia, Vancouver, BC V6T1Z4, Canada
e-mail: jesse.gillis@gmail.com

J.L. Perez Velazquez, R. Wennberg (eds.), *Coordinated Activity in the Brain*, Springer Series in Computational Neuroscience 2, DOI 10.1007/978-0-387-93797-7_11,

these three steps into analyses with a minimum of additional assumptions, i.e., a time–frequency transformation, followed by characterization of the time–frequency domain's local activity, and then mixture distribution analysis.

2 Mathematical Preliminaries and Wavelets

Convolution (or *faltung*) is a mathematical operation performed between two functions, $g(t)$ and $f(t)$, and defined by:

$$\int_0^t f(\tau)\, g\, (t-\tau) d\, \tau$$

In a discrete sense, convolution can be implemented as a sliding dot product, where the vectors are defined by the values f and g take with discrete time. The convolution of two Gaussians will also be a Gaussian. In signal processing, the beginning and end of signals present a problem in applying convolution since the signal within an interval [0, t] will be affected by the values outside of that interval. Zero-padding edges, making time periodic, or excluding regions that are affected are possible options. The area under a convolution is the product of the area under its constituent functions, variances add, as do means (or centroids). "Smoothing" generally takes the form of convolution with a Gaussian. A specific "filtering" of a signal consists of performing a convolution on the signal with the response of the filter to an impulse. One simple example of this would be to see that convolution with the vector (1, –1) will remove any changes in the signal of duration greater than the difference between adjacent time values. In fact, this is the discrete implementation of taking a derivative, and also defines the Haar function. The zero-padded dot product of this vector with a signal will give the difference between the points where it is own values are nonzero. The nonzero portion then shifts along the extent of the signal, giving differences as it goes, which determines the output signal. This translation of the function along the signal can be combined with dilations to give the wavelet transform discussed in the following section. If the function being translated (a mother wavelet) is doubled in extent (duration) it will now be taking differences between groups of points – a larger scale derivative. In this way the wavelet transform gives the variation of a signal at different timescales (dilations) over time (translation). The initial function that is dilated is generally known as the "mother wavelet" and can take many different forms so long as it satisfies the admissibility criterion. That is, if $h(t)$ is the wavelet used for analysis, the Fourier spectrum of $h(t)$ must satisfy $\int \mathrm{F(s)}^2/\mathrm{sds} < \infty$. If, for example, we took the negative derivative of our derivative function, and smoothed it, we would obtain the Mexican Hat wavelet, defined by

$$\sqrt{\tfrac{1}{2\pi}}(1-t^2)\exp(-t^2/2)$$

That is, the negative second derivative of a Gaussian, which will take regions of change (edges at various scales) in the input signal and convert them into zeroes in the output signal. The Mexican hat can be scaled, translated, or constructed in higher dimensions in exactly the same way by examining an n-dimensional Gaussian distribution.

Recently these methods have been applied to the investigation of the frequency of neuronal firing thereby broadening it to include spike correlations and more complex elements of the pattern in time and frequency (e.g., Rieke et al., 1997; Gray et al., 1989; Nicolelis et al., 1995). Specifically, those signal methods allowing the examination of the variability of rhythm in the temporal domain (nonstationarity) are called time–frequency techniques, of which the wavelet transform is only one example.

3 Time–Frequency Transforms

One of the chief tasks of signal processing is finding representations for a signal that allow it to be seen and understood simply. As we have seen, regions in the hippocampus, as well as behavioral states, have been shown to correlate with different frequencies of activity. Because of this, a historically useful technique has been to transform the signal so that the magnitude of its different component frequencies can be directly plotted. The most common technique for obtaining the distribution of signal frequency is the Fourier transform with its inverse, respectively defined by:

$$X(\omega) = F\left[x(t)\right] = \int_{-\infty}^{\infty} x(t)e^{-i\omega t}dt$$

$$x(t) = F^{-1}\left[X(\omega)\right] = \frac{1}{2\pi}\int_{-\infty}^{\infty} X(\omega)e^{i\omega t}d\omega$$

In general I will use $x(t)$ to refer to an arbitrary signal with forward Fourier transform $X(\omega)$. This transform separates a signal into its constituent sine and cosine components, each of which could be defined entirely by a weight, frequency, and phase, but without any reference to variation of those parameters in time. This requires these components to be unchanging or stationary in time. If frequency components do change in time, then the Fourier transform does not provide a useful representation of the signal, because the contribution (amplitude) of different frequencies to a signal does not alone characterize the signal; it would be easier to understand the signal as having a changing contribution of frequencies. A general approach to the problem of nonstationary components would be to change the limits of the Fourier integral so that frequency components were captured at specific times, a technique that forms the basis of the spectrogram. Although nonstationary signals have a frequency that changes with time, there will typically be a fixed duration over which the frequency can be taken to be roughly constant. Breaking a signal into pieces and then performing a Fourier transform upon each piece as if it were stationary gives us both the frequency information contained within the Fourier transform, and knowledge of which piece of the signal these frequencies are operating in. It is this technique which forms the basis for the short-time Fourier transform (STFT), defined by (integrals are in the interval $(-\infty, \infty)$ except where stated):

$$STFT_x(t,\omega) = \int x(\tau)h(\tau - t)e^{-i\omega\tau}d\tau$$

with $h(t)$ giving the windows to look at specific times in the signal (Akay, 1998). A rectangular window gives high overall resolution but induces a ripple (typically called the Gibbs ripple) in the frequency domain due to the lack of continuity in the time domain, so a smooth window choice is necessary to improve the visualization of the frequency and its useful resolution. Computationally, $h(t)$ serves to break the signal into pieces, with the STFT being performed by taking the Fourier transform within a piece or window and then shifting the window forward to repeat. Since similar windowing occurs in the frequency domain, the STFT also be thought of as a collection of band-pass filters plotted along the same time axis. A discrete form of the STFT is given by:

$$STFT(k,n) = \sum_{m=0}^{L-1} x(n+m)g(m)\exp\left(\frac{-2i\pi mk}{N}\right)$$

where k is the frequency, n the time going up to N, L the discrete window length, and $g(m)$ a Gaussian window:

$$g(m) = \exp\left[-\left(2 - \frac{4m}{L-1}\right)^2\right], \quad 0 \le m \le L-1$$

These methods can be thought of as band-pass filters performed over a range of frequencies. The squared magnitude of the STFT is the spectrogram, which has been put to extensive use in diverse areas due to its very well-defined properties (Cohen, 1989). The most negative of those properties is its inherent tradeoff between temporal and frequency resolutions, sometimes called the uncertainty principle. As the window lengthens, the frequency resolution improves, but this must decrease the temporal resolution proportionately, leaving their product as a constant. The variance is the most common way to characterize the localization of a probability distribution, $f(t)$, where the variance is described by

$$\|(t - t_0)\cdot x(t)\|^2 = \int (t - t_0)^2 \cdot |x(t)|^2 dt$$

$$t_0 = \frac{\int t\cdot |x(t)|\,dt}{\int |x(t)|\,dt}$$

and t_0 is the mean. This usage can be generalized from probability distributions, and we may take the same definition of variance to describe the resolution of a function,

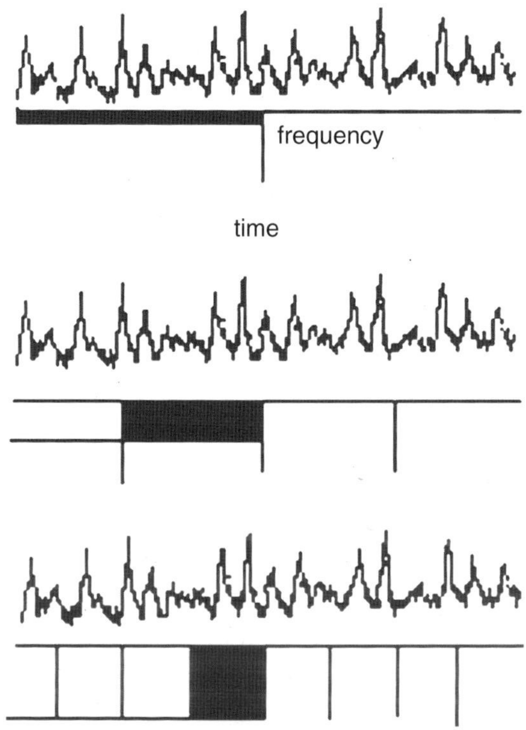

Fig. 1 A graphical representation of the uncertainty principle. An arbitrary signal is shown with the time and frequency resolutions below. The length of the rectangles represents the time resolution and the height of the frequency resolution with shorter lengths representing finer resolutions. In order to improve time resolution, shorter windows in time must be taken, but as the time window used to analyze the window decreases in length, the frequency content becomes less well resolved in each interval. This is reflected by the increasing width of the rectangle below each section with decreasing temporal length. The product of the two resolutions (the area of each rectangular) represents the joint time–frequency resolution, and the inverse relationship described creates a lower limit in the total possible resolution

where t_0 remains the first-order moment (Teolis, 1998). The uncertainty relation (shown graphically in Fig. 1) can be derived using the Cauchy–Schwarz inequality as follows

$$\left|\left\langle tx(t), x'(t)\right\rangle\right| \leq \|tx(t)\| \cdot \left\|x'(t)\right\| = \|tx(t)\| \cdot \|2\pi i\omega X(\omega)\|$$

$$\begin{aligned}\left\langle tx(t), x'(t)\right\rangle = \int tx(t)x'(t)dt = \int t \cdot \frac{1}{2}\partial_t x^2(t)dt = \frac{1}{2}\left(tx^2(t)\Big|_{-\infty}^{+\infty} - \int x^2(t)\right) \\ = \frac{1}{2}\|x\|^2\end{aligned}$$

Reorganizing, this then gives us the uncertainty relation

$$\|tx(t)\| \cdot \|\omega X(\omega)\| \geq \frac{1}{4\pi} \|x\|^2$$

Because the spectrogram is a bilinear signal energy distribution (the squared magnitude of the STFT), it is possible for cross-terms to be generated (Kadambe and Boudreaux-Bartels, 1992). This will take the form of false spatio-temporal patterns appearing in the time–frequency plot which are not reflective of the actual signal activity.

While the spectrogram is typically used, during the last 15 years more or less, beginning with the work of Choi and Williams (1989) and Zhao, Atlas and Marks (1990), there has been a substantial exploration and expansion of the techniques available for examining the rhythmic or time–frequency behavior of biological signals. The spectrogram is only one member of a more general class of time–frequency representation known as Cohen's class of distribution and defined by:

$$C_x(t,\omega;\phi) = \int\int\int e^{i(-\theta t - \tau\theta + \omega u)} \phi(\theta,\tau) x(u+\tau/2) x^*(u-\tau/2) du\, d\tau\, d\theta$$

with $x(t)$ being the signal to be analyzed, $x^*(t)$ the complex conjugate, and $\varphi(\theta,\tau)$ the kernel of the distribution (Williams, 1998). The shape of the kernel will determine the properties of the time–frequency distribution generated including resolutions, interference, and the specific uncertainty relation governing the distribution. For example, the kernel of the spectrogram will be a two-dimensional Gaussian, if the windowing is Gaussian.

Another very popular distribution is the Wigner distribution which has a kernel of unity as first used by Ville (1948) for signal processing. To improve on the STFT, the Wigner distribution is frequently used, defined by

$$Wigner(t,\omega) = \int x(t+\tau/2) x^*(t-\tau/2) \exp(-i\omega\tau) d\tau$$

That is, it is a transform from Cohen's class with a kernel of 1 (Wigner 1932; Claasen and Mecklenbrauker, 1980). The Wigner distribution offers excellent time and frequency resolutions, but suffers from substantial interference between cross-terms. That is, if the signal can be considered $x(t) = a(t)+b(t)$ then the magnitude of the Wigner distribution will take the form $W_{aa}(t,\omega)+W_{ab}(t,\omega)+W_{ba}(t,\omega)+W_{bb}(t,\omega)$, where the middle two terms are cross-terms that generate an artifact, representing time–frequency activity where the signal is inactive. In addition to generating additional components in the time–frequency domain, the Wigner distribution will similarly generate additional noise (in time) by causing any noise present in one piece of the signal to migrate throughout the signal (Cohen, 1989). There are two general approaches to dealing with these undesirable properties of the Wigner distribution, although for each there is a cost in terms of lost resolution. The first method is to use smoothing as a filter, and the second is to generate a kernel with properties to minimize interference. Each method has its defects. In particular, smoothing will generally reduce resolution, and producing a kernel optimized for a particular signal will tend to exaggerate artifacts within the signal since the method of analysis itself has incorporated the existence of the artifact (Baraniuk and Jones, 1993).

Choi and Williams (1989) developed the exponential distribution as an intermediate between the low-resolution, low-interference spectrogram, and the high-resolution, high-interference Wigner distribution. The exponential distribution (ED) diminishes interference while still maintaining a high overall time–frequency resolution (Williams and Jeong, 1992). The exponential distribution obtains its name from its kernel, given by $\exp(-\theta^2\tau^2/\sigma)$, σ determines the kernel′s breadth. Varying the value of σ allows the kernel shape to be altered to obtain better resolution (worse interference) or less interference (worse resolution). For very large σ, the exponential distribution becomes the Wigner distribution with a kernel of unity and obtains optimal resolution, but significant interference (Cohen and Posch, 1985). Although the uncertainty relation is usually understood to be specific to the spectrogram, for each time–frequency distribution there will be a corresponding uncertainty relation expressing the joint resolution possible. In every case there are two layers of tradeoffs, between resolution and interference, and then between time and frequency resolutions (Jones and Parks, 1992).

The tradeoff between time and frequency resolutions in signal processing has so far been unique for each time–frequency plot generated. The wavelet transform is intended to be an improvement over this, by tiling the time–frequency domain with resolutions particularly suitable to most signals, as shown in Fig. 2. The wavelet

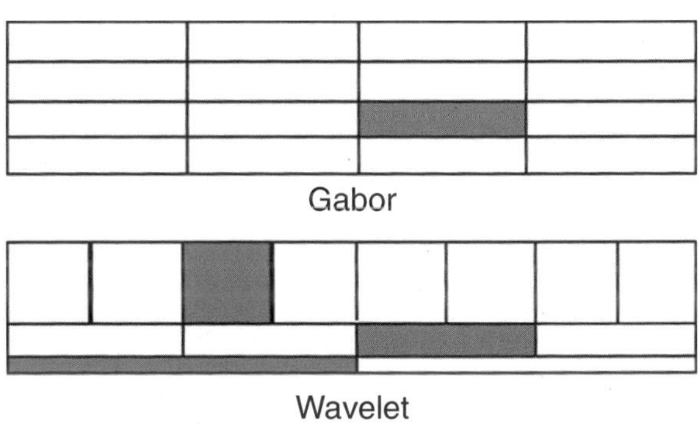

Fig. 2 The wavelet transform′s time–frequency resolution. As in Fig. 1, the tiling in the time–frequency domain is represented by rectangles over which signal properties are constant. That is, we only have one measurement in each rectangle with the x-axis representing time and the y-axis representing frequency. Thus, the Gabor transform has a measurement every fixed interval in time and every fixed interval in frequency. The wavelet transform trades frequency and temporal resolutions so that the intervals in time are longer for low frequencies and shorter for high frequencies. Because the joint frequency (area) remains the same, this allows for higher frequency resolution at low frequencies and lower frequency resolution at high frequencies, represented by the changing rectangle widths over the y-axis. This is sometimes a useful property since high-frequency signals may change rapidly in time and often do not require fine resolution in frequency

transform uses a window that varies in frequency, so that at low frequencies, the frequency resolution is very good, and at high frequencies, the temporal resolution is very good (Daubechies, 1990). The wavelet transform is not usually defined with reference to a frequency, and a scaling factor, s, proportional to $1/f$ is typically used:

$$WT(t,s) = \int x(\tau)\left(\frac{1}{s}\right)^{1/2} h^*\left(\frac{\tau - t}{s}\right) d\tau$$

and $h(t)$ is the wavelet used for analysis. Computationally, the wavelet transform consists of a series of translations, dilations, and convolutions of a prototype shape (the mother wavelet) over the extent of the signal, with the size of the wavelet determining the scale of the output. While techniques do exist (signal replication, specific constraints) for choosing the wavelet to be especially appropriate to the signal, this is typically a difficult to make (and usually unjustified) choice. A useful representation of wavelet data is the scalogram, consisting of the squared magnitude of the wavelet transform, thus paralleling the use of the spectrogram in the case of the STFT (Rioul and Flandrin, 1992). Just as in the case of Cohen's class, this will produce cross-terms creating the problem of interference (Kadambe and Boudreaux-Bartels, 1992). In fact, the similarity to Cohen's class can be extended further, since wavelet results can be obtained by using Cohen's class kernel that varies with frequency (Jeong and Williams, 1990).

4 Time–Frequency Properties and Hippocampal Rhythms

There is an extensive literature comparing the properties of the Wigner distribution, the exponential distribution, and the wavelet transform examined in this work (Jones and Parks, 1992; Loughlin et al., 1994; Papandreou and Boudreaux-Bartels, 1993). The general properties, that the Wigner will have a better resolution than the spectrogram, or that smoothing will reduce interference and resolution, are not dependent upon signal specifics although windowing is a factor. It is the specific qualitative tradeoffs that vary most from signal to signal. The spectrogram's joint resolution is often sufficient to examine features of the time–frequency domain within the dominant frequency band, using different window sizes where appropriate, it is only by comparison with higher resolution techniques like the Wigner distribution that any specific observations as to omitted time–frequency behavior can be made. Truly prominent interference patterns in the Wigner distribution can often be seen only in areas of the time–frequency domain where there is little physiological signal contribution. In hippocampal rhythms, when compared with the spectrogram with a variety of windows, some artifacts induced by interference can be observed in the Wigner. Smoothing does reduce these undesirable components, but some "false" features remained noticeable even for smoothing that gave a resultant time–frequency representation approaching the spectrogram in resolution. Over the range of parameter values that can be chosen, the exponential distribution shifts in appearance from spectrogram-like to Wigner-like (high parameter) – the differences being mostly minor between the spectrogram and exponential, and Wigner and Exponential in those cases. In the intermediate case, with parameter 1, the Exponential Distribution gives a very clear view of signal components outside the main band. In the scalogram representation, the wavelet transform also exhibits interference. Although the magnitude of the interference is in the same range as that of the Gaussian-windowed spectrogram, the effect is somewhat more pronounced due to the changes in resolution. That is, in the low-frequency range, the higher resolution produced an effect similar but opposite to the effect of smoothing in the Wigner distribution; resolution was traded for interference.

The tradeoff between interference and resolution is somewhat unusual in the case of hippocampal rhythms because they are frequently less likely to be multicomponent than other physiological rhythms, with interference contributing less to the signal than is often the case (e.g. Abeyeskera et al., 1986). A representative time–frequency hippocampal rhythm including seizure activity is shown in Fig. 3. Interference is due to the interaction of multiple components, but the multicomponent aspect of this signal was somewhat anomalous since it occupied a narrow range of frequencies and also related more to changes in frequency power (and therefore breadth of the dominant band) with time than to changes in mean frequency. The exponential distribution exhibited few advantages over other methods for this

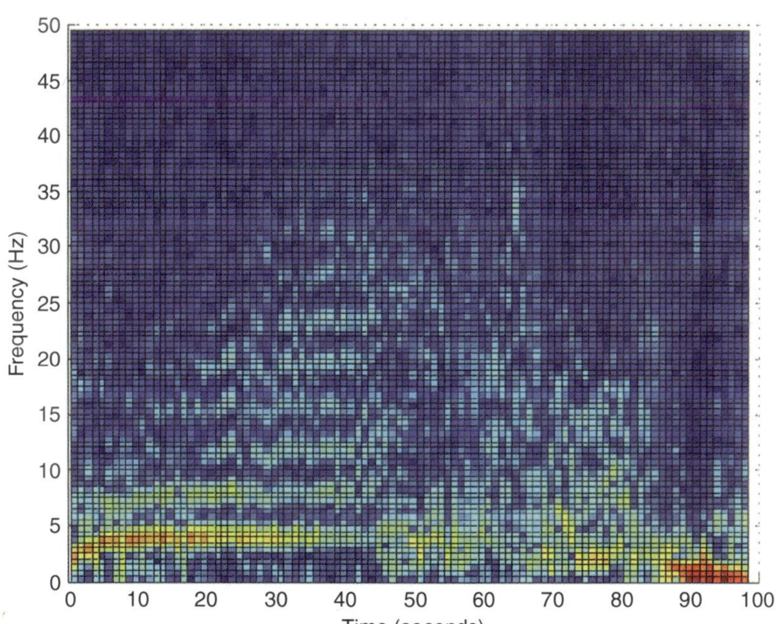

Fig. 3 The Gabor transform performed on a rhythmic hippocampal brain slice signal with seizure activity starting at 50 s. Note the disruption at onset of the seizure and the presence of harmonics created by deviation from a sinusoidal waveform in the clearly rhythmic signal's extent

signal, while the Wigner distribution exhibited surprisingly little interference compared to its norm (Claasen and Mecklenbrauker, 1980; Cohen, 1989). Because we are concerned with fine-scale time–frequency properties, even this less-than-usual interference was inhibiting, but for other applications, such as detecting prominent changes in frequency distribution, it will be less problematic.

The short-time Fourier transform will often be sufficient to examine rhythm properties, and is typically a more sensible choice than the conceptually more complicated wavelet transform (which has come close to becoming the modern default). In particular, it suffers less from having analytically sophisticated problems (interference, changing resolution) that investigators may not be interested in exploring. However, there remain a wide variety of other available time–frequency representations. Three recent innovations are the Zhao–Atlas–Marks (ZAM) cone kernel distribution, reduced interference distribution (RID), and multi-taper analysis. The RID is quite similar in its signal requirements to the Wigner, so it could be readily applied to analysis of this signal. It has a large advantage over the Wigner distribution in that it is designed to exhibit markedly less interference, with little loss in resolution. The ZAM distribution has also been found to perform particularly well when faced with near discontinuous signals (Devedeux and Duchene, 1994), which is one of the characteristics of the intact hippocampal preparation, as exhibited by the local highs in the time–frequency domain. Multi-taper analysis is used to provide greater statistical reliability in the case of stationary signals by averaging the signal over multiple independent window choices. While it certainly can be adapted to non-stationary signals, it does not provide an optimum windowing (optimums depend on the signal feature of interest), and the usual constraint of the uncertainty relation still apply.

5 Feature Extraction and Classification

Much of time–frequency analysis in electrophysiology consist of attempting to classify different signals in functionally significant ways on the basis of response to different stimuli, or on the basis of shared characteristics as in epilepsy (Williams et al., 1994). For these classification schemes, the best results are typically obtained when some form of feature detection is employed (Brown et al., 1994). Wavelet-based techniques are among the most commonly used for the extraction of features from images or signals (Brooks et al., 2001). In fact, wavelets are not only a technically common choice, but a biologically natural one as well. Speaking anthropomorphically, evolution has made prominent use of wavelets in designing our visual processing systems with activity in the human visual cortex representing the activity of wavelets (Whittaker and Siegfried, 1983). This parallel between quantitative signal processing and qualitative observation means that it is possible to use wavelet-based feature detection to identify the features that we also find visually arresting. One simple form of wavelet-based object detection is to use the coefficients derived from the wavelet transform to characterize different features (Brooks et al., 2001). The coefficients represent similarity to the mother wavelet at different timescales, so using them to characterize a spatiotemporal pattern means that the features can be understood as reflecting the contribution of a particular dilation of the wavelet. This type of feature identification is well suited to spike sorting because it allows a simple characterization of the shape of the spike if the wavelet choice is suitable (Quiroga et al., 2002). A popular choice for edge detection is the "Mexican Hat" or Ricker wavelet, which is a smoothed second derivative (or Laplacian in higher dimensions), and makes edges into zeroes (Navarro et al., 1995). Edges occur where the data is rapidly changing in the z-axis, and the perimeter's position can be calculated from the second derivative of intensity, $D^2(image) = 0$ calculated in the appropriate direction, which is defined by the orientation formed by zero crossings. In this sense, and consistent with ordinary experience, an edge occurs when the gradient is relatively high in magnitude. In smoothed or smooth images, the condition of linear variation near zero crossings hold, and the direction may be found by the Laplacian (Marr and Hildreth, 1980). Because it operates at multiple scales – effectively at different levels of smoothing – the wavelet transform can perform multiscale edge detection, and these boundaries can be considered perimeters at a particular scale.

6 Clustering and Associated Techniques

Once features defining distinct spatiotemporal patterns have been identified they can be sorted into distinct clusters or into a weighted sum of overlapping distributions. Separating data into more homogenous subgroupings is called *cluster analysis*, and it finds general application in spike sorting, where a discrete classification for the spikes is desirable (Quiroga et al., 2002). Clustering is a problem that is NP-hard. Algorithmic problems can be classified as P, NP, NP-complete, or NP-hard (Baase and Van Gelger, 2000). Problems sitting within the category "P" can be solved in polynomial time as a function of input side, for example, as input length cubed. Problems classified as "NP" are problems whose solutions can be verified within polynomial time. Clearly then, every P problem is an NP problem, but whether any NP problems exist which are not P problems has not yet been determined. NP-complete problems are those which are in NP and to which every other problem in NP is reducible. NP-complete problems are thought not to be in P but this remains unproven. NP-hard problems are typically optimization problems such as clustering where no available algorithms for the verification of solutions within polynomial time.

A clustering of n objects into K groupings requires testing k^n possible clusterings to determine an optimal solution and so algorithms that provide approximate solutions have been developed (Lingras and Huang, 2005). Two popular clustering techniques are hierarchical grouping and K-means. Hierarchical grouping (Sharma and Werner, 1981) operates by joining together groups of clusters according to minimum intra-group error. In this framework, the error corresponds to the distance between vectors representing the pattern being separated. K-means clustering operates by generating centroids for clusters randomly, assigning objects to these centroids based on proximity, and then re-evaluating the centroids, and so on. An alternative class of method to these clustering techniques is to only separate data in a probabilistic sense, thus yielding distributions which determined the likely contributions to a single measurement. This is known as a finite mixture distribution, and while dependent on the choice of distribution, a probabilistic separation of data will often yield statistically more elegant or simpler results (Flury, 1997).

A mixture distribution is a probability density function which can be written as a weighted sum of its constituent conditional probability functions. The weighting, or mixing coefficients, must sum to one, and the constituent distributions typically represent discrete classes of observation. For k clusters, a mixture density $f(y)$ has the form:

$$f(\vec{y}) = \sum_{j=1}^{k} p_j f_j(\vec{y})$$

with p_j being the weighting coefficient and f_j representing the component density for the jth distribution with random variables y.

One technique to fit a distribution to the data generated by a presumed mixture is to use the expectation–maximization (EM) algorithm (Dempster et al., 1977). The EM algorithm is a technique for dealing with incomplete data to determine the underlying distribution that generated it. As it operates, the EM algorithm will act to iteratively maximize the observed likelihood or log-likelihood data by an expectation (E) step followed by a maximization (M) step, and repeating. In the E step, the expectation of the log-likelihood for the particular case is determined conditionally from the observed data (given parameters used to determine the complete data in this step) and it is then substituted for the complete log-likelihood; In the M-step the likelihood equations are solved to maximize the complete data log-likelihood (Everitt and Hand, 1981; Flury, 1997). One standard case which is easily solved is to fit a mixture of Gaussian distributions using the EM algorithm to determine the mean and covariance matrices which define them. Let $\vec{X} = (\mathrm{X}_1 \ldots \mathrm{X}_k)^{\mathrm{T}}$ represent a multinomial random vector with exactly one X_j equal to 1 and the others zero, with $p_j = \Pr[X_j = 1], j = 1,\ldots, k$ and where each $p_j>0$ and all p_j's for j=1...k sums to one. The pdf of $\vec{X}$ is $f_x(x_1, \ldots, x_k) = \prod_{j=1}^{k} p_j^{x_j}$.

Now, let $\vec{Y} = (Y_1 \ldots Y_m)^T$ represent a random variable defined conditionally on $\vec{X}$. Then the marginal density on $\vec{Y}$ is $f_y(\vec{y}) = \sum \prod_{j=1}^{k} [p_j f_j(\vec{y})]^{x_j} = \sum_{j=1}^{k} p_j f_j(\vec{y})$, i.e., a finite mixture distribution. We define the posterior probabilities, p_{jy} as

$$p_{jy} = \Pr[X_j = 1][\vec{Y} = \vec{y}] = \frac{p_j f_j(\vec{y})}{\sum_{h=1}^{k} p_h f_h(\vec{y})} \qquad j = 1, \ldots, k$$

Consider the sample $\begin{pmatrix} \vec{X}_1 \\ \vec{Y}_1 \end{pmatrix}, \ldots, \begin{pmatrix} \vec{X}_N \\ \vec{Y}_N \end{pmatrix}$ from the joint distribution of $\vec{X}$ and $\vec{Y}$. We take $\vec{X}_i = \begin{pmatrix} X_{1i} \\ \ldots \\ X_{ki} \end{pmatrix}$ as the ith observation for the vector of variables. Then the complete log-likelihood function, l_c, for the unknown parameters, and N, $\vec{X}_i$, and $\vec{Y}_i$ becomes

$$\begin{aligned} l_c &= \sum_{i=1}^{N} \sum_{j=1}^{k} x_{ji} \log[p_j f_j(\vec{y}_i)] \\ &= \sum_{j=1}^{k} \sum_{x_{ji}=1} \log p_j + \sum_{j=1}^{k} \sum_{x_{ji}=1} \log f_j(\vec{y}_i) \end{aligned}$$

The two halves of the likelihood equation can thus be maximized separately and after some calculation the iterative steps of the EM algorithm for normal mixtures can be obtained:

$$\begin{aligned} \text{E}-\text{step}: \quad & \pi_{hi} = \frac{p_h f_h(\vec{y}_i)}{f(\vec{y}_i)} \qquad ; h = 1, \ldots, k;\ i = 1, \ldots, N. \\ \text{M}-\text{step}: \quad & p_h = \frac{1}{N} \sum\nolimits_{i=1}^{N} p_{hi} \qquad ; h = 1, \ldots, k \\ & \vec{\mu}_h = \frac{1}{N p_h} \sum\nolimits_{i=1}^{N} p_{hi} \vec{y}_i \quad ; h = 1, \ldots, k \\ & \psi_h = \frac{1}{N p_h} \sum\nolimits_{i=1}^{N} p_{hi} (\vec{y}_i - \vec{\mu}_h)(\vec{y}_i - \vec{\mu}_h)^T \qquad ; h = 1, \ldots, k \end{aligned}$$

where $\vec{\mu}$ is the mean and ψ the covariance matrix (Flury, 1997; Redner and Walker, 1984).

7 Conclusions

In the time–frequency domain, the signal recording constitutes a distribution of frequencies that are changing with time – a non-stationary rhythm. An unbiased classification of those distributions offers a way of reducing signal behavior into separable components drawn from the distribution of frequencies it exhibits over time. By way of analogy, this methodology resembles spike sorting, with three central aspects: (1) representation of key features in an appropriate domain (time–frequency); (2) feature extraction in that domain; and (3) feature classification by clustering or mixture distribution. For example, in Buzsáki et al. (2003)the three sorting steps are respectively, filtering (frequency cutoff), power cutoff, and clustering (grouping cutoff). Spike sorting can also be determined from the intrinsic character of the signal (e.g., Letelier and Weber, 2000; Snider and Bonds, 1998). In this chapter, we reviewed methods for the characterization of the complete signal by examining the time–frequency domain, characterizing local maxima by their time (duration) and frequency (mean frequency) parameters, and then grouping those parameters using the estimation–maximization (EM) algorithm. This provides a natural and relatively unbiased way to simply characterize the experimental recordings.

References

Abeyeskera R, Bolton R, Westphal L, and Boashash B (1986) Patterns in Hilbert transforms and Wigner-Ville distributions of electrocardiograms data. Proc IEEE ICASSP 1793–1796.

Achermann P, and Borbély AA (1998) Coherence analysis of the human sleep electroencephalogram. Neuroscience 85: 1195–208.

Akay M (Ed.), (1998) Time Frequency and Wavelets in Biomedical Signal Processing. New York: IEEE, Inc.

Baase S, Van Gelder S (2000) Computer Algorithms: Introduction to Design and Analysis. Reading, Massachusetts: Addison-Wesley, 3rd edn.

Baraniuk RG, Jones DJ (1993) Signal-dependent time-frequency representation: Optimal kernel design. IEEE Trans Signal Proc 41: 1589–1602.

Brooks RR, Grewe L, Lyengar SS (2001) Recognition in the wavelet domain: a survey. J Electron Imaging 10: 757–784.

Brown ML, Williams WJ, Hero III AO (1994) Nonorthogonal Gabor representations of biological signals. Proc Int Conf Acoustics, Speech, Signal Process 4: 305–308.

Buzsáki G, Buhl DL, Harris KD, Csicsvari J, Czh B, Morozov A (2003) Hippocampal network patterns of activity in the mouse. *Neuroscience* 116: 201–211.

Choi HI, Williams WJ (1989) Improved time-frequency representation of multicomponent signals using Exponential kernels. IEEE Trans Acoust, Speech, Signal Proc ASSP-37 6: 862–871.

Claasen TC, Mecklenbrauker WG (1980) The Wigner distribution – A tool for time-frequency signal analysis. Philips J Res 35: 276–300, 217–250.

Cohen, L (1989) Time-frequency distributions – A review. Proc IEEE 77: 941–981.

Cohen L, and Posch TE (1985) Generalized ambiguity function. IEEE Int Conf Acoust, Signal Process 1033–1036.

Daubechies I (1990) The wavelet transform, time-frequency localization and signal analysis. IEEE Trans Inf Theory 36: 961–1005.

Dempster AP, Laird NM, Rubin DB (1977) Maximum likelihood from incomplete data via the EM Algorithm. J R Stat Soc B 39:1–38.

Devedeux D, Duchene J (1994) Comparison of various time/frequency distributions applied to synthetic uterine EMG signals. IEEE Int Symp Time-Frequency Time-Scale Anal 572–575.

Everitt BS, and Hand DJ (1981) Finite Mixture Distributions. London, New York: Chapman and Hall.

Flury B (1997) A First Course in Multivariate Statistics. New York: Springer-Verlag.

Gray CM, Konig P, Engel AK, Singer W (1989) Oscillatory responses in cat visual cortex exhibit inter-columnar synchronization which reflects global stimulus properties. Nature 338: 334–337.

Jones DL, Parks TW (1992) A resolution comparison of several time-frequency representations. IEEE Signal Proc Mag 9: 21–67.

Jeong J, Williams WJ (1990) Variable windowed spectrograms: connecting Cohen's class and the wavelet transform. *IEEE ASSP Workshop on Spectrum Estimation and Modeling*, 270–273.

Kadambe S, Boudreaux-Bartels GF (1992) A comparison of the existence of "cross terms" in the Wigner distribution and the squared magnitude of the wavelet transform and the short time Fourier transform. IEEE Trans Signal Proc 40: 2498–2517.

Letelier JC, Weber P (2000) Spike sorting based on discrete wavelet transform coefficients. J Neurosci Meth 101: 93–106.

Lingras P, Huang X (2005) Statistical, evolutionary and neurocomputing clustering techniques: Cluster-based vs. object-based approaches. Artif Intell Rev 23: 3–29.

Loughlin P, Pitton J, Atlas L (1994) Construction of positive time-frequency distributions. IEEE Trans Signal Proc 42: 2697–2705.

Marr D, Hildreth E (1980) Theory of edge detection. Proc Roy Soc London B 207: 187–217.

Navarro R, Tabernero A, Cristobal G. (1995) Image representation with Gabor wavelets and its applications. In P. W. Hawkes (Ed.), *Advances in Imaging and Electron Physics*. Orlando, FL: Academic.

Nicolelis MA, Baccala LA, Lin RCS, Chapin JK (1995) Sensorimotor encoding by synchronous neural ensemble activity at multiple levels of the somatosensory system. Science 268: 1353–1358.

Papandreou A, Boudreaux-Bartles G (1993) Distributions for time-frequency analysis: A generalization of Choi-Williams and the Butterworth distribution. IEEE Trans Signal Proc 41: 463–472.

Quian Quiroga R, Sakowica O, Basar L, Schürmann M (2002) Wavelet Transform in the analysis of the frequency composition of evoked potentials. Brain Res Protoc 8: 16–24.

Redner RA, Walker HF (1984) Mixture densities, maximum likelihood and the EM algorithm. SIAM Rev 26: 195–239

Rieke F, Bialek W, Warland D (1997) Spikes: Exploring the Neural Code. Boston: MIT Press.

Rioul O, Flandrin P (1992) Time-scale energy distributions: A general class of extending wavelet transforms. IEEE Trans Signal Proc 40: 1746–1757.

Sharma SC, Werner Al (1981) Improved method of grouping provincewide permanent traffic counters. *Transp Res Rec* 815: 12–18.

Snider RK, and Bonds AB (1988) Classification of nonstationary neural signals. J Neurosci Meth 84: 155–166.

Teolis, A (1998) Computational Signal Processing with Wavelets. Boston, MA: Birkhauser.

Ville, J (1948) Theorie et applications de la notion de signal analytique. Cables et Transmissions 20AA: 61–74.

Whittaker SG, Siegfried JB (1983) Origins of the wavelet in the visual evoked potential. Electroencephalogr Clin Neurophysiol 55: 91–101.

Wigner E (1932) On the quantum correction for thermodynamic equilibrium. Phys Rev 40: 749–759.

Williams W (1998) Recent advance in time-frequency representations: Some theoretical foundations. In Akay M, (Ed.), Biomedical Signal Processing. Amsterdam: IEEE.

Williams WJ, Brown ML, Zaveri HP, Shevrin H (1994) Feature extraction from time-frequency distributions. In C. H. Dagli (Ed.), *Intelligent Engineering Systems through Artificial Neural Networks*. 4, pp. 823–829. New York: ASME Press.

Williams WJ, Jeong J (1992) Reduced interference time-frequency distributions. In B. Boashash (Ed.), *Time-Frequency Signal Analysis: Methods and Application.* Melbourne: Longman and Cheshire.

Zhao Y, Atlas LE, Marks R (1990) The use of cone shaped kernels for generalized time frequency representations of nonstationarity signals. IEEE Trans Acoust, Speech, Signal Proc 38: 1084–1091.

Complex Network Modeling: A New Approach to Neurosciences

Yamir Moreno

Abstact Recently, many seemingly diverse systems have been found to share unique structural properties that in their turn greatly influence the dynamical features and function of these systems. In this chapter, we focus on recent discoveries about the structure and function of complex networks to show how emergent properties in neuronal systems can be studied using the tools developed in the last several years. We first introduce basic notions to understand what a complex network is and how one can characterize its topology. Next, we discuss some topological properties recently revealed in the literature and their connections to the structural properties of other complex networks found in many fields of science. The second part of this chapter is devoted to revise the relationship between the structural properties of brain-like systems and their dynamical behavior. In particular, we pay attention to synchronization phenomena, quite relevant in many contexts of neurosciences. We round off the chapter by outlining some perspectives and discussing future and promising lines of research.

1 Introduction

Contemporary science has witnessed a movement toward research in complex systems. The aim is to develop models that capture the emergent behavior of many systems for which our current understanding of their isolated components does not provide answers to the observed collective behavior. The task is not easy, as these systems are made up of many non-identical interacting elements whose interactions determine the dynamical behavior of the systems at a global scale. On its turn, the dynamics also affects the way in which the elements interact, producing a sort of feedback loop between the structure and dynamics difficult to capture by

---

Y. Moreno (✉)
Institute for Biocomputation and Physics of Complex Systems (BIFI) and Department of Theoretical Physics, University of Zaragoza, Zaragoza 50009, Spain
e-mail: yamir.moreno@gmail.com

J.L. Perez Velazquez, R. Wennberg (eds.), *Coordinated Activity in the Brain*, Springer Series in Computational Neuroscience 2, DOI 10.1007/978-0-387-93797-7_12,

current state-of-the-art models. The first step to approach the problem is to study the backbone of these complex systems. This is precisely the goal of what has been recently coined as complex networks modeling.

The network approach is particularly suitable to explore several aspects of complexity. The massive and comparative analysis of networks from different fields, spurred by our current computation capabilities, has produced a series of unexpected results and has shown that previous models proposed in mathematical graph theory are very far from the real needs (Newman, 2003b; Boccaletti et al., 2006). The first issue that has been faced is certainly structural and consists of identifying a series of unifying principles and statistical properties common to most of real networks. Although new metrics and algorithms appear constantly, the progresses achieved in the last decade allow us to have at hand an important set of tools to unambiguously characterize, from a topological point of view, complex networked systems. Another important body of works has dealt with spreading and percolation-like processes on top of networks, addressing a variety of phenomena ranging from disease spreading to information flow and resilience to random failures and attacks. Finally, a third and promising branch of research has arisen in the last few years spurred by the new insights gained through network modeling. It consists of studying the dynamical behavior of large assemblies of (often nonlinear) dynamical systems interacting via complex topologies. Phenomena such as synchronization, the emergence of cooperation in social and biological systems, as well as signaling and gene regulatory dynamics and other biochemical processes could be now tackled with a fresh viewpoint by considering both sources of entangled complexity: the structure and the dynamics of the systems constituents.

In this contribution, we shall revise some recent results in the field of neurosciences. The reader should take this effort just as a glimpse of the already large body of research works on the subject. We first discuss the topological properties of complex networks in general and give an example of a brain-like system where the tools introduced can provide new and useful insights. Next, we address a particularly relevant phenomenon in brain-like systems, namely, synchronization of nonlinear dynamical elements. To be concise, we focus on the synchronization of phase oscillators on top of different complex topologies. Finally, we briefly discuss more realistic models and round off the chapter with some perspective remarks.

2 The Structure of Complex Networks

We first give a brief account of the most used topological metrics that allow the statistical characterization and modeling of the *structural properties* of a network. The aim is to make the rest of the chapter self-consistent so that we limit the exposition to those quantities that are relevant for the discussion that follows. The interested reader is referred to recent reviews on the subject of complex networks, where one can find much more complete texts about how to characterize and model the structural properties of complex networks (Newman, 2003b; Boccaletti et al., 2006). In those works, the reader could also find the many ways that exist in the literature to generate networks in the computer that mimic most of the properties observed in real world networks. We conclude this section by discussing the topological features of a network in the context of neurosciences, namely, the so-called brain functional networks.

2.1 Definitions and Notations

Formally, the interactions between the elements of a networked system can be represented as a graph. Graph theory (Bollobás, 1998) is the natural framework for the exact mathematical treatment of complex networks. A complex network (or graph) is made up by nodes, which are the system's components, and links or edges, which represent the interactions between them. As we shall discuss, the word *interaction* is intended in a wide sense, as it can stand for many diverse forms of relationships among the elements of the system. For instance, in a communication system such as the Internet, the physical entities represented as edges could be as diverse as phone lines, optical fibers, or even electromagnetic waves (wireless connections). The same happens for biological systems, where the interactions between nodes could be due to a physical interaction or a functional relationship. In the latter case, there are no physical links between the system's elements, but they are nonetheless *connected*.

In general, the topology of networks can be classified according to the way their links are defined. They can be *undirected* or *directed*, *weighted* or *unweighted*, and combinations of them. For instance, an undirected network is that in which the interaction of node i with node j also implies that j interacts with i. This is the case of many communication networks where the interactions are reciprocal – two routers exchange their routing tables without any preferred direction. On the contrary, in a directed network, the interaction takes place following a specific direction. In other words, the fact that node i interacts with node j does not necessarily means that node j also points to that node. An example of this kind of network is the World Wide Web (WWW), where one Web page may have a link pointing to another Web page, but the latter does not "reciprocate" the original one and does not contain a link to it. Additionally, the links of a network can also account for the "traffic" or strength of interactions among two nodes of the graph. This amounts to adding to the links more information about the way two nodes interact – in air transportation networks, for instance, it is not only important to know that two airports are connected through a commercial air flight, but also how many passengers travel from one point to the other. Finally, we mention here another feature that can be incorporated to the definition of links in a network and that, although per se is not identified by a tag as before, is of utmost relevance in the context of neural networks. Admittedly, the interactions in neuronal systems (but also in biological systems such as gene networks) can be of two different kinds: excitatory or inhibitory. In these cases, an additional label is ascribed to the links to account for the nature of the interaction between two nodes.

Formally, an *undirected (directed) graph* $G = (\mathscr{N}, \mathscr{L})$ consists of two sets $\mathscr{N}$ and $\mathscr{L}$, such that $\mathscr{N} \neq \emptyset$ and $\mathscr{L}$ is a set of unordered (ordered) pairs of elements of $\mathscr{N}$. The elements of $\mathscr{N} \equiv \{n_1, n_2, \ldots, n_N\}$ are the *nodes* (or *vertices*, or *points*) of the graph G, while the elements of $\mathscr{L} \equiv \{l_1, l_2, \ldots, l_K\}$ are its *links* (or *edges*, or *lines*). The number of elements in $\mathscr{N}$ and $\mathscr{L}$ are denoted by N and K, respectively.

We should refer to a node by its order i in the set $\mathscr{N}$. In the same way, if the graph is undirected, each of its links is defined by a couple of nodes i and j, and will be denoted as (i, j) or alternatively by l_{ij}. Moreover, the link is said to be *incident* in nodes i and j. Usually, the two nodes i and j are referred to as the *end-nodes* of link (i, j) and two nodes joined by a link are referred to as *adjacent* or *neighboring*. As discussed previously, in a directed graph, the order of the two nodes is important as there is a direction associated to the interaction: l_{ij} stands for a link from i to j, and $l_{ij} \neq l_{ji}$. The usual way to picture a graph is by drawing a dot for each node and joining two dots by a line if the two corresponding nodes are connected by a link. How these dots and lines are drawn is irrelevant, and the only thing that matters is which pairs of nodes form a link and which ones do not. We also stress that, in general, a network does not contain *self-loops*, i.e., links from a node to itself, nor *multiple edges*, i.e., couples of nodes connected by more than one link, since these elements are not allowed by the standard definition of graph given above. Graphs with either of these elements are called *multigraphs* (Boccaletti et al., 2006). In this chapter, the emphasis will be on graphs rather than multigraphs and, unless explicitly mentioned, the term graph will be used to refer to an undirected graph.

Another useful distinction is given by the relation between the number of nodes of a graph and the number of links between its nodes. For a graph G of size N, the number of edges K is at least 0 and at most $N(N-1)/2$ (when all the nodes are pairwise adjacent). This latter graph is what we call a *complete* or *all-to-all* network. On the other hand, G is said to be *sparse* if $K \ll N^2$ and *dense* if $K = \mathscr{O}(N^2)$. Additionally, a topological configuration of particular interest in many biological and social networks is the complete graph made up by three nodes, denoted by K_3, and called a *triangle* (also known as *cycles* or *loops* of length 3, see below). In general, not only triangles are relevant but also other kinds of subgraphs and cycles. We define a *subgraph* $G' = (\mathscr{N}', \mathscr{L}')$ of $G = (\mathscr{N}, \mathscr{L})$ as a graph such that $\mathscr{N}' \subseteq \mathscr{N}$ and $\mathscr{L}' \subseteq \mathscr{L}$. If G' contains all the links of G that join two nodes in $\mathscr{N}'$, then G' is said to be the *subgraph induced* by $\mathscr{N}'$ and is denoted as $G' = G[\mathscr{N}']$. A subgraph is said to be *maximal* with respect to a given property if it cannot be extended without losing that property.

From a topological point of view, a central concept in network (or graph) theory is that of reachability of two different nodes of the graph representing the topology of the system. A given graph is said to be *connected* if, for every pair of distinct nodes i and j, there is a path from i to j, otherwise it is said *unconnected* or *disconnected*. A *component* of the graph is a maximally connected induced subgraph. A *giant component* is a component whose size is of the same order as N. From the previous definitions, it follows that two nodes that are not adjacent may nevertheless be reachable by passing through other nodes in the graph – provided that the two nodes belong to the same component of the graph, i.e, as far as they do not belong to two disconnected set of nodes. We refer to a *walk* from node i to node j as an alternating sequence of nodes and edges (a sequence of adjacent nodes) that begins with i and ends with j. In this way, the length of the walk is given by the number of edges in this sequence. We can also distinguish a *trail*, a walk in which no edge is repeated, from a walk, in which no node is visited more than once, referred to as a *path*. The most important property related to the above definitions is the walk of minimal length between two nodes, known as *shortest path* or *geodesic*. Alternatively, one can also speak in terms of a *cycle* as a closed walk, of at least three nodes, in which no edge is repeated. Consequently, a cycle of length k is usually said a k-cycle and denoted as C_k. As we have previously stated, C_3 is a triangle, and C_4 is called a quadrilateral, C_5 a pentagon, and so on.We shall come again to this kind of structures later on when motifs will be discussed.

Finally, when a real network is characterized or a graph is generated by numerical means, one needs a way to represent the interactions that take place between the nodes that made up the systems. This is usually achieved by considering the matricial representation of a graph. A graph G can be completely described by giving the *adjacency* (often called *connectivity*) matrix $\mathscr{A}$, a $N \times N$ square matrix whose entry a_{ij} $(i, j = 1, \ldots, N)$ is equal to 1 when the link l_{ij} exists, and zero otherwise. This is thus a symmetric matrix for undirected graphs. We note here that this representation is particularly useful for analytic calculations, but from a practical point of view, it may pose a serious constraint for numerical simulations when the system sizes are very large. One way to reduce the size of the adjacency matrix is to try alternative representations where the number of zeros in the matrix is significantly reduced. One possibility is to store the same information contained in the adjacency matrix in lists or vectors, one for each node, whose lengths or sizes are given by the number of neighbors of a given node. Note that a "safe" way is to impose the same size for all such vectors equal to the largest connectivity of the graph – although it is not optimal, the number of zeros is significantly reduced particularly for sparse graphs.

2.2 Topological Properties

The *degree* (or *connectivity*) k_i of a node i is the number of edges incident with the node, and is defined in terms of the adjacency matrix $\mathscr{A}$ as

$$k_i = \sum_{j \in \mathscr{N}} a_{ij}. \tag{1}$$

The degree of a node has two components when the graph is directed, namely, the number of outgoing links $k_i^{out} = \sum_j a_{ij}$, which is referred to as the out-degree of the node, and the number of ingoing links $k_i^{in} = \sum_j a_{ji}$, referred to as the in-degree of the node. For these graphs, the total degree is thus defined as $k_i = k_i^{out} + k_i^{in}$. Finally, a list of the node degrees of a network is called the *degree sequence*.

The probability that a node chosen uniformly at random has degree k or, equivalently, the fraction of nodes in the graph having degree k, is the most basic topological characterization of a graph G and is known as the *degree distribution* $P(k)$. It is denoted as P_k, or p_k, to indicate that the variable k assumes non-negative integer values, although sometimes for very large system sizes, a continuous approximation for k is usually taken and convenient. The degree distribution allows to distinguish between two important classes of real world networks: homogeneous and heterogeneous (*scale-free*) networks. The first kind of networks is characterized by a Poisson degree distribution, which means that the fluctuations around the mean value of connectivity decays exponentially fast for large values of k. This is the reason why such networks are also called *exponential* networks. Conversely, the latter class is defined by those networks that show a power law degree distribution of the form $P_k \sim k^{-\gamma}$. Interestingly enough, this kind of networks are ubiquitous in Nature (Newman, 2003b; Boccaletti et al., 2006), and most of them are characterized by an exponent γ which takes on values between 2 and 3. Although we will not discuss in detail directed networks, let us just mention here for the sake of completeness that when the network is directed one needs to consider two distributions, $P(k^{in})$ and $P(k^{out})$.

Information on how the degree is distributed among the nodes of an undirected network can be obtained either by a plot of $P(k)$, or by the calculation of the moments of the distribution. The n-moment of $P(k)$ is defined as

$$\langle k^n \rangle = \sum_k k^n P(k). \tag{2}$$

The first moment $\langle k \rangle$ is the mean degree of G. The second moment measures the fluctuations of the connectivity distribution. The fact mentioned above that almost all real world networks are power law distributed and that the exponent $2 < \gamma < 3$ results in that the second moment of the degree distribution is not bounded (i.e., diverges) when the size of the system goes to infinity. Even for finite systems, the deviation around the mean value of the distribution is so large that statements about a "typical" degree value are meaningless. This mathematical property has radical consequences for the dynamical processes taking place on top of such networks. In fact, the divergence of $\langle k^2 \rangle$ in the limit of infinite graph size, has been shown to be at the root of drastic changes in the behavior of epidemic spreading in heterogeneous, scale-free distributions (Newman, 2003b; Boccaletti et al., 2006).

The degree distribution completely determines the statistical properties of uncorrelated networks. However, a large number of real networks are *correlated* in the sense that the probability that a node of degree k is connected to another node of degree, say k', depends on k. In these cases, it is necessary to introduce a *conditional* probability $P(k'|k)$, being defined as the probability that a link from a node of degree k points to a node of degree k'. $P(k'|k)$ satisfies the normalization $\sum_{k'} P(k'|k) = 1$, and the degree detailed balance condition $kP(k'|k)P(k) = k'P(k|k')P(k')$ (Boguñá et al., 2003). For uncorrelated graphs, in which $P(k'|k)$ does not depend on k, the detailed balance condition and the normalization give $P(k'|k) = k'P(k')/\langle k \rangle$.

Another useful way to characterize the degree correlations of a given network is to measure the *average nearest neighbors degree* of a node i as

$$k_{nn,i} = \frac{1}{k_i} \sum_{j \in \mathcal{N}_i} k_j = \frac{1}{k_i} \sum_{j=1}^{N} a_{ij} k_j, \tag{3}$$

where the sum runs on the nodes belonging to $\mathcal{N}_i$, the set of first neighbors of i. By using the definition (3), one can calculate the average degree of the nearest neighbors of nodes with degree k, denoted as $k_{nn}(k)$, obtaining an expression that implicitly incorporates the dependence on k (Pastor-Satorras et al., 2001). Such a quantity can be, indeed, expressed in terms of the conditional probability as

$$k_{nn}(k) = \sum_{k'} k' P(k'|k). \tag{4}$$

If there are no degree correlations, Eq. (4) gives $k_{nn}(k) = \langle k^2 \rangle / \langle k \rangle$, i.e., $k_{nn}(k)$ is independent of k. Correlated graphs are classified as *assortative* if $k_{nn}(k)$ is an increasing function of k, whereas they are referred to as *disassortative* when $k_{nn}(k)$ is a decreasing function of k (Newman, 2002). In other words, in assortative networks the nodes tend to connect to their connectivity peers, while in disassortative networks nodes with low degree are more likely connected to highly connected ones. Degree correlation are usually quantified by reporting the numerical value of the slope of $k_{nn}(k)$ as a function of k, or by calculating the Pearson correlation coefficient of the degrees at either ends of a link (Newman, 2002, 2003a).

Shortest paths play an important role in the transport and communication within a network. Suppose one needs to send a data packet from one computer to another through the Internet: the geodesic provides an optimal pathway, since one would achieve a fast transfer and save system resources (Pastor-Satorras and Vespignani, 2004). For such a reason, shortest paths have also played an important role in the characterization of the internal structure of a graph (Wesserman and Faust, 1994; Scott, 2000). It is useful to represent all the shortest path lengths of a graph G as a matrix $\mathcal{D}$ in which the entry d_{ij} is the length of the geodesic from node i to node j. The maximum value of d_{ij} is called the *diameter* of the graph. A measure of the typical separation between two nodes in the graph is given by the *average shortest path length*, also known as *characteristic path length*, defined as the mean of geodesic lengths over all couples of nodes (Watts, 1999; Watts and Strogatz, 1998):

$$L = \frac{1}{N(N-1)} \sum_{i,j \in \mathcal{N}, i \neq j} d_{ij}. \tag{5}$$

In real world networks, the value of L is very small when compared to the system size. In fact, L scales with the number of nodes at most as $log(N)$. This property, a signature of all so-called complex networks regardless of their degree distributions, is what is known as the *small-world* property (the reader may remember the popular term "six degrees of separation" often used in the media to account for the fact that two people in the world are quite close to each other if one follows their acquaintances). Formally, the previous definition of L diverges if there are disconnected components in the graph. One possibility to avoid the divergence is to limit the summation in formula (5) only to couples of nodes belonging to the largest connected component (Watts and Strogatz, 1998). Alternatively, one may consider the harmonic mean (Marchiori and Latora, 2000) of geodesic lengths, and define the so-called *efficiency* of G as (Latora and Marchiori, 2001, 2003)

$$E = \frac{1}{N(N-1)} \sum_{i,j \in \mathcal{N}, i \neq j} \frac{1}{d_{ij}}. \tag{6}$$

Such a quantity is an indicator of the traffic capacity of a network, and avoids the divergence of formula (5), since any couple of nodes belonging to disconnected components of the graph yields a contribution equal to zero to the summation in formula (6).

The communication of two non-adjacent nodes, say j and k, depends on the nodes belonging to the paths connecting j and k. Consequently, a measure of the relevance of a given node can be obtained by counting the number of geodesics going through it, and defining the so-called *node betweenness*. Together with the degree, the betweenness is one of the standard *measures of node centrality*, originally introduced to quantify the importance of an individual in a social network (Wasserman and Faust, 1994; Scott, 2000; Latora and Marchiori). More precisely, the *betweenness* b_i of a node i, sometimes referred to also as *load*, is defined as (Wasserman and Faust, 1994; Scott, 2000; Freeman, 1977, 1979)

$$b_i = \sum_{j,k \in \mathcal{N}, j \neq k} \frac{n_{jk}(i)}{n_{jk}}, \tag{7}$$

where n_{jk} is the number of shortest paths connecting i and j, while $n_{jk}(i)$ is the number of such paths passing through i.

An important local property of complex networks is the *clustering coefficient* C, a measure introduced by Watts and Strogatz (1998). The clustering, also known as transitivity, is a typical property of acquaintance networks, where two individuals with a common friend are likely to know each other (Wasserman and Faust, 1994). In terms of a generic graph G, transitivity means the presence of a high number of triangles. This can be quantified by introducing first a quantity c_i (the local clustering coefficient of node i), expressing how likely $a_{jm} = 1$ for two neighbors j and m of node i. Its value is obtained by counting the actual number of edges (denoted by e_i) in G_i (the subgraph of neighbors of i defined as in Sect. 2.1). Notice that G_i can be, in some cases, unconnected. The local clustering coefficient is defined as the ratio between e_i and $k_i(k_i - 1)/2$, the maximum possible number of edges in G_i (Watts, 1999; Watts and Strogatz, 1998)

$$c_i = \frac{2e_i}{k_i(k_i - 1)} = \frac{\sum_{j,m} a_{ij}a_{jm}a_{mi}}{k_i(k_i - 1)}. \tag{8}$$

The clustering coefficient of the graph is then given by the average of c_i over all the nodes in G:

$$C = \langle c \rangle = \frac{1}{N}\sum_{i \in \mathcal{N}} c_i. \tag{9}$$

By definition, $0 \leq c_i \leq 1$, and $0 \leq C \leq 1$. In some situations, it is also useful to consider $c(k)$, the clustering coefficient of a connectivity class k, which is defined as the average of c_i taken over all nodes with a given degree k.

Finally, we discuss two topological properties that are not associated to the smallest scale of a graph (that of a node), but to the mesoscale. They have to do with characterizing groups of nodes. The first of such properties is the notion of a motif. A motif M is a pattern of interconnections occurring either in an undirected or in a directed graph G at a number significantly higher than in randomized versions of the graph, i.e., in graphs with the same number of nodes, links, and degree distribution as the original one, but where the links are distributed at random. As a pattern of interconnections, M is usually meant as a connected (undirected or directed) n-node graph which is a subgraph of G.

The concept of motifs was originally introduced in Shen-Orr et al. (2002), Milo et al. (2002), Mangan and Alon (2003), Milo et al. (2004), and Kashtan et al. (2004) by U. Alon and coworkers, who studied small n motifs in biological and other networks. The research of the significant motifs in a graph G is based on matching algorithms counting the total number of occurrences of each n-node subgraph M in the original graph and in the randomized ones. The statistical significance of M is then described by the *Z-score*, defined as (Milo et al., 2002, 2004)

$$Z_M = \frac{n_M - \langle n_M^{rand} \rangle}{\sigma_{n_M}^{rand}}, \tag{10}$$

where n_M is the number of times the subgraph M appears in G, and $\langle n_M^{rand} \rangle$ and $\sigma_{n_M}^{rand}$ are, respectively, the mean and standard deviation of the number of appearances in the randomized network ensemble.

To round off this section, we next discuss the second of the graph properties that deal with groups of nodes. At a larger scale, but still in between the micro (nodes) and macro (system-wide) scales, graphs can also exhibit other non-random structures, namely, community structures. As it is typical for many metrics in graph theory, the notion of community and the first network formalizations of the concept were historically proposed in the context of social sciences (Wasserman and Faust, 1994). Given a graph $G(\mathcal{N}, \mathcal{L})$, a *community* is a subgraph $G'(\mathcal{N}', \mathcal{L}')$

whose nodes are tightly connected, i.e., cohesive. Since the structural cohesion of the nodes of G' can be quantified in several different ways, there are different formal definitions of community structures.

The strongest definition requires that all pairs of community members choose each other. Such a requirement leads to the definition of a clique. A *clique* is a maximal complete subgraph of three or more nodes, i.e., a subset of nodes all of which are adjacent to each other, and such that no other nodes exist adjacent to all of them. This definition can be extended by weakening the requirement of adjacency into a requirement of reachability: a *n-clique* is a maximal subgraph in which the largest geodesic distance between any two nodes is no greater than n. When $n = 1$, this definition coincides with that of a clique. 2-cliques are subgraphs in which all nodes need not to be adjacent but are reachable through at most one intermediary. In 3-cliques all nodes are reachable through at most two intermediaries, and so on. Whereas the concept of n-clique involves increasing the permissible path lengths, an alternative possibility to relax the strong assumption of cliques involves reducing the number of other nodes to which each node must be connected. A *k-plex* is a maximal subgraph containing n nodes, in which each node is adjacent to no fewer than $n - k$ nodes in the subgraph. A different class of definitions is based on the relative frequency of links. In this case communities are seen as groups of nodes within which connections are dense, and between which connections are sparser (Alba, 1973; Flake et al., 2002).

While the simplest formal definitions in this class have been proposed in Seidman (1983) and Borgatti et al. (1990), a less stringent definition is the following: G' is a community if the sum of all degrees within G' is larger than the sum of all degrees toward the rest of the graph (Radicchi et al., 2004). The definitions presented here are not the only possible choices. Several other definitions, possibly more appropriate in some cases, can be found in Wasserman and Faust (1994) and Danon et al. (2005). The reader can also find in Danon et al. (2005) a short review of current methods and algorithms for community detection.

2.3 Topology of Functional Brain Networks

The tools of complex networks have been recently used to address the problem of how one can characterize the topology of different areas of the brain that are connected by a *functional* relationship. By using functional magnetic resonance imaging (FMRI), Eguiluz et al. (2005) extracted functional networks connecting correlated human brain areas. The experiments consisted of exposing several healthy individuals to different tasks, while recording their magnetic resonance brain activity in a number of brain sites (or voxels in the terminology of Eguiluz et al. (2005)). With the data recorded, it is possible to construct a correlation matrix that is later filtered by a positive threshold, r_c, to keep only those voxels that show the highest levels of correlation. Two sites of this filtered correlation matrix are considered to be functionally connected, which means that the activity recorded at site x_i at time t is concurrent to the activity measured at site x_j at the same time. Finally, by counting the number of sites (nodes) in the thresholded matrix and the links between them, a complex functional brain network can be drawn and its topological properties studied.

Figure 1 shows the connectivity distributions of the functional networks extracted. It is apparent from the figure that these networks are heavy tailed and described by a degree distribution that follows a power law of the form $P(k) \sim k^{-\gamma}$, being γ around 2, regardless of the threshold value used to filter the original correlation matrix. The observation of scale-freeness in the distributions is remarkable as it means that most areas of the brain are functionally connected to a few other areas, while just a tiny fraction of the voxels are linked to many others. The authors also showed that if one does not take into account the time correlations between different voxels, the resulting network would be gaussian distributed (see the inset). Finally, they additionally pointed out that the scale-free character of the networks obtained remains unaltered even for tasks engaging different brain regions.

As for other topological properties, Eguiluz et al. (2005) also reports that the clustering coefficient of these functional networks remains several orders of magnitude larger than that of the equivalent randomized networks. This feature, together with the low average path length of the networks is indicative of the small-world structure in the sense of Watts and Strogatz (1998). More importantly, this work showed that the scaling law for the degree distribution is robust in front of different parameters and task conditions, which points to invariant topological properties of an underlying dynamical network. Whether brain networks' scaling properties are altered under various pathologies remains an open question. However, as we shall see in the next sections, the quantitative features already revealed by Eguiluz et al. (2005) are of utmost importance for dynamical processes that takes place on top of these networks, and in particular, for the fast transmission of signals and synchronization phenomena.

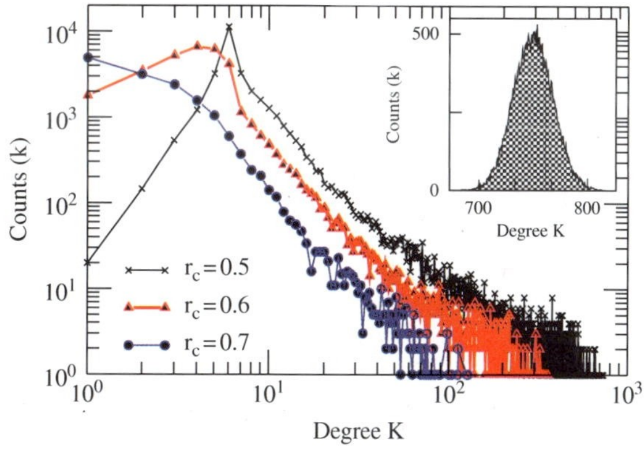

Fig. 1 Connectivity distributions of functional brain networks constructed following the methodology introduced in Eguiluz et al. (2005). The inset shows their equivalent (same size and number of links) randomly connected networks. Figure reprinted from Eguiluz et al. (2005)

3 Brain-Like Dynamical Processes

Once we have the tools to characterize the topology of a given networked system, we turn our attention to the dynamics taking place on the nodes of that system. One of the main theoretical challenges that we face today is to understand what is the relation between the structure and the function of complex systems. We certainly know that the dynamical properties of a complex system are not merely the sum of the individual dynamical features, but rather the result of a collective behavior. As many recent studies have shown, topology affects function and in many cases the other way around as well. Therefore, let us now discuss some dynamical processes that are relevant in neuronal systems and how different underlying topologies give rise to diverse dynamical behaviors of the system as a whole. We note that the processes that we are going to discuss next are still under scrutiny and that although they are sound, future progresses along these lines of research might shed new light or even change some of the partial conclusions we have arrived to up to now.

3.1 Synchronization Processes

Synchronization has been shown to be of special relevance in neural systems. The brain is composed of billions of neurons coupled in a hierarchy of complex network connectivity. In the last several years, significant progresses have been made in unraveling the detailed interconnections of different types of neurons at the level of cellular circuits (Binzegger et al., 2004; Silberberg et al., 2005; Markram et al., 2004). Nowadays, it is known that at this level, the neuronal networks of mammalian cortex also possess a complex structure, sharing small-world and scale-free features. This system is composed by two basic neuron types: excitatory principal cells and inhibitory interneurons. In contrast to the more homogeneous principal cell population, interneurons are very diverse in terms of morphology and function. It appears that there is an apparently inverse relationship between the number of neurons in various interneuron classes and the spatial extent of their axon trees – most of the neurons have only local connections, while a small number of neurons have long-range axons (Buzsáki et al., 2004). These properties of neuronal networks reflect a compromise between computational needs and wiring economy (Chklovskii et al., 2002, Karbowski, 2001). The latter is because in order to establish and maintain neuronal connections, the system is subjected to a significant metabolic cost. If we assume that this cost should be reduced, it follows that the wiring length should be globally minimized. Indeed, the wiring economy is apparent in the distributions of projection length in neural systems, which show that most neuronal projections are short (Sik et al., 1995; Braitenberg and Schüz, 1998). However, there also exists a significant number of long-distance projections (Sik et al., 1994; Kaiser and Hilgetag, 2006). This long-range connections are essential from the functional point of view.

Admittedly, large-scale synchronization of oscillatory neural activity is believed to play a crucial role in the information and cognitive processing (Fries, 2005). At the level of cellular circuits, oscillatory timing can transform unconnected principal cell groups into temporal coalitions, providing maximal flexibility and economic use of their spikes (Konig et al., 1996). Brains have developed mechanisms for keeping time by inhibitory interneuron networks (Buzsáki and Chrobak, 1995). The wiring will be the most economic if the connections were all local. However, in this case physically distant neurons are not connected, and synaptic path length and synaptic delays become exceedingly long for synchronization in large networks.

The study of synchronization phenomena in complex networks can help us to understand the relation between the structure of the system (wiring economy) and its function (dynamical synchronization properties). In fact, some recent studies of synchronization in random networks have shown that the stability of the synchronized state is optimal in fully random networks with a uniform connectivity per node, independent of the network size (Zhou et al., 2006). The same happens if interneuronal oscillators are coupled (Buzsáki et al., 2004). However, fully random connections irrespective of the physical distance are not economic if wiring cost is taken into account. Along this line of research, Buzsáki et al. (2004) studied a model of interneuronal networks containing local neurons (Gaussian distribution of projection length) and a fraction of long-range neurons (power law distribution of projection length), and showed that the ratio of synchrony to wiring length is optimized in the small-world regime with a small fraction of long-range neurons. Thus, most wiring is local and neurons with long-range connectivity and large global impact are rare, as consistent with observations. The authors argued that the complex wiring of diverse interneuron classes could represent an economic solution for supporting global synchrony and oscillations at multiple timescales with minimum axon length (Buzsáki et al., 2004). While such mathematical consideration can predict the scaling relationship among the interneuron classes in brain structures of varying sizes, understanding the role of complex neuronal connectivity, most likely mediated by synchronization, is still one of the main challenges in neuroscience. The theory reviewed here will surely contribute to their understanding when more systematic information of neuronal connectivity becomes available in the future.

3.2 *Synchronization of Kuramoto Oscillators*

Let us discuss in more detail one of the simplest systems where synchronization can be achieved. The idea is to study the general principles ruling the relationship between the structure and the dynamics of a system made up of oscillators and later on, with the insights gained with such simple system, go on by studying more realistic models (i.e., with many more degrees of freedom) of neuronal activity.

In order to study how topology influences collective dynamics, we assume that each network's component is an oscillator and that each interacts with the others following the Kuramoto model (Kuramoto, 1975, 1984; Acebron et al., 2005).

This choice is due to the seniority, elegance, and universality of this model. Studies on synchronization in complex topologies where each node is considered as a Kuramoto oscillator, were first reported for Watts-Strogatz (WS) networks (Watts, 1999; Hong et al., 2002) and Barabási-Albert (BA) scale-free graphs (Moreno and Pacheco, 2004; Moreno et al., 2004). The Kuramoto model (KM) on top of complex topologies is described by the equations of motion

$$\frac{d\theta_i}{dt} = \omega_i + \sum_j \sigma_{ij} a_{ij} \sin(\theta_j - \theta_i) \quad (i = 1, \ldots, N), \tag{11}$$

where σ_{ij} is the coupling strength between pairs of connected oscillators and a_{ij} are the elements of the connectivity matrix ($a_{ij} = 1$ if i is linked to j and 0 otherwise). The original Kuramoto model is recovered by letting $a_{ij} = 1, \forall i \neq j$ (all-to-all, complete graph) and $\sigma_{ij} = K/N, \forall i, j$. The intrinsic or natural frequencies ω_i are distributed according to some distribution $g(\omega)$, that is usually assumed to be unimodal and symmetric about its mean frequency Ω.

The global dynamics of the system Eq. (11) turns out to be qualitatively the same as for the original Kuramoto model for the case in which we are interested, namely, for networks whose degree distribution follows a power law $P(k) \sim k^{-\gamma}$, with $2 < \gamma < 3$, as the functional brain network previously discussed. The coherence of the set of N oscillators can be characterized by the order parameter $0 < r(t)\exp(\mathrm{i}\phi(t)) = \frac{1}{N}\sum_{j=1}^{N}\exp(\mathrm{i}\theta_j(t)) < 1$. For small values of the coupling ($\sigma_{ij} = \sigma\forall i$), the strength of the interactions is not enough to break the incoherence produced by the individual dynamics of the oscillators. This behavior persists until a certain critical value σ_c is attained, where some elements lock their relative phase and a cluster of synchronized nodes comes up. When $\sigma > \sigma_c$, the population of oscillators splits into a partially synchronized state made up of oscillators locked in phase that adds to r and a group of nodes whose natural frequencies are too spread as to be part of the coherent pack. Finally, for large enough values of σ, almost all nodes get entrained around the mean phase ϕ and the system settles into a completely synchronized state where $r \approx 1$.

The very existence of a transition point (also referred to as a critical point) separating the incoherent and the coherent regimes for the Kuramoto model on top of SF networks is surprising, as most of previous studies on dynamical processes taking place on top of these networks revealed the lack of critical points when $\gamma \leq 3$ (Newman, 2003b; Boccaletti et al., 2006). This is the case, for instance, of the spreading of a disease, which always pervades the system no matter the value of the epidemic spreading rate (Newman, 2003b; Boccaletti et al., 2006). On the other hand, the KM on top of homogeneous-like networks such as a random graph shows a critical point greater than that obtained in SF nets. As the equations of motion are functionally the same and only the second term on the right hand of Eq. (11) contains the topology of both kinds of networks, the differences in the collective behavior should come from the microscopic organization of the groups of coherent oscillators. Therefore, the dynamical organization of these groups (driven by the underlying substrate) is at the very root of the mechanisms behind the differences observed. Recently, Gómez-Gardeñes et al. (2007a,b) proposed a new local parameter that captures and quantifies the way in which clusters of locked oscillators emerge. The main difference with respect to r is that one measures the degree of synchronization of nodes (r) with respect to the average phase ϕ and the other (r_{link}) the degree of synchronization between every pair of connected nodes.

This new parameter, r_{link}, gives the fraction of all possible links that are synchronized in the network and is defined as

$$r_{link} = \frac{1}{2N_l} \sum_i \sum_{j \in \Gamma_i} \left| \lim_{\Delta t \to \infty} \frac{1}{\Delta t} \int_{t_r}^{t_r+\Delta t} e^{i[\theta_i(t)-\theta_j(t)]} dt \right|, \tag{12}$$

being Γ_i the set of neighbors of node i, t_r the time the system needs to settle into the stationary state, and Δt a large averaging time. By computing the extend to which pairs of connected oscillators are synchronized in terms of a symmetric matrix whose entries corresponds to every possible pair of links, the clusters of synchrony for any value of σ can be identified (Gómez-Gardeñes et al., 2007a,b). From the microscopic analysis, it turns out that for homogeneous topologies (Erdös-Rényi graphs (ER)), many small clusters of synchronized pairs of oscillators are spread over the graph and merge together to form a giant synchronized cluster (GC) when the effective coupling is increased. On the contrary, in heterogeneous graphs, a central core containing the hubs first comes up driving the evolution of synchronization patterns by absorbing small clusters. Moreover, the evolution of r_{link} as σ grows explains why the transition is sharper for ER networks: nodes are added first to the GC and latter on the links among these nodes that were missing in the original clusters of synchrony, while in SF graphs oscillators are added to the largest synchronized component together with most of their links, resulting in a much slower growth of r_{link}. Finally, the probability that a node with degree k belongs to the largest synchronized cluster is an increasing function of k for every σ, namely, the more connected a node is, the more likely it takes part in the cluster of synchronized links. Moreover, the highly connected nodes are also more robust to perturbations once the fully synchronized state is attained, that is, the average time $\langle \tau \rangle$ a node needs to get back into the fully synchronized state is inversely proportional to its degree, i.e., $\langle \tau \rangle \sim k^{-1}$ (Moreno and Pacheco, 2004).

The observed differences in the behavior at a local scale, which are depicted in Fig. 2, are rooted in the growth of the *GC*. That is, in one case (ER-like networks), almost all the nodes of the network takes part of the giant component from the beginning and later on, when σ is increased, what is added to the *GC* are the links among these nodes that were missing in the original cluster of synchronized nodes. For SF-like networks, the mechanism is the opposite. Nodes are added to the *GC together* with most of their links, resulting in a growth of r_{link} much slower than for the homogeneous topologies. This study about the patterns of self-organization toward synchronization reveals that the quantitative difference about the macroscopic behavior, shown by the computation of the evolution of the global coherence r for ER and SF

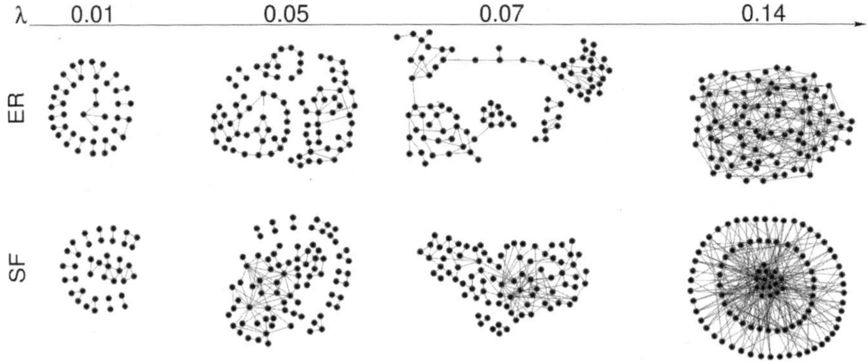

Fig. 2 Giant synchronized components for several values of λ and two network topologies (ER and SF). The size of the underlying networks is $N = 100$ nodes. See the text for further details. Figure reprinted with permission from Gómez-Gardeñes et al. (2007a)

networks, has its roots on a qualitatively different route at the microscopic level of description. The use of the new parameter r_{link} which involves the computation of the degree of coherence between each pair of linked nodes is a useful tool for describing such differences. Moreover, the results suggest that the degree of heterogeneity of the network is the key ingredient to explain the two different routes observed. We stress that the dynamical organization of synchronized clusters and the key role of hubs is not the result of the very simple model discussed so far. As we will see in the next section, hubs also play a key (and not always advantageous) role in more realistic, albeit complicated, models of neuronal activity.

Another important feature coming from the synchronization of Kuramoto oscillators is that structural differences appear when the control parameter σ grows. Imagine that one measures the topological properties of a network by measuring the degree of local synchrony in the system. Therefore, the structure of the network will be given by the emerging topology of oscillators beating at the same pace. In order words, even if the underlying network is of size N, unless the system is completely synchronized, the observer would only detect the topology resulting from the emerging collective behavior, the rest of the nodes being invisible for the measuring system. Strikingly, if the substrate network is a SF graph, one would be able to distinguish it from the homogeneous case. A clear evidence of this effect is given by the synchronization of networks with community structure. It has been shown that the community structure is progressively unveiled when the coupling σ drives the system's dynamics toward the coherent state. Figure 3 shows this process. In particular, the nodes belonging to the first community level are the first to get synchronized, subsequently the second level nodes achieve the frequency entrainment and finally the whole system shows global synchronization. Thus, one can conclude that the inner the link is the faster it gets synchronized and that different values of σ would reveal different levels and correspondingly distinct topological features.

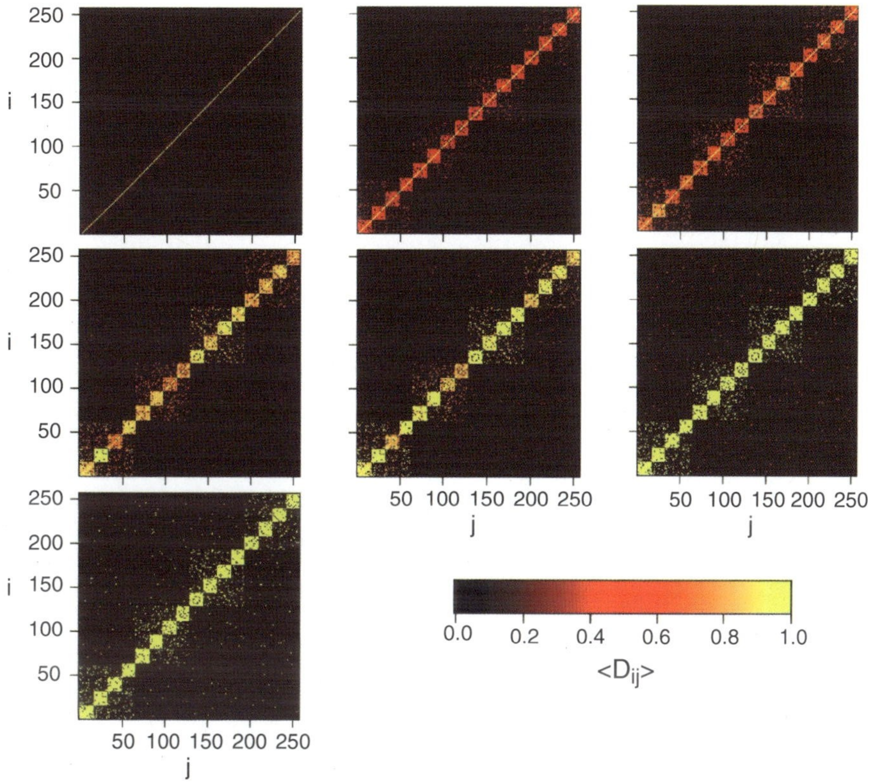

Fig. 3 We represent the degree of synchronization between pairs of connected nodes for several values of the coupling λ in a modular network (with two organizational levels) of $N = 256$ nodes. The color code is a measure of the degree of synchronization. From left to right and top to bottom, λ has been set to 0.011, 0.026, 0.032, 0.035, 0.038, 0.046, 0.210 (corresponding to full synchronization). See Gómez-Gardeñes et al. (2007b) for further details. Figure reprinted with permission from Gómez-Gardeñes et al. (2007b)

3.3 More Realistic Dynamical Models

Recent simulations of synchronization dynamics of brain-like networks have gone one step forward. In a series of papers (Zhou et al., 2006; Zemanová et al., 2006; Zhou et al., 2007) the dynamics of a realistic cortico-cortical projection network of the cat has been modeled at the level of functional areas (Scannell et al., 1999). At this level, the network displays a hierarchical cluster organization (Hilgetag and kaiser, 2004). There are four prominent clusters that agree broadly with the four functional subsystems: visual (V), auditory (A), somatomotor (SM), and frontolimbic (FL). These works simulate the network dynamics by a 2-level model: each node (cortical area) is represented by a SW subnetwork of neurons (network of networks). It was shown that the model possesses two distinguished regimes, weak and strong synchronization. In the weak synchronization regime, the model displays biologically plausible dynamical clusters. The functional connectivity, obtained by passing the correlation matrix through various thresholds, exhibits various levels of organization. The clusters with the highest levels of synchronization are from respective functional subsystems and are related to specialized functions of the subsystems. The specialized clusters are integrated into larger clusters through brain areas having many inter-community connections. As a whole, the functional connectivity reveals the hierarchical organization of the structural connectivity (Zhou et al., 2006). The dynamics forms four major clusters, in excellent agreement with the four functional subsystems (Zemanová et al., 2006). Furthermore, brain areas that bridge different dynamical clusters are found to be the areas involved in multisensory associations. Zhou et al., (2007) pushed in this direction and performed a comparative study in which the authors showed that representing the brain areas with a periodic, low-dimensional neuronal mass oscillator describing alpha waves (da Silva et al., 1974) cannot resolve the four clusters. In this case, the detailed network topology becomes rather irrelevant to the dynamical patterns, which essentially do not change when the network is randomized. This result can be understood by recalling previous analyses based on random networks where it is shown that synchronization is mainly determined by the node intensity (Zhou et al., 2006; Zhou and kurths, 2006). Furthermore, the transition from the weak to the strong synchronization regime is similar to that found for Kuramoto oscillators in complex networks already discussed (Gómez-Gardeñes et al., 2007a,b).

Excitable Integrate-and-Fire Oscillators (IFOs) models in SW networks have also been revisited (Roxin et al., 2004) to study the possibility of self-sustained activity induced by the topology itself. Roxin et al. (2004) considered a unidirectional ring of IFOs with density p of random long-range directed connections and showed that periodical patterns persist at low values of p, while long transients of disordered activity patterns are observed for high values of p. The rationale behind the observed behavior is rooted at the trade-off between the average path length and the speed of activity propagation. For low p configurations, the distances in the networks decrease logarithmically with size, while the superposition of activities is almost the same than in the regular configuration, i.e., the same activity occurs but in a "smaller" network able to self-sustain its excitation. However, for large values of p, the superposition of activity between excited domains plays also an important role. The consequence is that both effects make the synchronized self-sustained activity collapse, leading to disordered patterns.

More realistic models were also studied shortly after the first works on complex networks appeared. In particular, Lago-Fernández et al. (2000) deals with a network of nonidentical Hodgkin and Huxley (1952) elements coupled by excitatory synapses in random, regular, and SW topologies. The parameters of the model neurons were kept to stay below the bifurcation point, until the input arrives and forces the system to undergo a saddle–node bifurcation on a limit cycle. The dynamics of the system ends up into a coherent oscillation or in the activation of asynchronous states. In absence of a detailed analysis of the mechanism that generates coherence, the simulations showed several effects of the topology on the dynamics, the most interesting of which is that achieving synchronization in regular networks takes longer compared to SW, where the existence of shortcuts favors faster synchronization. The results obtained in all cases show that the randomness of the topology has strong effects on the dynamics of these models, in particular the average connectivity is a control parameter for the transition between asynchronous and synchronous states.

Large-scale, highly detailed neural network computer models incorporating the latest advances in complex network modeling are becoming tractable due to the current computation capabilities. Although these cellularly detailed models are computationally expensive, they have the great advantage of being physiologically realistic. Therefore, they can provide new insights about how structural properties (such as the network topology) at the inner (cellular) level shape the dynamical behavior of both normal and diseased brains, thus allowing to understand several brain disorders such as epilepsy through network modeling. One of such models has been recently proposed in Morgan and Soltesz (2008). The authors were able to develop a computer model of the dentate gyrus, a brain structure that plays a crucial role in temporal lobe epilepsy. The results of this study are of great interest because the authors convincingly argued that the appearance of a small number of highly connected neurons after brain damage is enough for the development of several disorders. In fact, what have showed is that, in their model, structures with highly connected nodes are harmful to network dynamics (in this case given by the propensity of the system to synchrony). In other words, highly connected nodes promote the abnormal synchronization of neural activity, just as revealed by more stylized models (such as the Kuramoto model previously discussed, see Fig. 2). However, in contrast to naive models and owing to the very detailed neuronal model in Morgan and Soltesz (2008), this kind of modeling studies can also allow meaningful "computer experiments", for instance, by manipulating several aspects of network connectivity patterns, and might even provide guidelines for treatments aimed at preventing diseases such as epilepsy after brain injury. This is the real value of such a huge but praiseworthy effort.

4 Outlook

Many dynamical processes in networks have been widely studied during the last decade (Newman, 2003b; Boccaletti et al., 2006). These include diffusion processes in social and technological systems, evolutionary dynamics in biological and social contexts, biochemical regulation in metabolic and genetic circuits, as well as neuronal-like dynamics. The ultimate goal is to unveil the relationship between the form and function of the complex system under study, i.e., how the dynamical patterns displayed by the system are affected by the structure of interactions among its constituents. It has become apparent in the last several years, that the new tools provided by complex networks modeling can help understand both the dynamics and the structural organization of neuronal systems. The first steps that have been taken are certainly insufficient, as many models are still too stylized to describe the dynamical complexity of the brain. However, we believe that such simple models can guide our search for more realistic models and can provide a basic understanding of what is going on in the system. In fact, many of the observed dynamical features of naive models are later verified, pointing to some degree of universality of the phenomenon under scrutiny, which allows to extract meaningful conclusions even from simplified models. Future research on neuronal systems will also benefit from the ever increasing data availability of brain activity patterns obtained through electroencephalography (EEG), positron-emission tomography (PET), and functional magnetic resonance (fMRI) techniques. The large amount of data have provided information about both the organization and the dynamical properties of large ensembles of neurons in the brain, and certainly will continue to do so in the near future.

One of the most fascinating theoretical problems nowadays is the design of realistic synthetic brain networks through optimization of single features (such as wiring) and a more accurate description of their developmental principles. Recent studies have shown that topological properties of neural systems include the small-world property, a scale-free architecture, and a modular and hierarchical organization in the form of multiple clusters at different levels. This picture of large-scale brain complex networks reveals the functional organization of the brain, being the connections between these functional clusters the long-range connections between neurons of different cortical areas. Additionally, at the most internal level, each functional area is viewed as a densely interconnected population of neurons. Surprisingly, the hierarchical compartmentalization of brain networks does not minimize the total wiring of the network (in contrast with previous ideas in theoretical neuroscience). On the other hand, these hierarchical modular layouts help to reduce the number of signal processing steps across the network while preventing the large-scale activation of the system (as observed during epileptic seizures). How this trade-off between different properties of the brain network is achieved remains elusive.

From a dynamical point of view, the neuronal activity patterns observed experimentally are characterized by oscillations over a broad spectrum and by synchronization phenomena over a wide range of spatial and temporal scales. This synchronized activity plays an important role in the information processing of the neural systems. In the case of neuronal networks it is of particular interest to study the transition from weak to strong synchronization regimes. The weak synchronization regime is seen as the biologically plausible dynamical state of the system: coherent activity is localized inside the most internal clusters or modules of the network. These local synchronized states are at the roots of the emergence of cerebral higher functions. On the other hand, in the strong synchronization regime, the coherent activity spreads over different modular levels and the behavior of the brain network resembles that of an epileptic seizure. The synchronization transition from weak to strong coherent dynamics is therefore more interesting than in generic complex networks since the stability of the weak synchronized state can be measured as an indicator of the biological reliability of the brain network. Understanding the transition from weak to strong synchronization is another challenging problem that many scientists are currently tackling using the tools of complex network modeling.

In summary, current efforts are aimed at developing a theoretical framework, based on the available experimental data, that allows one to go from the local interactions at the most internal level of the hierarchical structure (i.e. the neuronal level) up to the global scale at the level of functional or cortical networks from purely functional principles. The goal is therefore to construct a multi-optimization algorithm that, considering several structural and functional issues of brain dynamics (such as robustness, synaptic plasticity, fast information processing, and dynamical multistability), gives rise to graph layouts similar to those found in brain networks. A second step should also consider dynamical models on top of complex brain networks. The expectancy is that by using methods borrowed from nonlinear dynamics and complexity theory we will be able to capture those characteristic patterns that signal the sharp transition from weak (local) to global (strong) synchronization in modular brain networks. If we are able to achieve such a goal, we might have at hand dynamical indicators of the degree of local synchronization that can be used and tested experimentally, for instance, when epileptic seizures occur.

Acknowledgments I am grateful to my collaborators on the subjects discussed here and would also like to thank A. Arenas, S. Boccaletti, A. Díaz-Guilera, L. M. Floria, J. Gómez-Gardeñes, and V. Latora for many helpful discussions on the structure and dynamics of complex networks during the last several years. The author is supported by MEC through the Ramón y Cajal Program. This work has been partially supported by the Spanish DGICYT Projects FIS2006-12781-C02-01 and FIS2005-00337 and by the European NEST Pathfinder project GABA under contract 043309.

References

J. A. Acebron, L. L. Bonilla, C. J. Perez Vicente, F. Ritort, and R. Spigler. The Kuramoto model: A simple paradigm for synchronization phenomena. Rev. Mod. Phys. **77**, 137–185 (2005).

R. D. Alba. A graph-theoretic definition of a sociometric clique. J. Math. Social. **3**, 113 (1973).

T. Binzegger, R. J. Douglas, and K. A. C. Martin. A quantitative map of the circuit of cat primary visual cortex. J. Neurosci. **24**, 8441 (2004).

S. Boccaletti, V. Latora, Y. Moreno, M. Chavez, and D.-U. Hwang. Complex Networks: Structure and Dynamics. Phys. Rep. **424,** 175 (2006).

M. Boguñá, R. Pastor-Satorras, and A. Vespignani. Epidemic spreading in complex nets with degree correlations. Lect. Notes Phys. **625**, 127 (2003).

B. Bollobás, *Modern Graph Theory* (Graduate Texts in Mathematics, Springer-Verlag, New York, 1998).

S. P. Borgatti, M. G. Everett, and P. R. Shirey. LS sets, lambda sets and other cohesive subsets. Social Networks **12**, 337 (1990).

V. Braitenberg and A. Schüz, *Cortex: Statistics and Geometry of Neuronal Connectivity* (Springer-Verlag, Heidelberg, Germany, 2nd ed., 1998).

G. Buzsáki and J. J. Chrobak. Temporal structure in spatially organized neuronal ensembles: a role for interneuronal networks. Curr. Opin. Neurobiol. **5**, 504 (1995).

G. Buzsáki, C. Geisler, D. A. Henze, and X. J. Wang. Interneuron diversity series: circuit complexity and axon wiring economy of cortical interneurons. Trends Neurosci. **27**, 186 (2004).

D. B. Chklovskii, T. Schikorski, and C. F. Stevens. Wiring optimization in cortical circuits. Neuron **34**, 341 (2002).

F. L. da Silva, A. Hoek, H. Smith, and L. Zetterberg. Model of brain rythmic activity. Kybernetik **15**, 27 (1974).

L. Danon, A. Diaz-Guilera, J. Duch, and A. Arenas. Comparing community structure identification. J. Stat. Mech. P09008 (2005).
V. M. Eguiluz, D. R. Chialvo, G. A. Cecchi, M. Baliki, and V. V. Apkarian. Scale-free brain functional networks. Phys. Rev. Lett. **94**, 018102 (2005).
G. W. Flake, S. R. Lawrence, C. L. Giles, and F. M. Coetzee. Self-organization and identification of Web communities. IEEE Comput. **35**, 66 (2002).
C. L. Freeman. A set of measures of centrality based on betweenness. Sociometry **40**, 35 (1977).
L. C. Freeman. Centrality in social networks: I. Conceptual clarification. Social Networks **1,** 215 (1979).
P. Fries. A mechanism for cognitive dynamics: neuronal communication through neuronal coherence. Trends Cogn. Sci. **9**, 474 (2005).
J. Gómez-Gardeñes, Y. Moreno, and A. Arenas. Paths to synchronization on complex networks. Phys. Rev. Lett. **98**, 034101 (2007a).
J. Gómez-Gardeñes, Y. Moreno, and A. Arenas. Synchronizability determined by coupling strengths and topology on Complex Networks. Phys. Rev. E **75**, 066106 (2007b).
C. Hilgetag and M. Kaiser. Clustered organization of cortical connectivity. Neuroinformatics **2**, 353 (2004).
A. L. Hodgkin and A. F. Huxley. A quantitative description of membrane current and its application to conduction and excitation in nerve. J. Physiol. **117**, 500 (1952).
H. Hong, M. Y. Choi, and B. J. Kim. Synchronization on small-world networks. Phys. Rev. E **65**, 026139 (2002).
M. Kaiser and C. C. Hilgetag. Nonoptimal Component Placement, but Short Processing Paths, due to Long-Distance Projections in Neural Systems. PloS Comput. Biol. **2**, 95 (2006).
J. Karbowski. Optimal wiring principle and plateaus in the degree of separation for cortical neurons. Phys. Rev. Lett. **86**, 3674 (2001).
N. Kashtan, S. Itzkovitz, R. Milo, and U. Alon. Efficient sampling algorithm for estimating subgraph concentrations and detecting network motifs. Bioinformatics **20**, 1746 (2004).
P. Konig, A. K. Engel, and W. Singer. Integrator or coincidence detector? the role of the cortical neuron revisited. Trends Neurosci. **19**, 130 (1996).
Y. Kuramoto. Self-entrainment of a population of coupled nonlinear oscillators. Lect. Notes Phys. **30**, 420 (1975).
Y. Kuramoto, *Chemical Oscillations, Waves, and Turbulence* (Springer-Verlag, New York, 1984).
L. F. Lago-Fernández, R. Huerta, F. Corbacho, and J. A. Sigüenza. Fast Response and Temporal Coherent Oscillations in Small-World Networks. Phys. Rev. Lett. **84**, 2758 (2000).
V. Latora and M. Marchiori. Efficient Behavior of Small-World Networks. Phys. Rev. Lett. **87**, 198701 (2001).
V. Latora and M. Marchiori. Economic Small-World Behavior in Weighted Networks. Eur. Phys. J. **B32**, 249 (2003).
V. Latora and M. Marchiori. A measure of centrality based on the network efficiency. Preprint cond-mat/0402050.
S. Mangan and U. Alon. Structure and function of the feed-forward loop network motif. Proc. Natl. Acad. Sci. USA **100**, 11980 (2003).
M. Marchiori and V. Latora. Harmony in the small-world. Physica **A285**, 539 (2000).
H. Markram, M. Toledo-Rodriguez, Y. Wang, A. Gupta, G. Silberberg, and C. Wu. Interneurons of the neocortical inhibitory system. Nat. Rev. Neurosci. **5**, 793 (2004).
R. Milo, S. Itzkovitz, N. Kashtan, R. Levitt, S. Shen-Orr, I. Ayzenshtat, M. Sheffer, and U. Alon. Superfamilies of Evolved and Designed Networks. Science **303**, 1538 (2004).
R. Milo, S. Shen-Orr, S. Itzkovitz, N. Kashan, D. Chklovskii, and U. Alon. Network Motifs: Simple Building Blocks of Complex Networks. Science **298**, 824 (2002).
Y. Moreno and A. F. Pacheco. Synchronization of Kuramoto oscillators in scale-free networks. Europhys. Lett. **68**, 603 (2004).
Y. Moreno, M. Vázquez-Prada, and A. F. Pacheco. Fitness for synchronization of network motifs. Physica A **343**, 279 (2004).

R. J. Morgan and I Soltesz. Nonrandom connectivity of the epileptic dentate gyrus predicts a major role for neuronal hubs in seizures. Proc. Natl. Acad. Sci. USA **105**, 6179 (2008).

M. E. J. Newman. Assortative mixing in networks. Phys. Rev. Lett. **89**, 208701 (2002).

M. E. J. Newman. Mixing patterns in networks. Phys. Rev. **E67**, 026126 (2003a).

M. E. J. Newman. The structure and function of complex networks. SIAM Rev. **45**, 167 (2003b).

R. Pastor-Satorras, A. Vázquez, and A. Vespignani. Dynamical and correlation properties of the Internet. Phys. Rev. Lett. **87**, 258701 (2001).

R. Pastor-Satorras, A. Vespignani, *Evolution and Structure of the Internet: A Statistical Physics Approach* (Cambridge University Press, Cambridge, 2004).

F. Radicchi, C. Castellano, F. Cecconi, V. Loreto, and D. Parisi. Defining and identifying communities in networks. Proc. Natl. Acad. Sci. USA **101**, 2658 (2004).

A. Roxin, H. Riecke, and S. A. Solla. Self-Sustained Activity in a Small-World Network of Excitable Neurons. Phyas. Rev. Lett. **92**, 198101 (2004).

J. Scannell, G. Burns, C. Hilgetag, M. ONeil, and M. Young. The connectional organization of the cortico-thalamic system of the cat. Cereb. Cortex **9**, 277 (1999).

J. Scott, *Social Network Analysis: A Handbook* (Sage Publications, London, 2nd ed., 2000).

S. B. Seidman. Internal cohesion of LS sets in graphs. Social Networks **5**, 97 (1983).

S. Shen-Orr, R. Milo, S. Mangan, and U. Alon. Motifs in the transcriptional regulation network of E. coli. Nat. Genet. **31**, 64 (2002).

A. Sik, M. Penttonen, A. Ylinen, and G. Buzsaki. Hippocampal CA1 interneurons: an in vivo intracellular labeling study. J. Neurosci. **15**, 6651 (1995).

A. Sik, A. Ylinen, M. Penttonen, and G. Buzsáki. Inhibitory ca1-ca3-hilar region feedback in the hippocampus. Science **265**, 1722 (1994).

G. Silberberg, S. Grillner, F. E. Lebeau, R. Maex, and H. Markram. Synaptic pathways in neural microcircuits. Trends Neurosci. **28**, 541 (2005).

S. Wasserman and K. Faust, *Social Networks Analysis* (Cambridge University Press, Cambridge, 1994).

D. J. Watts, *Small Worlds: The Dynamics of Networks Between Order and Randomness* (Princeton University Press, Princeton, NJ, 1999).

D. J. Watts and S. H. Strogatz. Collective dynamics of small-world networks. Nature **393**, 440 (1998).

L. Zemanová, C. Zhou, and J. Kurths. Structural and functional clusters of complex brain networks. Physica D **224**, 202 (2006).

C. Zhou and J. Kurths. Dynamical Weights and Enhanced Synchronization in Adaptive Complex Networks. Phys. Rev. Lett. **96**, 164102 (2006).

C. Zhou, A. E. Motter, and J. Kurths. Universality in the Synchronization of Weighted Random Networks. Phys. Rev. Lett. **96**, 034101 (2006).

C. Zhou, L. Zemanová, G. Zamora, C. C. Hilgetag, and J. Kurths. Hierarchical organization unveiled by functional connectivity in complex brain networks. Phys. Rev. Lett. **97**, 238103 (2006).

C. Zhou, L. Zemanová, G. Zamora, C. C. Hilgetag, and J. Kurths. Structure function relationship in complex brain networks expressed by hierarchical synchronization. New J. Phys. **9**, 178 (2007).

Index

C